Computational Fluid Mechanics

Computational Fluid Mechanics

Editors

Peng Du
Abdellatif Ouahsine
Haibao Hu
Xiaopeng Chen

MDPI • Basel • Beijing • Wuhan • Barcelona • Belgrade • Manchester • Tokyo • Cluj • Tianjin

Editors

Peng Du
School of Marine Science
and Technology
Northwestern Polytechnical
University
Xi'an
China

Abdellatif Ouahsine
Roberval Laboratory
UT Compiègne-Sorbonne
University
Compiègne
France

Haibao Hu
School of Marine Science
and Technology
Northwestern Polytechnical
University
Xi'an
China

Xiaopeng Chen
School of Marine Science
and Technology
Northwestern Polytechnical
University
Xi'an
China

Editorial Office
MDPI
St. Alban-Anlage 66
4052 Basel, Switzerland

This is a reprint of articles from the Special Issue published online in the open access journal *Journal of Marine Science and Engineering* (ISSN 2077-1312) (available at: www.mdpi.com/journal/jmse/special_issues/computational_fluid_mechanics).

For citation purposes, cite each article independently as indicated on the article page online and as indicated below:

LastName, A.A.; LastName, B.B.; LastName, C.C. Article Title. *Journal Name* **Year**, *Volume Number*, Page Range.

ISBN 978-3-0365-7171-3 (Hbk)
ISBN 978-3-0365-7170-6 (PDF)

Contents

About the Editors vii

Preface to "Computational Fluid Mechanics" ix

Peng Du, Abdellatif Ouahsine, Haibao Hu and Xiaopeng Chen
Computational Fluid Mechanics Methods and Applications in Marine Engineering
Reprinted from: *J. Mar. Sci. Eng.* **2023**, *11*, 606, doi:10.3390/jmse11030606 1

Peng Du, Abdellatif Ouahsine, Philippe Sergent, Yannick Hoarau and Haibao Hu
Investigation on Resistance, Squat and Ship-Generated Waves of Inland Convoy Passing Bridge Piers in a Confined Waterway
Reprinted from: *J. Mar. Sci. Eng.* **2021**, *9*, 1125, doi:10.3390/jmse9101125 5

Jianxun Kuang, Guodong Chen, Zhoulong Yuan, Xiajun Qi, Qianhao Yu and Zhen Liu
Dynamic Interactions of a Cable-Laying Vessel with a Submarine Cable during Its Landing Process
Reprinted from: *J. Mar. Sci. Eng.* **2022**, *10*, 774, doi:10.3390/jmse10060774 21

Shuo Huang, Weiqi Liu, Wanzhen Luo and Kai Wang
Numerical Simulation of the Motion of a Large Scale Unmanned Surface Vessel in High Sea State Waves
Reprinted from: *J. Mar. Sci. Eng.* **2021**, *9*, 982, doi:10.3390/jmse9090982 37

Qiong Zhang, Bo Shui and Hanhua Zhu
Study on Sloshing Characteristics in a Liquid Cargo Tank under Combination Excitation
Reprinted from: *J. Mar. Sci. Eng.* **2022**, *10*, 1100, doi:10.3390/jmse10081100 59

Yunlong Ma, Qiaogao Huang, Guang Pan and Pengcheng Gao
Investigation of the Hydrodynamic Characteristics of Two Manta Rays Tandem Gliding
Reprinted from: *J. Mar. Sci. Eng.* **2022**, *10*, 1186, doi:10.3390/jmse10091186 77

Xing He, Qiaogao Huang, Guocang Sun and Xihui Wang
Numerical Research of the Pressure Fluctuation of the Bow of the Submarine at Different Velocities
Reprinted from: *J. Mar. Sci. Eng.* **2022**, *10*, 1188, doi:10.3390/jmse10091188 91

Yao Shi, Jinyi Ren, Shan Gao and Guang Pan
Numerical Investigation on Air Film Fusion of Pressure-Equalizing Exhaust around Shoulder Ventilation of Submarine-Launched Vehicle
Reprinted from: *J. Mar. Sci. Eng.* **2021**, *10*, 39, doi:10.3390/jmse10010039 109

Rui Lu, Guang Pan, Kun Tan and Shaoping Yin
Numerical Simulation of Cavitation and Damping Force Characteristics for a High-Speed Supercavitation Vehicle
Reprinted from: *J. Mar. Sci. Eng.* **2021**, *9*, 1171, doi:10.3390/jmse9111171 125

Pengcheng Gao, Qiaogao Huang and Guang Pan
Propulsion Performance and Wake Dynamics of Heaving Foils under Different Waveform Input Perturbations
Reprinted from: *J. Mar. Sci. Eng.* **2021**, *9*, 1271, doi:10.3390/jmse9111271 141

Han Li, Qiaogao Huang, Guang Pan, Xinguo Dong and Fuzheng Li
Effects of Blade Number on the Propulsion and Vortical Structures of Pre-Swirl Stator Pump-Jet Propulsors
Reprinted from: *J. Mar. Sci. Eng.* **2021**, *9*, 1406, doi:10.3390/jmse9121406 . . . . **153**

Fuzheng Li, Qiaogao Huang, Guang Pan, Denghui Qin and Han Li
Influence of Various Stator Parameters on the Open-Water Performance of Pump-Jet Propulsion
Reprinted from: *J. Mar. Sci. Eng.* **2021**, *9*, 1396, doi:10.3390/jmse9121396 . . . . **175**

Ho-Sang Jang, Se-Yun Hwang and Jang-Hyun Lee
Numerical Prediction of Convective Heat Flux on the Flight Deck of Naval Vessel Subjected to a High-Speed Jet Flame from VTOL Aircraft
Reprinted from: *J. Mar. Sci. Eng.* **2022**, *10*, 260, doi:10.3390/jmse10020260 . . . . **199**

Bo Hu, Chuan Wang, Hui Wang, Qian Yu, Jinhua Liu and Yong Zhu et al.
Numerical Simulation Study of the Horizontal Submerged Jet Based on the Wray–Agarwal Turbulence Model
Reprinted from: *J. Mar. Sci. Eng.* **2022**, *10*, 1217, doi:10.3390/jmse10091217 . . . . **221**

Seyed Ahmad Reza Saeidi Hosseini, Abdolmajid Mohammadian, Philip J. W. Roberts and Ozeair Abessi
Numerical Study on the Effect of Port Orientation on Multiple Inclined Dense Jets
Reprinted from: *J. Mar. Sci. Eng.* **2022**, *10*, 590, doi:10.3390/jmse10050590 . . . . **243**

About the Editors

Peng Du

Dr. Peng DU received his M.Sc. degree from Northwestern Polytechnical University in 2015 and a Ph.D. from UT Compiègne-Sorbonne University of France in 2019. His expertise is in the fields of hydrodynamics, naval architecture, computational and experimental fluid mechanics etc. He has published more than 30 articles in the journals including Ocean Engineering, Physics of Fluids, etc., and 6 books till 2022. He obtained an associate professor position from Northwestern Polytechnical University directly after Ph.D.

Abdellatif Ouahsine

Prof. Abdellatif Ouaisne is at the University of Technology of Compiegne. The research activities concern the numerical simulation of problems related to fluid-structure interactions, free-surface fluid flows, and numerical modeling of hydrodynamic processes and transport in natural environments.

Haibao Hu

Prof. Haibao Hu is at the Northwestern Polytechnical University. He is the manager of the "High speed water tunnel Laboratory ". He has long been committed to the research of hydrodynamics and bionic drag reduction technology, presided over more than 10 projects funded by the National Natural Science Foundation and Shaanxi province, and published more than 60 academic papers.

Xiaopeng Chen

Prof. Xiaopeng Chen is at the Northwestern Polytechnical University. His research is mainly in the area of experimental and computational fluid mechanics. He has published more than 50 academic papers.

Preface to "Computational Fluid Mechanics"

Ocean flows and their interactions with marine structures, vehicles, etc., are of great importance for the research of marine science and technology. The revealing of their mechanisms inevitably involves fluid mechanics. Computational Fluid Mechanics (CFD) techniques are developing to a higher level nowadays with the flourishing of computer science. CFD is able to predict ocean flow details and hydrodynamic interactions with increasing accuracy.

This reprint collected the cutting-edge developments of CFD in the area of marine science and technology. CFD methods including novel numerical algorithms, advanced treatments of solving procedures, parallel acceleration techniques, and the utilization of CFD for fluid–structure interaction, wave modeling, ship dynamics, etc., are included. The audience of this book can be researchers and engineers in the area of marine science and technology.

This reprint is organized by Associate Professor Peng DU, Professor Abdellatif Ouahsine, Haibao HU, and Xiaopeng Chen. The authors, including Zhen Liu, Wanzhen Luo, Hanhua Zhu, Qiaogao Huang, Yao Shi, Guang Pan, Jang-Hyun Lee, Yang Yang, Abdolmajid Mohammadian, etc., are acknowledged for their contributions.

Peng Du, Abdellatif Ouahsine, Haibao Hu, and Xiaopeng Chen
Editors

Journal of
Marine Science and Engineering

MDPI

Editorial

Computational Fluid Mechanics Methods and Applications in Marine Engineering

Peng Du [1,*], Abdellatif Ouahsine [2], Haibao Hu [1] and Xiaopeng Chen [1]

1 School of Marine Science and Technology, Northwestern Polytechnical University, Xi'an 710072, China
2 Laboratoire Roberval, UT Compiègne-Sorbonne Université, 60203 Compiègne, France
* Correspondence: dupeng@nwpu.edu.cn

Citation: Du, P.; Ouahsine, A.; Hu, H.; Chen, X. Computational Fluid Mechanics Methods and Applications in Marine Engineering. *J. Mar. Sci. Eng.* **2023**, *11*, 606. https://doi.org/10.3390/jmse11030606

Received: 28 February 2023
Accepted: 1 March 2023
Published: 13 March 2023

Ocean flows and their interactions with marine structures, vehicles, etc., are of great importance for the research of marine science and technology. The revealing of their mechanisms inevitably involves fluid mechanics. Computational fluid mechanics techniques are developing to a higher level nowadays, with the flourishing of computer science. It is possible to predict ocean flow details and hydrodynamic interactions with increasing accuracy.

This book has collected cutting-edge developments in computational fluid mechanics in the area of marine science and technology. Certain numerical methods and their engineering applications have been covered by the 14 collected works. They can be divided into five groups according to the following subjects:

(1) Hydrodynamics of surface vehicles

The dynamic interactions of a cable-laying vessel with a submarine cable during its landing process were investigated in [1]. The effects of the cable length, the current velocity, the incident wave, and the wind direction on vessel stability and the tensions in the mooring lines and cable were investigated. The most unfavorable environmental conditions for submarine cable laying were determined under different environmental conditions. The results are critical to the safety of the cable. The resistance, squat, and ship-generated waves of inland convoy passing bridge piers in a confined waterway were studied in [2]. The existence of piers was found to only influence the transient hydrodynamics of the convoy, but not the averaged properties. Ship-generated waves, especially wave profiles at a specific lateral position, were characterized. The motion of a large-scale unmanned surface vessel in high sea-state waves was studied in [3]. The effects of different sea states, as well as different wave directions, on the motion of a USV (unmanned surface vessel) were compared. Sloshing characteristics in a liquid cargo tank under combination excitation were investigated in [4]. The pressure distribution characteristics at different positions of the cargo tank were discussed, along with the influence of different excitation conditions on the pressure of the cargo tank. This is very significant for the stability and safety of ship navigation.

(2) Hydrodynamics of underwater vehicles

The hydrodynamic characteristics of the tandem gliding of two manta rays were investigated in [5]. A numerical method was used to explore the influence of the front-to-back distance and the angle of attack on the overall and individual hydrodynamic performance of a pair of manta rays gliding. This provides a theoretical basis for understanding the biological habits of manta rays and for the design of an underwater bionic robot group system. Pressure fluctuations in the bow of a submarine at different velocities were studied in [6], and this can provide important clues for the hydrodynamic noise source of underwater vehicles.

(3) Hydrodynamics of trans-media vehicles

The air film fusion of a pressure-equalizing exhaust around the shoulder ventilation of submarine-launched vehicles was investigated in [7]. A numerical calculation method based on the VOF (volume of fluid) multiphase flow model was established to study the fusion characteristics of the air film at the shoulder of the underwater vehicle, as well as the distribution of surface pressure along the vehicle's axial direction. The cavitation and damping force characteristics for a high-speed supercavitation vehicle were studied in [8]. A homogeneous equilibrium flow model and a Schnerr–Sauer model based on the Reynolds-averaged Navier–Stokes method were used. The effect of the navigation speed and angle of attack on the cavitation morphology and dynamic characteristics was investigated. The propulsion performance and wake dynamics of heaving foils under different waveform input perturbations were studied in [9]. This work further explains the effect of different waveform perturbation signals on the base sinusoidal signal and provides a new control idea for underwater vehicles.

(4) Hydrodynamics of propellers

The effects of blade number on the propulsion and vortical structures of pre-swirl stator pump-jet propulsors (PJP) were investigated in [10]. It was found that the blade number was also vital for further PJP design, particularly when the main concerns were exciting force and noise performance. The influence of various stator parameters on the open-water performance of pump-jet propulsion was studied in [11], in order to improve the hydrodynamic performance of pump-jet propulsion. The torque balance locations under various parameters were compared, and each component force was analyzed in detail to explain the reason for performance variation.

(5) Heat and flow characteristics of jets

Convective heat flux on the flight deck of a naval vessel subjected to a high-speed jet flame from VTOL (vertical take-off and landing) aircraft was numerically predicted in [12]. A procedure for analyzing the convective heat transfer imposed on the deck by the high-temperature and high-velocity impact of a VTOL jet was described. A horizontal submerged jet (HSJ) based on the Wray–Agarwal turbulence model was studied in [13], in order to further understand the mechanism of a HSJ. The results have a guiding significance for engineering practice and academic research. The effect of port orientation on multiple inclined dense jets was investigated in [14]. The outcomes may be favorable for outfall system applications involving dilution.

In conclusion, this book presents 14 works regarding numerical methods and their application to five different areas of marine science and technology, including the hydrodynamics of surface vehicles, underwater vehicles, trans-media vehicles, propellers, and the physics of jets. We hope these works provide some clues for related fundamental research and engineering applications.

Author Contributions: Conceptualization, P.D. and A.O.; resources, P.D., A.O., H.H. and X.C.; writing—original draft preparation, P.D.; writing—review and editing, A.O., H.H. and X.C.; supervision, P.D. All authors have read and agreed to the published version of the manuscript.

Funding: This research was supported by the National Natural Science Foundation of China (Grant Nos. 52201380).

Acknowledgments: The authors wish to thank all contributors to this Special Issue.

Conflicts of Interest: The authors declare no conflict of interest.

References

1. Kuang, J.; Chen, G.; Yuan, Z.; Qi, X.; Yu, Q.; Liu, Z. Dynamic Interactions of a Cable-Laying Vessel with a Submarine Cable during Its Landing Process. *J. Mar. Sci. Eng.* **2022**, *10*, 774. [CrossRef]
2. Du, P.; Ouahsine, A.; Sergent, P.; Hoarau, Y.; Hu, H. Investigation on Resistance, Squat and Ship-Generated Waves of Inland Convoy Passing Bridge Piers in a Confined Waterway. *J. Mar. Sci. Eng.* **2021**, *9*, 1125. [CrossRef]
3. Huang, S.; Liu, W.; Luo, W.; Wang, K. Numerical Simulation of the Motion of a Large Scale Unmanned Surface Vessel in High Sea State Waves. *J. Mar. Sci. Eng.* **2021**, *9*, 982. [CrossRef]
4. Zhang, Q.; Shui, B.; Zhu, H. Study on Sloshing Characteristics in a Liquid Cargo Tank under Combination Excitation. *J. Mar. Sci. Eng.* **2022**, *10*, 1100. [CrossRef]
5. Ma, Y.; Huang, Q.; Pan, G.; Gao, P. Investigation of the Hydrodynamic Characteristics of Two Manta Rays Tandem Gliding. *J. Mar. Sci. Eng.* **2022**, *10*, 1186. [CrossRef]
6. He, X.; Huang, Q.; Sun, G.; Wang, X. Numerical Research of the Pressure Fluctuation of the Bow of the Submarine at Different Velocities. *J. Mar. Sci. Eng.* **2022**, *10*, 1188. [CrossRef]
7. Shi, Y.; Ren, J.; Gao, S.; Pan, G. Numerical Investigation on Air Film Fusion of Pressure-Equalizing Exhaust around Shoulder Ventilation of Submarine-Launched Vehicle. *J. Mar. Sci. Eng.* **2022**, *10*, 39. [CrossRef]
8. Lu, R.; Pan, G.; Tan, K.; Yin, S. Numerical Simulation of Cavitation and Damping Force Characteristics for a High-Speed Supercavitation Vehicle. *J. Mar. Sci. Eng.* **2021**, *9*, 1171. [CrossRef]
9. Gao, P.; Huang, Q.; Pan, G. Propulsion Performance and Wake Dynamics of Heaving Foils under Different Waveform Input Perturbations. *J. Mar. Sci. Eng.* **2021**, *9*, 1271. [CrossRef]
10. Li, H.; Huang, Q.; Pan, G.; Dong, X.; Li, F. Effects of Blade Number on the Propulsion and Vortical Structures of Pre-Swirl Stator Pump-Jet Propulsors. *J. Mar. Sci. Eng.* **2021**, *9*, 1406. [CrossRef]
11. Li, F.; Huang, Q.; Pan, G.; Qin, D.; Li, H. Influence of Various Stator Parameters on the Open-Water Performance of Pump-Jet Propulsion. *J. Mar. Sci. Eng.* **2021**, *9*, 1396. [CrossRef]
12. Jang, H.-S.; Hwang, S.-Y.; Lee, J.-H. Numerical Prediction of Convective Heat Flux on the Flight Deck of Naval Vessel Subjected to a High-Speed Jet Flame from VTOL Aircraft. *J. Mar. Sci. Eng.* **2022**, *10*, 260. [CrossRef]
13. Hu, B.; Wang, C.; Wang, H.; Yu, Q.; Liu, J.; Zhu, Y.; Ge, J.; Chen, X.; Yang, Y. Numerical Simulation Study of the Horizontal Submerged Jet Based on the Wray–Agarwal Turbulence Model. *J. Mar. Sci. Eng.* **2022**, *10*, 1217. [CrossRef]
14. Saeidi Hosseini, S.A.R.; Mohammadian, A.; Roberts, P.J.W.; Abessi, O. Numerical Study on the Effect of Port Orientation on Multiple Inclined Dense Jets. *J. Mar. Sci. Eng.* **2022**, *10*, 590. [CrossRef]

Journal of
Marine Science and Engineering

MDPI

Article

Investigation on Resistance, Squat and Ship-Generated Waves of Inland Convoy Passing Bridge Piers in a Confined Waterway

Peng Du [1,2,3], Abdellatif Ouahsine [3], Philippe Sergent [4], Yannick Hoarau [5] and Haibao Hu [2,*]

1 Research & Development Institute, Northwestern Polytechnical University, Shenzhen 518000, China; dupeng@nwpu.edu.cn
2 School of Marine Science and Technology, Northwestern Polytechnical University, Xi'an 710072, China
3 Roberval Laboratory, UT Compiègne-Sorbonne University, CS 60319, 60203 Compiègne, France; ouahsine@utc.fr
4 Cerema, CS 60039, 60200 Compiègne, France; philippe.sergent@cerema.fr
5 ICUBE Laboratory, UMR 7357 CNRS, University of Strasbourg, 67000 Strasbourg, France; hoarau@unistra.fr
* Correspondence: huhaibao@nwpu.edu.cn

Abstract: The average and unsteady hydrodynamics of an inland convoy passing bridge piers in a confined waterway were investigated using both numerical and experimental approaches. The numerical simulations are realized by solving the RANS (Reynolds-averaged Navier–Stokes) equations accounting for the solid body motion using the sliding mesh technique, while the experiments were carried out in the towing tank. The advancing resistance, trim, sinkage and ship-generated waves were analyzed as functions of the water depth, distance between bridge piers, draught and velocity. The existence of the piers is found to only influence the transient hydrodynamics of the convoy, but not the averaged properties. The ship-generated waves, especially the wave profiles at a specific lateral position, were characterized. Two wave crests exist at the pier position because of the additional reflections, creating a very complex wave pattern in the confined waterway.

Keywords: hydrodynamics; inland convoy; bridge pier; confined waterway; resistance

Citation: Du, P.; Ouahsine, A.; Sergent, P.; Hoarau, Y.; Hu, H. Investigation on Resistance, Squat and Ship-Generated Waves of Inland Convoy Passing Bridge Piers in a Confined Waterway. *J. Mar. Sci. Eng.* **2021**, *9*, 1125. https://doi.org/10.3390/jmse9101125

Academic Editors: Kostas Belibassakis and Md Jahir Rizvi

Received: 11 September 2021
Accepted: 11 October 2021
Published: 14 October 2021

1. Introduction

The interactions between ships and structures in water are of great importance, and include ship overtaking, passing moored vessels, crossing bridge piers, etc. They can influence the kinematic and kinetic properties of ships, and have to be considered during ship design and manufacture. When ships are maneuvered near these structures, special attention should be given to the rudder action and propulsion of the ship to avoid instabilities and possible collisions. Among these phenomena, the first three (ship encountering, overtaking, the passage of moored vessels) have been widely investigated [1–6], either by theoretical analysis, experiments or numerical simulations. However, the passage of bridge piers has not been studied in depth, but is equally important for ship navigation, especially in confined waters. Normally, there exist multiple bridge piers in water, which increase the restriction level of the waterway. When the distance between piers is very small (less than 4 times the ship breadth), this interaction becomes evident. Accidents are prone to happen at the pier position if the vessel is not properly controlled, causing damage to both the ship and bridge, especially in shallow waters. The hydrodynamics during the passage are therefore extremely significant to avoid accidents under these conditions.

Some work has been done to investigate the impact during a ship-bridge collision, with emphasis on the impact force of the collision and the damage [7–16]. However, the hydrodynamics during the passage are rarely investigated [17–21]. Tan and Gan [22] mathematically studied the influence of water level on ship-bridge collision. They concluded that the turbulence area near the bridge pier has a negative impact on safe passage. Li et al. [17] developed a 3-D Rankine boundary element method and simulated the unsteady hydrodynamic interaction during the process. A free surface re-meshing algorithm was proposed to treat the relevant moving boundary problem. The characteristics of the hydrodynamic loads were specified and the critical positions for the peak values have been identified. Zhang et al. [18] used a 3-D Rankine source boundary element method fully based on NURBS (non-uniform rational B-spline). The interaction forces and moments were numerically predicted. Zhang et al. [23] used CFD (computational fluid dynamics) techniques for this problem, where the unsteady viscous effect was considered. The transient forces and moments were predicted and the influence of the distance between the ship and the pier was analyzed. Xiang et al. [24] extended a method for evaluating the calm water-interacting loads with a submerged floating tube bridge up to the point of collision. Guo et al. [20] proposed an empirical formula relating to the minimum distance between the navigation ship and the bridge, which demonstrated that this distance increases logarithmically with flow velocity, and linearly with the pier's diameter in the channel. Zheng [25] found that pier scouring made the bending moments of pier bodies increase under the same ship impact force and the reduction effect of the ship impact resistance force on piers by scouring cannot be ignored. Khangaonkar et al. [21] presented an assessment of zone of influence (ZOI) from the Hood Canal Bridge in the Salish Sea, Washington. The results confirmed that this bridge obstructs the brackish outflow surface layer, which induces increased local mixing near the bridge, causes pooling of water (up-current) during ebb and flood, and results in shadow/sheltering of water (down-current). The study of this phenomenon is still worthy, which is promising to uncover the unsteady hydrodynamic process and give useful insights and suggestions for ship maneuvering actions.

In this study, an inland convoy passing bridge piers was simulated using CFD methods to investigate the influence of the piers on the inland convoy in a confined waterway. The advancing resistance, squat and wave patterns are analyzed and characterized.

2. Numerical and Experimental Details

2.1. Governing Equations Accounting for Solid Body Motions

The governing equations of fluid flow (RANS) are as follows [26,27]:

$$\nabla \cdot \mathbf{u} = 0, \tag{1}$$

$$\frac{\partial \mathbf{u}}{\partial t} + \nabla \cdot (\mathbf{u}\mathbf{u}) = -\frac{\nabla p}{\rho} + \nabla \cdot (\nu_{eff} \nabla \mathbf{u}) + \mathbf{g} + \mathbf{f}_{\sigma}, \tag{2}$$

where $\mathbf{u}$, ρ and p are the velocity, density and pressure of the fluid. The fluid is treated as incompressible (ρ = const). The open-source code OpenFOAM is used throughout this work. The pressure and velocity are decoupled using the merged-PISO-SIMPLE scheme, which allows larger time steps and thus accelerates the simulations. $\nu_{eff} = \nu + \nu_t$ is the effective dynamic viscosity, consisting of the fluid viscosity (ν) and the eddy viscosity (ν_t) calculated from a certain turbulence model. The $SST\ k-\omega$ model is adopted in this study. It is a two-equation eddy-viscosity model which combines the advantages of the $k-\epsilon$ and $k-\omega$ models. The $k-\omega$ model is used in the boundary layer and the $k-\epsilon$ model is used in the free stream flow. Thereby it is less sensitive to free stream conditions and has better performance predicting the flow separation and reattachment. Different turbulence models, including the $k-\varepsilon$, $k-w$ and $SST\ k-\omega$ models, were all tested as in Figure 1. It can be observed that the $SST\ k-\omega$ model produces the most accurate resistance coefficient among them. The $SST\ k-\omega$ model was ultimately selected, on account of its precision performance.

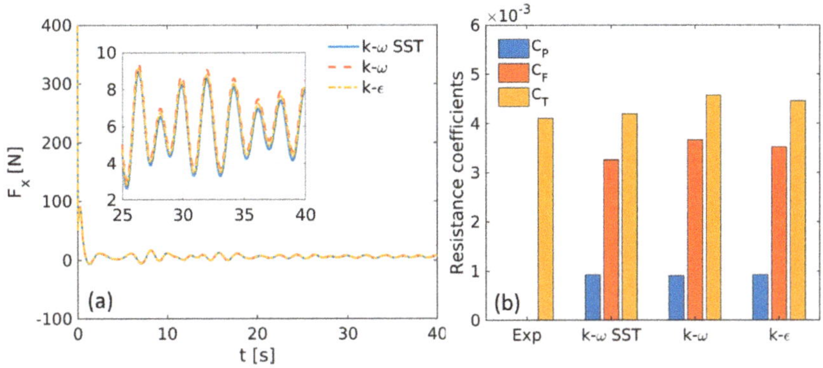

Figure 1. (**a**) Convergence histories of the drag (F_x) and (**b**) comparison of resistance coefficients with different turbulence models of a KVLCC2 model. C_T, C_F, C_P are the total, frictional and pressure resistance coefficients, respectively. The experimental data can be found in [28].

The VOF (volume-of-fluid) model is adopted to capture the free surface between water and air:

$$\frac{\partial \alpha}{\partial t} + \nabla \cdot (\alpha \mathbf{u}) + \nabla \cdot [\alpha(1-\alpha)\mathbf{u}_r] = 0, \quad (3)$$

where $\mathbf{u}_r$ is the compressive velocity field acting in the normal direction towards the interface. $\alpha = V_w/V$ is the phase fraction, also called the indicator function, where V_w is the volume of water inside a control volume V. The VOF method is conservative because α is bounded between 0 and 1. $\alpha = 0$ and $\alpha = 1$ correspond with air and water. The last term of Equation (3) can prevent excessive smearing of the free surface.

When the solid in the computational domain moves, the surrounding cells should also be moved to maintain their shapes, i.e., the mesh quality. The grid velocity $\mathbf{u}_g$ should thus be added into the momentum Equation (2) through the ALE (arbitrary Lagrangian–Eulerian) approach:

$$\frac{\partial \mathbf{u}}{\partial t} + \nabla \cdot ((\mathbf{u} - \mathbf{u}_g)\mathbf{u}) = -\frac{\nabla p}{\rho} + \nabla \cdot (\nu_{eff} \nabla \mathbf{u}) + \mathbf{g} + \mathbf{f}_\sigma, \quad (4)$$

where the grid velocity $\mathbf{u}_g$ can be determined by the space conservation law:

$$\frac{\partial}{\partial t}\int_V dV - \int_S \mathbf{u}_g \cdot d\mathbf{S} = 0, \quad (5)$$

During the simulation, the sliding mesh technique is utilized to account for the process of the convoy passing the bridge piers. Interfaces are created between the domains of the convoy and the piers. The adjacent mesh domains exchange information through interpolation at the interfaces. Each cell face of the local patch (slave patch) accepts contributions from the overlapping faces of the neighbor patch (master patch) at the sliding interface, with the weights defined as a fraction of the intersecting areas. For each face, the sum of the weights should be 1. Interpolation weights at the interface are constructed such that the interpolation results in conservation. Conservation errors are introduced as the sum of weights deviates from 1 where the patch geometries are not well matched. Consistent and conservative discretization across the interface is achieved using weighted interpolation [29,30].

For the flow variables from the master patch to the slave patch [31]:

$$\phi_{S_i} = \sum_n W_{M_n - S_i} \phi_{M_n}, \quad (6)$$

where ϕ and W denote the flow variable and the weight. The subscripts 'M' and 'S' represent the master and slave patches, respectively. It can be observed that the value of the flow variable at the slave patch is calculated from the weighted sum of that at the master patch.

For the ones from the slave patch to the master patch:

$$\phi_{M_j} = \sum_m W_{S_m - M_j} \phi_{S_m}, \tag{7}$$

in order for the interface discretization to remain conservative, the following three constraints are obeyed:

$$\sum W_{M_n - S_i} = 1.0, \tag{8}$$

$$\sum W_{S_m - M_j} = 1.0, \tag{9}$$

$$W_{M_n - S_i} \left| S_{M_n} \right| = W_{S_m - M_j} \left| S_{M_j} \right| = \left| S_{M \cap S} \right|, \tag{10}$$

with the additional symmetry constraint:

$$W_{M_n - S_i} > 0 \Rightarrow W_{S_i - M_n} > 0, \tag{11}$$

where ϕ_S and ϕ_M are the variables of the slave and master patch, respectively. i and j denote the ith and jth facets at the slave and master patches. n is the number of master facets neighboring the slave facet i, and m is the number of the slave facets neighboring the master facet j. W_{M-S} and W_{S-M} are the master-to-slave and slave-to-master weighting factors, respectively. $|S_M|$, $|S_S|$ and $|S_{M \cap S}|$ are the surface area of the master facet, the surface area of the slave facet and the intersection between the master and slave facets, respectively.

The value of the weighting factors is basically the percentage of the surface intersection between two overlapping facets deduced from Equation (10):

$$W_{M - S_i} = \frac{\left| S_{M \cap S_i} \right|}{\left| S_{M_n} \right|} \in [0.0, 1.0], \tag{12}$$

$$W_{S - M_i} = \frac{\left| S_{S \cap M_j} \right|}{\left| S_{S_n} \right|} \in [0.0, 1.0], \tag{13}$$

Because of the domain discretization, the two meshes may not overlap with each other, making the weighting factors underestimated (smaller than 1.0), i.e., the simulation will not be conservative. It is therefore suggested to use enough mesh resolution at the interface and use similar mesh sizes between the master and slave patches for the sake of accuracy. If this cannot be satisfied, remedies can be adopted. Corrections for low weight are used here, which rescale the face weighting factors so they will sum up to 1.0. For each neighboring face involved, the rescaling will be proportional to the initial value of the neighbor face weighting factor, so the overall correction for each face ends up being weight interpolated

2.2. Testing Setups

The experiments were carried out in the ANAST laboratory of the University of Liège. The testing setup and parameter definitions can be found in Figure 2. The towing tank was adjusted with trapezoidal banks with a slope of 2:1, resembling the real conditions of an inland waterway. Two bridge piers were designed and placed symmetrically in the channel An inland convoy model with one pusher and one barge was used with the scale ratio 1/25, according to the dimension of the towing tank. The length of the convoy in model scale was 7.0 m. This convoy is widely-used during inland waterway transport in Europe. In the experiments, the convoy model was towed by the carriage with a specific velocity. It passed through the space between the piers, which are composed of two concrete blocks with the dimensions 40 cm × 16 cm. A six-component dynamometer balance was used to measure the forces and moments on the vessel. The system was calibrated before the experiments, with reported uncertainty of 8% for the force. More details about the experiments can be found in [32,33].

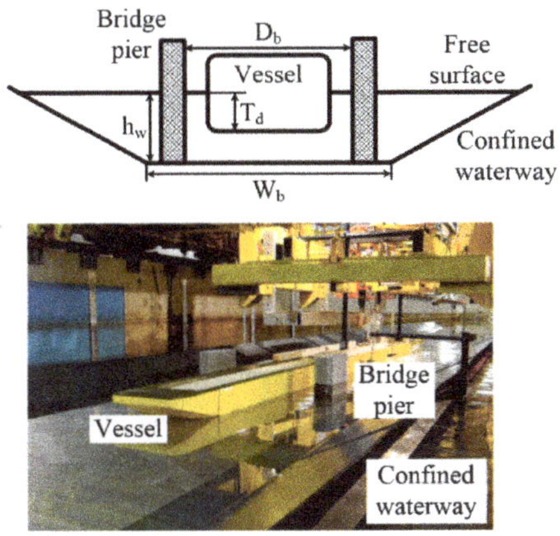

Figure 2. Schematic and photo of the inland convoy passing the bridge piers.

The computational domain was designed as the experiments in Figure 2. The length of the channel was 14.5 m. Since the domain is symmetrical in the y direction, only one-half was simulated with the symmetry boundary condition to reduce the grid number. The boundary conditions and principal dimensions of the domain are denoted in the figure. The advective boundary condition was adopted to avoid wave reflections at both ends of the channel. The sliding interfaces were designed between the convoy and the bridge pier/channel bottom [6]. The area near the pier is zoomed in Figure 3b to demonstrate their relative position. The computational mesh is shown in Figure 3c. The areas of the convoy, free surface and the bridge pier are clearly refined to capture the complex flow and accurately calculate the forces. Boundary layers were added near the hull with the dimensionless wall distance y^+ being about 60. The converged solution of a towed case was used as an initial condition to reduce the influence of the impulsive start.

The channel dimensions and testing conditions were designed as in Table 1. All tests were performed with the channel bottom width 1.44 m. Two pier distances, two draughts, three water depths and several velocities were tested during the simulations and experiments. The maximum Froude number was 0.738, which means that all our tests were in sub-critical conditions ($Fr_h < 1$). The initial position of the convoy was placed 6.5 m away from the pier to guarantee the fully-developed wakes generated by the vessel. The vessel took about 16~18 s to finish the whole passing process in the current domain. The dimensions of the channel and pier resemble real conditions during inland waterway transport. The mesh was generated using the snappyHexMesh utility in OpenFOAM. The mesh numbers can be found in Table 1.

Verification and Validation (V & V) studies were carried out in advance using the standard hull forms KCS and KVLCC2. A grid convergence study with three mesh resolutions and a time step convergence study with four timesteps were performed to select appropriate mesh generation strategy and time steps. The resistance was also validated with the ITTC benchmarks. Details of the grid and the time step independence study can be found in our earlier work [34]. The accuracy of the simulation can thus be guaranteed. A fixed time step of 0.001 s was selected in this work.

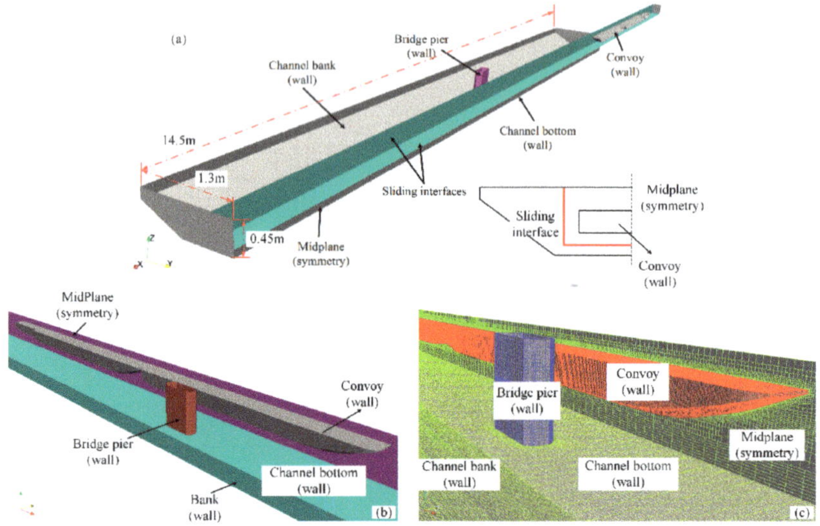

Figure 3. (**a**) Computational domain, (**b**) zoom view and (**c**) meshes near the bridge pier. The boundary conditions are denoted next to the boundary names.

Table 1. Test configurations and mesh numbers of the convoy passing the bridge piers. The units of the channel bottom width W_b, distance between bridge piers D_b, draught T_d and water depth h_w are (m). The unit of the maximum velocity V_{max} is (m/s). Fr_h^{max} is the maximum depth-Froude number in each case, corresponding with V_{max}. N_{mesh} is the mesh number.

Case	W_b	D_b	T_d	h_w	V_{max}	Fr_h^{max}	N_{mesh}
1	1.44	0.7	0.04	0.12	0.80	0.738	1,284,531
2				0.18	0.91	0.685	1,423,626
3				0.24	0.91	0.593	1,426,806
4			0.1	0.18	0.80	0.602	1,340,223
5				0.24	0.91	0.593	1,437,711
6		0.8	0.04	0.12	0.80	0.738	1,278,281
7				0.18	0.91	0.685	1,419,086
8				0.24	0.91	0.593	1,534,598
9			0.1	0.18	0.80	0.602	1,331,889
10				0.24	0.91	0.593	1,471,276

During the solution of the RANS equations, the time derivative term was discretized using the implicit Euler scheme. The convection terms in the momentum and VOF equations were discretized with the second-order linear upwind scheme and the second-order TVD (total variation diminishing) scheme with van Leer's flux limiter. The diffusion term was discretized by a second order central differencing scheme with non-orthogonal correction. Five outer corrections and two pressure corrections were employed for the velocity–pressure coupling. All equations were solved to the tolerance of $\mathcal{O}(-8)$ [35,36]. Each case used 52 processors and takes about 12 CPU hours to finish. The actual physical time was about the same level.

To characterize the process of the convoy passing bridge piers, a parameter ξ_{sb} similar to that of ship–ship interaction is defined in this study.

$$\xi_{sb} = x_{sb}/L_m, \tag{14}$$

where ξ_{sb} is the longitudinal distance between the convoy and the bridge pier. $L_m = (L_s + L_b)/2$ is the average length of the convoy and the pier. The unsteady forces and wave changes can then be characterized as functions of the relative position between the amidships of the convoy and the pier. Figure 4 demonstrates the three most important positions of the passing process. When $\xi_{sb} = -1$, the convoy meets the piers. When $\xi_{sb} = 0$, the centers of the convoy and the pier are located at the same longitudinal position. When $\xi_{sb} = 1$, the passing ends. The positions within $-1 < \xi_{sb} < 1$ are important for the transient properties of the convoy.

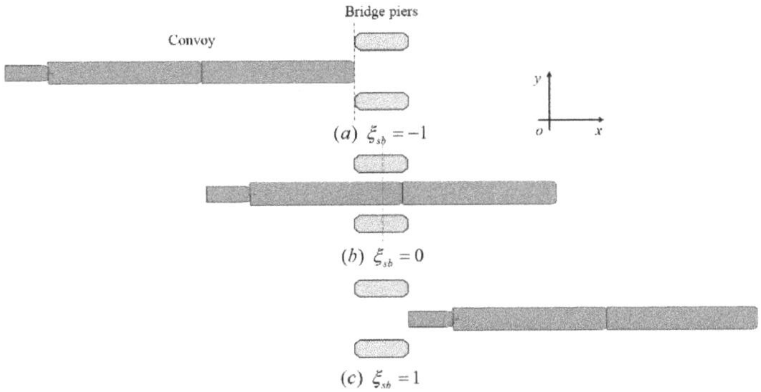

Figure 4. Relative position (ξ_{sb}) between the convoy and bridge piers.

3. Results and Discussions

3.1. Advancing Resistance during the Convoy Passing Bridge Piers

The transient effects while the convoy is passing bridge piers are significant since they may change the unsteady hydrodynamics and maneuverability during the crossing. The instantaneous resistance coefficient ($C_x = \frac{R_x}{\frac{1}{2}\rho v^2 BT}$, B and T are the width and draught of the vessel) is plotted against the relative position (ξ_{sb}) in Figure 5 as a function of the pier distance, with two smaller windows showing a shorter period of time. To concentrate on the transient process, the average value is removed from the resistance directly obtained from the simulations. With a smaller pier distance, the oscillation of the resistance is stronger since the space restriction is increased. The force oscillation mainly lies in the range of $-1 < \xi_{sb} < 1$. The resistance changes intensively near the position $\xi_{sb} = 0$ where the centers of the convoy and the piers coincide during the passing process. This means that the convoy experiences the strongest time-varying interactions with the piers at this position, to which attention should be paid during real maneuvering. Similar trends can be found in Figure 6 as a function of the water depth. The oscillation is higher with a smaller water depth. However, slight unsteady effects can also be observed when $\xi_{sb} < -1$ and $\xi_{sb} > 1$. This can be caused by the disturbances introduced by the evolution of the system, because the initial conditions may not be realistic at the beginning.

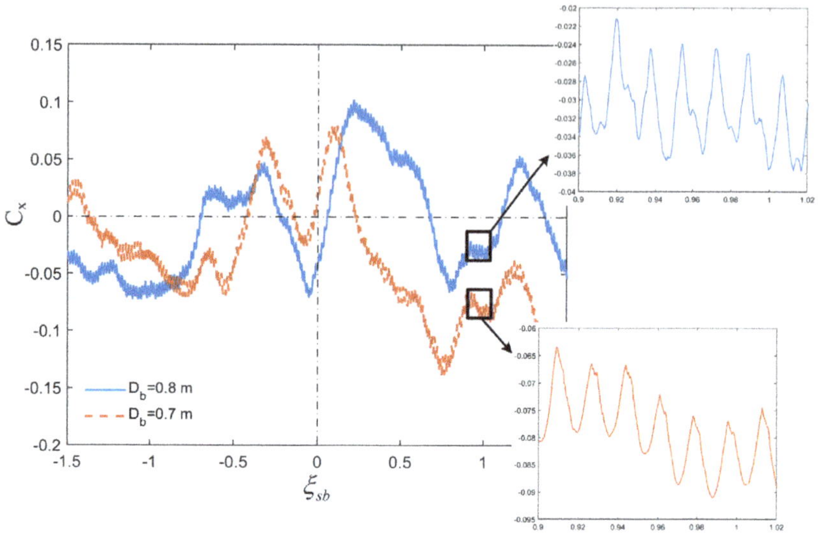

Figure 5. Resistance coefficient (C_x) with the relative position (ξ_{sb}) under different pier distances (D_b). The velocity, draught and water depth of the convoy are 0.91 m/s, 0.04 m and 0.24 m, respectively.

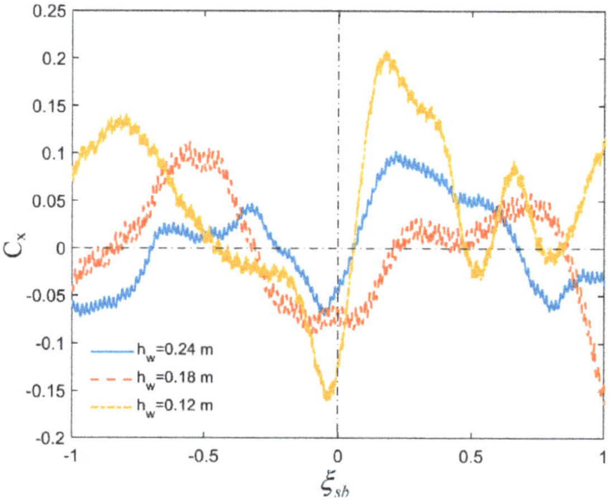

Figure 6. Resistance coefficient (C_x) with the relative position (ξ_{sb}) under different water depths (h_w). The velocity and draught of the convoy are 0.91 m/s and 0.04 m, respectively. The pier distance is 0.8 m.

The average resistance during the passing process is measured in the experiment. In Figure 7a, the simulated values are compared with the experiments. The direction of the convoy advancement is defined as positive, which is the reason why the resistance values are all negative in the results. The simulations agree with the experimental data, but the resistance was underestimated. The largest error between the two was about 7%. This means that our simulations can predict the unsteady process with acceptable accuracy.

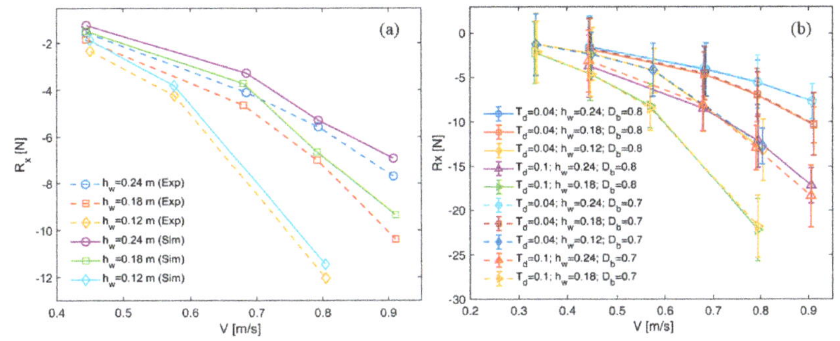

Figure 7. (**a**) Comparison between experimental and numerical resistances of the convoy passing the bridge piers; (**b**) Advances vs. water depths, pier distances, draughts and velocities. R_x, h_w, V, D_b, T_d are the resistance, water depth, convoy speed, distance between the bridge piers and the draught, respectively. The units of h_w, D_b, T_d are (m). The error bar denotes the error between the experimental and simulation results.

All the advancing resistances with various water depths, pier distances, draughts and velocities are demonstrated in Figure 7b. Clearly, the resistance increases with the convoy velocity and draught. With the decrease of the water depth, the advancing resistance increases because of the enhancement of the confinement level, which makes the convoy consume more fuel and increases the difficulty of ship maneuvering. When the water depth is too small, the possibility of grounding also increases. This relates to the squat effect, a combination of trim and sinkage, which will be analyzed later. Similar to the water depth, the resistance is reckoned to increase with the decrease of the pier distance, which also increases the confinement of the waterway. However, this is not evident in the figure. The distance between the bridge piers does not seem to influence the average resistance of the convoy. The bridge piers do increase the confinement, however, because of their small size, their influences are temporary and instantaneous. The unsteady effect of the passing process will counteract it. Therefore, the appearance of structures near the convoy only influences transient properties, but not averaged ones, which means that the overall fuel consumption is not affected by the passing process.

3.2. Trim and Sinkage Analysis

Squat is caused by the overall reduction in under keel clearance (UKC) forward or aft between a convoy at rest and underway due to the accelerated flow around the moving body [37–40]. It greatly influences resistance and maneuverability. In shallow water, a large squat can lead to grounding, posing potential dangers to the ship navigation. Squat is actually the combination of the trim and sinkage of the convoy. As the resistance, averaged trim and sinkage values are plotted against the water depth h_w, distance between bridge piers D_b, draught T_d and velocity V. Their definitions can be observed in Figures 8 and 9. The trim is defined as the pitch angle underway. It can be seen that the trim increases with the augmentation of the draught and velocity, and the reduction of the water depth. The same trend can be observed for the sinkage. However, the distance between the bridge piers does not influence the averaged properties. The trim and sinkage are caused by the return current around the convoy. When the space is restricted, the flow there will be accelerated, which leads to pressure reduction. Thus, an attracting force to the nearest wall is generated, causing the so-called sinkage. Because of the flow direction and the hull form, the motions of the bow and stern are different, a pitching moment is also generated; the trim then appears.

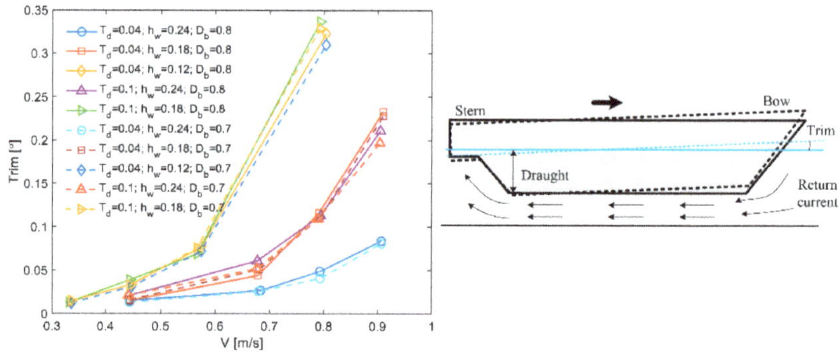

Figure 8. Trim as functions of the water depth h_w, distance between bridge piers D_b, draught T_d, ship velocity V.

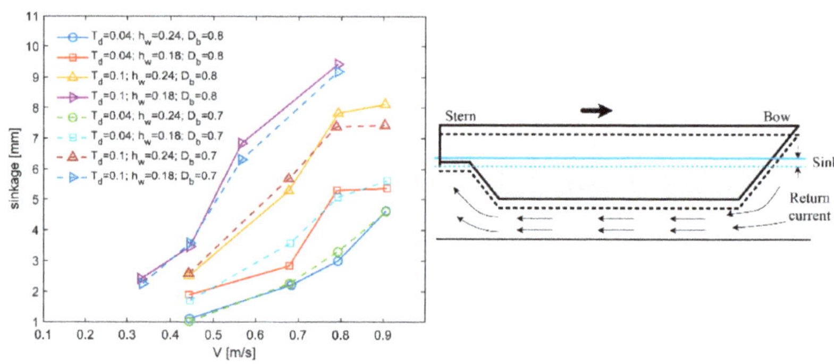

Figure 9. Sinkage as functions of the water depth h_w, distance between bridge piers D_b, draught T_d and ship velocity V.

3.3. Ship-Generated Waves Influenced by Bridge Piers

The appearance of the bridge piers will inevitably influence the wave pattern generated by the convoy. The ship waves with different speeds are shown in Figure 10, and were well captured by our simulations. Two important positions with $\xi_{sb} = 0$ and $\xi_{sb} = 1$ are demonstrated. The traditional Kelvin wave pattern can be clearly observed [41]. The waves generated by the bows of the two barges and the pusher are the most evident. Because of the bank, the waves are reflected and then superposes with the original ones, creating a very complicated wave pattern. With a higher velocity, the wave change becomes more intensive, which can be observed by the isolines in the figure. It is clear that the wave fields are disturbed by the bridge piers. To quantitatively characterize the wave properties, the wave profiles at the lateral position $y/L = 0.037$ (the center between the convoy and the bridge pier) are extracted for comparison in Figure 11. The overall ship waves are lower than the initial water level. The wave crests originate from the bows and sterns. A larger speed makes the wave profile higher in front of the convoy and lower in the middle. The wave change is more violent with a higher convoy velocity. At the position of the pier, two wave crests were observed. The waves there are zoomed in Figure 11b. It can be seen that this is caused by the wave reflection of the pier, which further superimposed with the original wave pattern. Because of the restricted space, the wave change becomes more intensive.

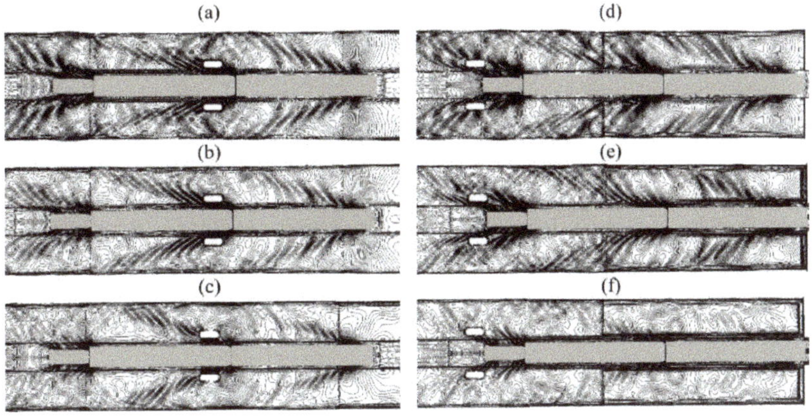

Figure 10. Ship-generated waves with different speeds. $\xi_{sb} = 0$ for (**a**–**c**). $\xi_{sb} = 1$ for (**d**–**f**). $V = 0.907$ m/s for (**a**,**d**). $V = 0.793$ m/s for (**b**,**e**). $V = 0.684$ m/s for (**c**,**f**). The draught, water depth and pier distance are 0.04, 0.24 and 0.8 m, respectively.

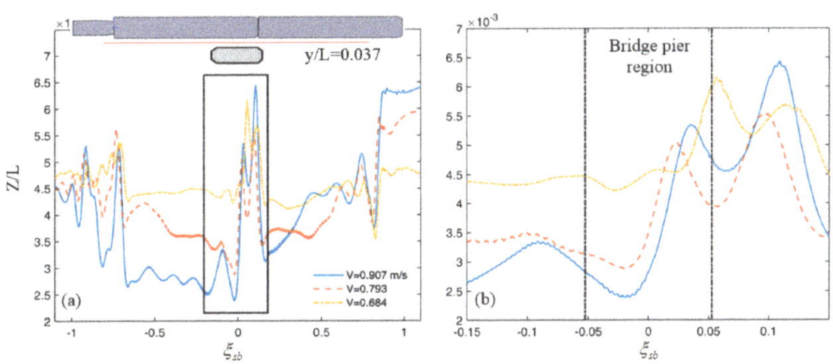

Figure 11. Wave profiles at the lateral position $y/L = 0.037$ (the center between the convoy and the bridge pier) with different speeds. L is the ship length. The red line denotes the sampling position of the wave. (**b**) is the zoomed view of the box in (**a**). The draught, water depth and pier distance are 0.04, 0.24 and 0.8 m, respectively. The dotted lines in (**b**) represent the position of the pier.

Figure 12 demonstrates the influence of water depth. The wave contours are different for the two water depths, especially near the pusher. The waves created by the two barges finally superimpose with the pusher waves and form a very complex wave pattern. The influence becomes clearer in the extracted wave profiles in Figure 13. The positions of the wave crests remain the same. With a smaller water depth and pier distance, the wave elevation becomes higher because of the space restriction. The water level is more easily influenced by the channel bottom when the water depth is lower, creating a more deformed wave pattern.

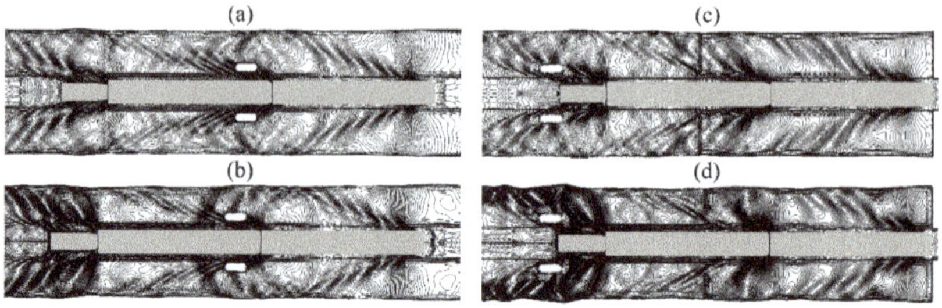

Figure 12. Ship-generated waves with different water depths. $\xi_{sb} = 0$ for (**a**,**b**). $\xi_{sb} = 1$ for (**c**,**d**). $h_w = 0.24\,m$ for (**a**,**c**). $h_w = 0.18\,m$ for (**b**,**d**). The convoy speed, draught and pier distance are 0.91 m/s, 0.04 m and 0.8 m, respectively.

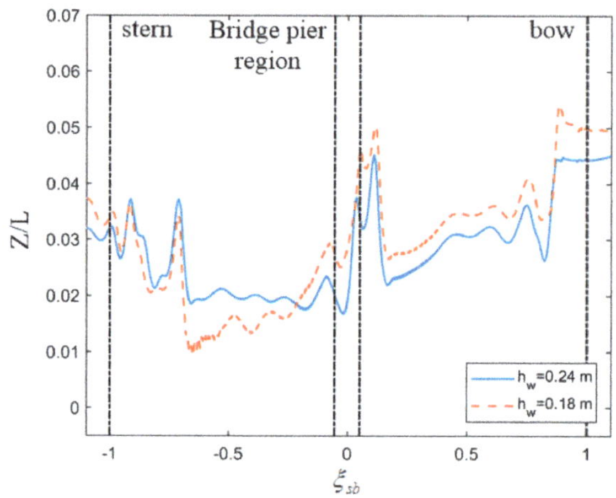

Figure 13. Ship wave profiles at the lateral position $y/L = 0.037$ (the center between the convoy and the bridge pier) with different water depths. L is the ship length. The water levels are adjusted using the initial value for comparison. The convoy speed, draught and pier distance are 0.91 m/s, 0.04 m and 0.8 m, respectively. The dotted lines denote the positions of the convoy and pier.

The influence of the pier distance can be found in Figure 14. Since the pier distances are small, no significant differences can be observed. However, they are evident enough in the extracted profiles (Figure 15). The pier will create a more confined condition beside the channel, and the water level will be elevated. A smaller pier distance will create a more confined condition, which will make the water level higher. Nevertheless, the pier distance will not influence the overall wave pattern because the wave crests and troughs are at the same positions. This is also the reason why no significant differences can be seen in Figure 14.

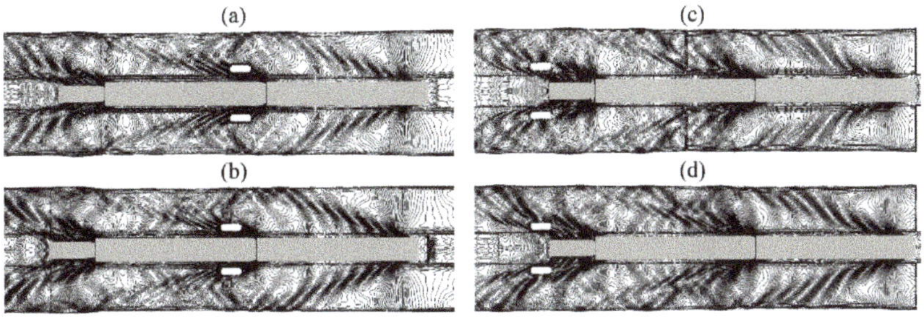

Figure 14. Ship-generated waves with different distances of the bridge piers. $\xi_{sb} = 0$ for (**a**,**b**). $\xi_{sb} = 1$ for (**c**,**d**). $D_b = 0.8\,m$ for (**a**,**c**). $D_b = 0.7\,m$ for (**b**,**d**). The convoy speed, draught and water depth are 0.91 m/s, 0.04 m and 0.24 m, respectively.

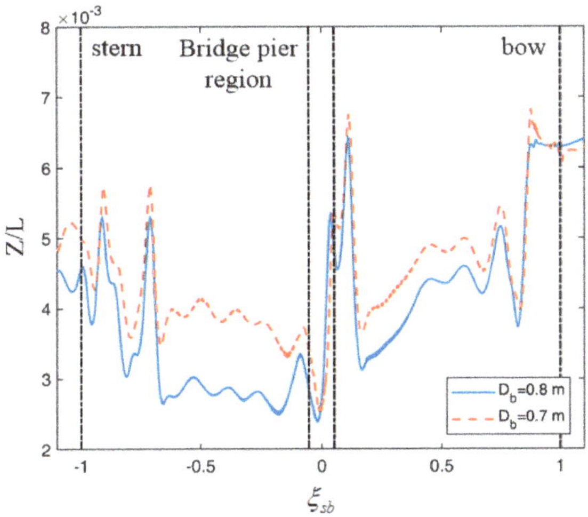

Figure 15. Ship wave profiles at the lateral position $y/L = 0.037$ (the center between the convoy and the bridge pier) with different pier distances. L is the ship length. The convoy speed, draught and water depth are 0.91 m/s, 0.04 m and 0.24 m, respectively. The dotted lines denote the convoy and pier positions.

4. Conclusions

During inland shipping, vessels will inevitably pass bridges, where the bridge piers exert stronger confinement apart from the channel banks. The ship dynamics and maneuverability change during this process. In this study, the hydrodynamics including the advancing resistance, trim, sinkage and ship-generated waves of the inland convoy passing the bridge piers in a confined waterway were investigated using the CFD and experimental approaches. The RANS equations were solved during the numerical simulations, using the sliding mesh technique for the motion of the convoy. The experiments were carried out in the towing tank of the ANAST laboratory in the University of Liège.

The averaged and transient resistances were both analyzed as functions of water depth, pier distance, draught and velocity. The averaged values agree well between simulations and experiments. The averaged resistance was found to increase with the increase of the draught and velocity and the decrease of the water depth. The influence of the bridge piers on the mean properties is not evident, since the averaged resistance does not change with the distance between the piers according to our results. In reality, this means that the fuel consumption remains the same with or without the bridge piers. However, the appearance of the piers has great influence on the transient properties of the convoy. A parameter of the relative position between the convoy and the pier was defined to characterize the crossing process. The fluctuation of the resistance during the procedure clearly increased when the space restriction of the piers increased.

The increase of the squat may lead to grounding. To characterize the squat in a confined waterway, the trim and sinkage were first analyzed as functions of the water depth, pier distance, draught and velocity. The averaged characteristics were emphasized here. Similar conclusions with the resistance can be drawn for the averaged squat.

The ship-generated waves were successfully captured by our simulations. Wave crests were found to originate from the bow and stern. In a confined waterway where both sides and the bottom are restricted, the ship waves are reflected at these positions, and further superpose with the original ones, creating a very complex wave pattern in the waterway. When influenced by the bridge piers, the waves will reflect once more, leading to the appearance of two wave crests near the pier.

As a result, the averaged and transient parameters were analyzed for a convoy passing bridge piers. The wave pattern near the bridge pier was analyzed to characterize the flow fields. These data provide insights into this phenomenon and can be used for their further predictions.

Author Contributions: Conceptualization, P.D. and A.O.; methodology, P.D.; software, P.D.; validation, P.D.; formal analysis, P.D., A.O., Y.H. and H.H.; investigation, P.D., A.O. and H.H.; resources, P.S.; data curation, P.D.; writing—original draft preparation, P.D.; writing—review and editing, P.D., Y.H. and H.H.; visualization, P.D.; supervision, A.O. and P.S.; project administration, A.O. and P.S.; funding acquisition, P.D. and A.O. All authors have read and agreed to the published version of the manuscript.

Funding: This research was funded by the Guangdong Basic and Applied Basic Research Foundation (Grant No. 2019A1515110863), Fundamental Research Funds for the Central Universities (Grant No. 3102020HHZY030004, 3102019JC006), National Natural Science Foundation of China (Grant No. 51979226, 52171324), Natural Science Basic Research Program of Shaanxi (Program No.2020JC-18) and Shaanxi Provincial Key R&D Program (2021KW-38).

Conflicts of Interest: The authors declare no conflict of interest.

References

1. Xie, N.; Iglesias, G.; Hann, M.; Pemberton, R.; Greaves, D. Experimental study of wave loads on a small vehicle in close proximity to a large vessel. *Appl. Ocean Res.* **2019**, *83*, 77–87. [CrossRef]
2. Xu, H.F.; Zou, L.; Zou, Z.J.; Yuan, Z.M. Numerical study on hydrodynamic interaction between two tankers in shallow water based on high-order panel method. *Eur. J. Mech. B Fluids* **2019**, *74*, 139–151. [CrossRef]
3. Gourlay, T. Sinkage and trim of two ships passing each other on parallel courses. *Ocean Eng.* **2009**, *36*, 1119–1127. [CrossRef]
4. Wang, J.; Zou, L.; Wan, D. Numerical simulations of zigzag maneuver of free running ship in waves by rans-overset grid method. *Ocean Eng.* **2018**, *162*, 55–79. [CrossRef]
5. Mousaviraad, S.M.; Sadat-Hosseini, S.H.; Carrica, P.M.; Stern, F. Ship–Ship interactions in calm water and waves. Part 2: URANS validation in replenishment and overtaking conditions. *Ocean Eng.* **2016**, *111*, 627–638. [CrossRef]
6. Wang, H.Z.; Zou, Z.J. Numerical study on hydrodynamic interaction between a berthed ship and a ship passing through a lock. *Ocean Eng.* **2014**, *88*, 409–425. [CrossRef]
7. Wuttrich, R.; Wekezer, J.; Yazdani, N.; Wilson, C. Performance Evaluation of Existing Bridge Fenders for Ship Impact. *J. Perform. Constr. Facil.* **2001**, *15*, 17–23. [CrossRef]
8. Wang, L.; Yang, L.; Tang, C.; Zhang, Z.; Chen, G.; Lu, Z. On the Impact Force and Energy Transformation in Ship-Bridge Collisions. *Int. J. Prot. Struct.* **2012**, *3*, 105–120. [CrossRef]
9. Wang, L.; Yang, L.; Huang, D.; Zhang, Z.; Chen, G. An impact dynamics analysis on a new crashworthy device against ship-bridge collision. *Int. J. Impact Eng.* **2008**, *35*, 895–904. [CrossRef]
10. Svensson, H. Protection of bridge piers against ship collision. *Steel Constr.* **2009**, *2*, 21–32. [CrossRef]
11. Zhi-Qiang, H.; Yong-Ning, G.; Zhen, G.; Ya-Ning, L. Fast evaluation of ship-bridge collision force based on nonlinear numerical simulation. *J. Mar. Sci. Appl.* **2005**, *4*, 8–14. [CrossRef]
12. Fan, W.; Liu, Y.; Liu, B.; Guo, W. Dynamic Ship-Impact Load on Bridge Structures Emphasizing Shock Spectrum Approximation. *J. Bridg. Eng.* **2016**, *21*, 04016057. [CrossRef]
13. Proske, D.; Curbach, M. Risk to historical bridges due to ship impact on German inland waterways. *Reliab. Eng. Syst. Saf.* **2005**, *90*, 261–270. [CrossRef]

14. Chu, L.M.; Zhang, L.M. Centrifuge Modeling of Ship Impact Loads on Bridge Pile Foundations. *J. Geotech. Geoenvironmental Eng.* **2011**, *137*, 405–420. [CrossRef]
15. Xie, Z.; Zhang, Y.; Zhou, J.; Zhu, W. Theoretical and experimental research on the micro interface lubrication regime of water lubricated bearing. *Mech. Syst. Signal Process.* **2021**, *151*, 107422. [CrossRef]
16. Xie, Z.; Zhu, W. An investigation on the lubrication characteristics of floating ring bearing with consideration of multi-coupling factors. *Mech. Syst. Signal Process.* **2022**, *162*, 108086. [CrossRef]
17. Li, L.; Yuan, Z.-M.; Ji, C.; Li, M.-X.; Gao, Y. Investigation on the unsteady hydrodynamic loads of ship passing by bridge piers by a 3-D boundary element method. *Eng. Anal. Bound. Elem.* **2018**, *94*, 122–133. [CrossRef]
18. Zhang, X.; Teng, B.; Liu, Z.; Zhang, L.W. Unsteady computation of hydrodynamic interaction forces and moments between ship hull and pier based on NURBS. *J. Ship Mech.* **2003**, *7*, 47–53.
19. Li, Z.; Du, P.; Ouahsine, A.; Hu, H. Ship Hydrodynamics of Several Typical Scenes During Inland Waterway Transport. *IOP Conf. Ser. Earth Environ. Sci.* **2021**, *697*, 012003. [CrossRef]
20. Guo, J.; Ai, W.Z.; Wang, J.H. The minimum distance between navigation ship and bridge. In *Design, Manufacturing and Mechatronics: Proceedings of the 2015 International Conference on Design, Manufacturing and Mechatronics (ICDMM2015)*; World Scientific: Singapore, 2016; pp. 1447–1453.
21. Khangaonkar, T.; Nugraha, A.; Wang, T. Hydrodynamic zone of influence due to a floating structure in a Fjordal Estuary-Hood Canal Bridge Impact Assessment. *J. Mar. Sci. Eng.* **2018**, *6*, 119. [CrossRef]
22. Tan, Z.; Gan, L. Risk Assessment of Water Level Effect on Ship-Bridge Collision. In Proceedings of the 2008 International Symposium on Safety Science and Technology, Shanghai, China, 6–9 August 2008; pp. 290–293.
23. Zhang, C.X.; Zou, Z.J.; Wang, H.M. Numerical prediction of the unsteady hydrodynamic interaction between a ship and a bridge pier. *Chin. J. Hydrodynam.* **2012**, *27*, 359–364.
24. Xiang, X.; Eidem, M.E.; Sekse, J.H.; Minoretti, A. Hydrodynamic loads on a submerged floating tube bridge induced by a passing ship or two ships in maneuver in calm water. In *International Conference on Offshore Mechanics and Arctic Engineering*; American Society of Mechanical Engineers: New York, NY, USA, 2016; Volume 49989, p. V007T06A047.
25. Zheng, G.D. Cross-sectional Curvature Method in Bridge Ship Impact Resistance Force Calculation. In Proceedings of the International Conference on Mechanics, Building Material and Civil Engineering (MBMCE), Guilin, China, 15–16 August 2015; pp. 893–898.
26. Rusche, H. Computational Fluid Dynamics of Dispersed Two-Phase Flows at High Phase Fractions. Ph.D. Thesis, Imperial College London (University of London), London, UK, 2003.
27. Jasak, H. Error Analysis and Estimation for the Finite Volume Method with Applications to Fluid Flows. Ph.D. Thesis, Imperial College London (University of London), London, UK, 1996.
28. Pereira, F.; Eça, L.; Vaz, G. Verification and Validation exercises for the flow around the KVLCC2 tanker at model and full-scale Reynolds numbers. *Ocean Eng.* **2017**, *129*, 133–148. [CrossRef]
29. Beaudoin, M.; Jasak, H. Development of a generalized grid interface for turbomachinery simulations with OpenFOAM. In Proceedings of the Open Source CFD International Conference, Berlin, Germany, 4 December 2008; Volume 2.
30. Darwish, M.; Geahchan, W.; Moukalled, F. Fully implicit method for coupling multiblock meshes with nonmatching interface grids. *Numer. Heat Transf. B Fundam.* **2017**, *71*, 109–132. [CrossRef]
31. Farrell, P.E.; Maddison, J.R. Conservative interpolation between volume meshes by local Galerkin projection. *Comput. Methods Appl. Mech. Eng.* **2011**, *200*, 89–100. [CrossRef]
32. Linde, F.; Ouahsine, A.; Huybrechts, N.; Sergent, P. Three-dimensional numerical simulation of ship resistance in restricted waterways: Effect of ship sinkage and channel restriction. *J. Waterw. Port Coast. Ocean Eng.* **2017**, *143*, 06016003. [CrossRef]
33. Kaidi, S.; Smaoui, H.; Sergent, P. Numerical estimation of bank-propeller-hull interaction effect on ship manoeuvring using CFD method. *J. Hydrodyn. B* **2017**, *29*, 154–167. [CrossRef]
34. Du, P.; Ouahsine, A.; Hoarau, Y. Solid body motion prediction using a unit quaternion-based solver with actuator disk. *Comptes Rendus Mécanique* **2018**, *346*, 1136–1152. [CrossRef]
35. Jiao, J.; Huang, S.; Soares, C.G. Numerical investigation of ship motions in cross waves using CFD. *Ocean Eng.* **2021**, *223*, 108711. [CrossRef]
36. Jiao, J.; Huang, S. CFD simulation of ship seakeeping performance and slamming loads in bi-directional cross wave. *J. Mar. Sci. Eng.* **2020**, *8*, 312. [CrossRef]
37. Yuan, Z.M.; Zhang, X.; Ji, C.Y.; Jia, L.; Wang, H.; Incecik, A. Side wall effects on ship model testing in a towing tank. *Ocean Eng.* **2018**, *147*, 447–457. [CrossRef]
38. Pompée, P.J. About modelling inland vessels resistance and propulsion and interaction vessel-waterway key parameters driving restricted/shallow water effects. In Proceedings of the Smart Rivers 2015, Buenos Aires, Argentina, 7–11 September 2015.
39. Barrass, C.B. The phenomena of ship squat. *Int. Shipbuild. Prog.* **1979**, *26*, 44. [CrossRef]
40. Gourlay, T.P. A brief history of mathematical ship-squat prediction, focusing on the contributions of EO Tuck. *J. Eng. Math.* **2011**, *70*, 5–16. [CrossRef]
41. Wang, J.; Wan, D. Cfd study of ship stopping maneuver by overset grid technique. *Ocean Eng.* **2020**, *197*, 106895. [CrossRef]

Journal of
Marine Science and Engineering

MDPI

Article

Dynamic Interactions of a Cable-Laying Vessel with a Submarine Cable during Its Landing Process

Jianxun Kuang [1], Guodong Chen [1], Zhoulong Yuan [1], Xiajun Qi [1], Qianhao Yu [1] and Zhen Liu [2,*]

[1] Zhejiang Qiming Electric Power Engineering Co., Ltd., Zhoushan 316000, China; zsxm2017@126.com (J.K.); chenguodong662702@126.com (G.C.); 13868209580@163.com (Z.Y.); xiajun_7@163.com (X.Q.); yuqianhao665219@126.com (Q.Y.)
[2] Department of Ocean Engineering, College of Engineering, Ocean University of China, Qingdao 266100, China
* Correspondence: liuzhen@ouc.edu.cn

Abstract: The rapid development of offshore electricity grid construction has led to a great demand for submarine cable deployment. In this study, a numerical model is established based on the commercial software ANSYS-AQWA to investigate the dynamic interactions between a cable-laying vessel and a submarine cable during its landing process, which has not yet been reported and is critical to the safety of the cable. The numerical model was validated by an experimental test on the mooring stability of a vessel conducted in a wave tank. The effects of the cable length, the current velocity, the incident wave, and the wind direction on vessel stability and the tensions in the mooring lines and cable were investigated. When the cable length is short, the submarine cable acts as a mooring cable that can stabilize the hull, but it is not safe to apply force to the submarine cable. At the same time, an increase in the current speed also increases the tensile force of the submarine cable. The influence of different incident wave directions and wind directions on the stability and tension of ships in mooring lines and cables was studied, and the most unfavorable environmental conditions for submarine cable laying were determined under different environmental conditions.

Keywords: cable-laying vessel; submarine cable; dynamic interactions; numerical simulation

Citation: Kuang, J.; Chen, G.; Yuan, Z.; Qi, X.; Yu, Q.; Liu, Z. Dynamic Interactions of a Cable-Laying Vessel with a Submarine Cable during Its Landing Process. *J. Mar. Sci. Eng.* **2022**, *10*, 774. https://doi.org/10.3390/jmse10060774

Academic Editor: Spyros A. Mavrakos

Received: 27 April 2022
Accepted: 31 May 2022
Published: 3 June 2022

1. Introduction

The rapid development of offshore resources, including islands, underwater minerals, wind and marine energies, and aquatic products, necessitates more human activities that involve electricity supply and transmission. Electricity is mainly transmitted by submarine cables in offshore circumstances. The offshore power grid is the same as the onshore grid, which needs more submarine cables to be laid to interconnect the networks [1,2]. Consequently, laying submarine cables safely has become an important mission for offshore electricity transmission engineering.

Landing cables is the most difficult part of the entire process of cable laying. In addition, a critical issue to prevent cable damage is to control the tension in the cable during laying [3,4]. Compared to the traditional landing methodology, floaters could be employed to provide additional buoyancy to the floating cable and prevent possible overloaded friction and bending stress concentration [5]. On the other hand, a reasonable design and deployment of the mooring lines could enhance the stability of a cable-laying vessel in complicated sea states and reduce the tension in the cable from the vessel motion. Therefore, the dynamic interactions between the cable and floaters, vessel, and mooring lines play a critical role in the safety and reliability of cable-laying engineering.

For the stability of the ships and offshore platforms, due to the large structures and multiple external and internal loads, numerical simulations have become a popular methodology to study the effects of environmental forces and structural parameters on the motion of ships and platforms. Hu et al. investigated the kinematic response of a liquified natural gas ship under the joint effects of waves, streams, and winds [6]. Shigunov et al. reported the dynamic stability of an offshore service vessel under various operating conditions [7]. Liu et al. preliminarily estimated the intensity response of a deep-water ship in still water and under regular wave conditions [8]. Putra et al. evaluated the tilting stability of a flat shell ship under different environmental conditions [9]. Roy et al. conducted an integrated numerical analysis on the motion and structural responses of an offshore spar platform to irregular waves [10]. Banik et al. investigated the effects of incident wave direction on the dynamic responses of a spar offshore platform [11]. Wang et al. proposed a motion equation and used it to study the influences of wave group parameters on the ship motions in random wave groups [12]. Su et al. investigated the seakeeping performance of a variable-structure ship under regular and irregular waves [13].

The mooring line is also important for the stability of the ships. Liu conducted a dynamic analysis of mooring lines and investigated their damping effects on floating platforms [14]. Sarkar et al. proposed a dynamic stiffness method to handle the kinematic problems of mooring lines [15] and considered seabed friction in a linear analysis under irregular wave conditions [16]. Gao et al. studied the mooring performance of a multi-body floating system under the joint impact of the wind, waves, and waterflow [17]. Nie et al. refined the mooring forces using a time-domain method under environmental forces [18]. Zheng et al. investigated the tensions of the bow–stern mooring system of a single ship under the combination impact of external loads [19]. Pham proposed a computational fluid dynamic model to simulate the tensions of mooring lines under various operating conditions [20]. Based on the experimental and numerical results, Nguyen et al. proposed an empirical formula to calculate the tension of mooring lines for oil ships and studied the effects of waterflow on mooring line tension [21].

For the stability and dynamic positioning of a cable-laying vessel, Fu et al. proposed a self-adapted surface tracking control method [22]. Yang et al. analyzed the dynamic responses of a submarine cable during the motion of a laying vessel and the kinematic characteristics of a cable during its laying [23,24]. Under extremely shallow water conditions, Cavefors et al. studied the motion effects of a cable-laying vessel on the tensions of the mooring lines under various wave conditions [25]. Zhang et al. and Wang et al. analyzed the effects of the vessel velocity, water depth, and length of the cable using models that coupled the vessel with the submarine cable [26,27]. In addition, related scholars carried out research related to cable structure health monitoring [28,29]. Drissi-Habti M et al. simulated the real strain of copper wire in high-voltage electric transport phases using an optical fiber sensor (FOS) [30].

From the literature review, it can be seen that most previous studies individually focused on the vessel stability and tensions of submarine cables, while the dynamic interactions between the moored cable-laying vessel and submarine cables during the cable landing process have not yet been reported. In this study, due to the engineering demands of the Qifan No. 9 vessel, a numerical model was established based on the commercial computational fluid dynamic software Ansys-AQWA and validated by the corresponding experimental data. The effects of the submarine cable length, current velocity, incident wave, and wind direction on vessel stability and tensions in the mooring lines and cables were investigated.

2. Numerical Model

2.1. Governing Equations

In this study, air and water are idealized and assumed to be incompressible, irrotational, and non-viscous. Based on the three-dimensional potential flow theory, the velocity potential ϕ satisfies the Laplace equation in the flow field [31]:

$$\nabla^2 \phi = 0 \tag{1}$$

Introducing the complex form of the velocity potential ϕ:

$$\phi = Re\left(\varphi e^{-i\,\omega t}\right) \tag{2}$$

where ω is the frequency of incident waves and t is the time.

The velocity potential can be divided into three parts: the incident potential ϕ_I, the diffraction potential ϕ_D, and the radiation potential ϕ_R, which all satisfy the Laplace equation. In addition, the incident potential can be written as follows:

$$\phi_I = \frac{A_i g}{\omega} \frac{\cos hk_i(z+d)}{\cosh kd} e^{[ik(x\cos\beta + y\sin\beta)]} \tag{3}$$

where A_i is the incident wave amplitude, k_i is the wave number, h is the water depth, g is the gravitational acceleration, and β is the intersection angle between the incident wave direction and the positive direction of the x-axis.

The governing equation of the diffraction potential; boundary conditions for the free surface, seabed, and rigid body surface; and definite condition at infinity can be written as follows:

$$\nabla^2 \varphi_D = 0 \tag{4}$$

$$\frac{\partial \varphi_D}{\partial z} - \frac{\omega^2}{g} \varphi_D = 0\ (z=0) \tag{5}$$

$$\frac{\partial \varphi_D}{\partial z} = 0\ (z=-d) \tag{6}$$

$$\frac{\partial \varphi_D}{\partial n} = -\frac{\partial \varphi_I}{\partial n} \tag{7}$$

$$\lim_{R\to\infty} \sqrt{R}\left(\frac{\partial \varphi_D}{\partial R} - iK\varphi_D\right) = 0 \tag{8}$$

If six degrees of freedom (DOFs) are considered, the radiation potential can be expressed as follows:

$$\varphi_R = i\omega \sum_{j=1}^{6} \varepsilon_j \varphi_j \tag{9}$$

The governing equation of the radiation potential; the boundary conditions for the free surface, seabed, and rigid body surface; and definite condition at infinity can be written as follows:

$$\nabla^2 \varphi_R = 0 \tag{10}$$

$$\frac{\partial \varphi_R}{\partial z} - \frac{\omega^2}{g} \varphi_R = 0\ (z=0) \tag{11}$$

$$\frac{\partial \varphi_R}{\partial z} = 0\ (z=-d) \tag{12}$$

$$\frac{\partial \varphi_j}{\partial n} = i\omega n_j (j = 1,\ 2,\ \ldots,\ 6) \tag{13}$$

$$\lim_{R\to\infty} \sqrt{R}\left(\frac{\partial \phi_R}{\partial R} - ik\phi_R\right) = 0 \tag{14}$$

where ε_j represents the motion at the jth degree of freedom, and φ_j represents the radiation potential caused by the motion at the jth degree of freedom. n is the unit vector, which is perpendicular to the floater boundary and points to the floater.

As the velocity potential and the surface pressure are determined, the wave force acting on the body can be calculated as follows:

$$(m_{ij} + \mu_{ij})\ddot{x}_j + \lambda_{ij}x_j + k_{ij}x = f_i(i = 1,2\ldots 6) \tag{15}$$

where m is the quality matrix, μ is the added mass matrix, λ is the damping coefficient matrix, and k is the restoring force matrix. f_i is the first-order wave force acting on the body.

Following the API codes [32], the wind force F_w and the water current force F_{cs} are estimated as follows:

$$F_w = C_w \sum (C_s C_h A) V_w{}^2 \tag{16}$$

$$F_{cs} = C_{ss} C_d A V_c^2 \tag{17}$$

where C_w is the wind force coefficient, which is commonly defined as 0.615 Ns^2/m^4. C_s and C_h are the ship shape coefficient and ship height coefficient, respectively. A is the projected area of the ship in the direction perpendicular to the incident wind direction, and V_w is the wind velocity. C_{ss} is the water current force coefficient, which is commonly defined as 515.62 Ns^2/m^4. C_d is the drag coefficient of the water flow, which is 0.5 for a cylinder and 1.5 for a flat plate. A is the projected area of the ship in the direction perpendicular to the incident water current, and V_c is the water current velocity.

2.2. Numerical Model Setup

The ship model is established based on the cable-laying vessel Qifan No. 9, designed and owned by Zhejiang Qiming Electric Power Group Co. Ltd. (Zhoushan, China). As shown in Figure 1, it is the first 5000 t professional vessel for submarine cable engineering [33]. The vessel is equipped with an advanced cable tray and a clean room for cable connection on site. The single construction distance for 220 kV voltage level cables can be enhanced from 30.0 km to 60.0 km. The eight-point mooring system enables the vessel to resist force 10 winds.

Figure 1. Qifan No. 9 cable-laying vessel.

The governing equations are solved in the commercial software Ansys-AQWA. The cable-laying vessel is modeled in Design Modeler integrated in Workbench, as shown in Figure 2. As the model structure affects the complexity and quality of the generated grids, and subsequently the computational time and accuracy, the vessel structure is simplified, and the extraction is conducted. The vessel draught is preset with centers of mass and gravity. As the positions of anchor points and fairleads on the vessel are fixed, the material and properties of the mooring lines are set in the model. The floating submarine cable for the landing process is also set in the related modules following the same procedure, as shown in Figure 2a. According to the vessel shape and size, the greatest length of the grid is 1.5 m. The primary grid shape is quadrangular with a total number of 11,371, as shown in Figure 2b.

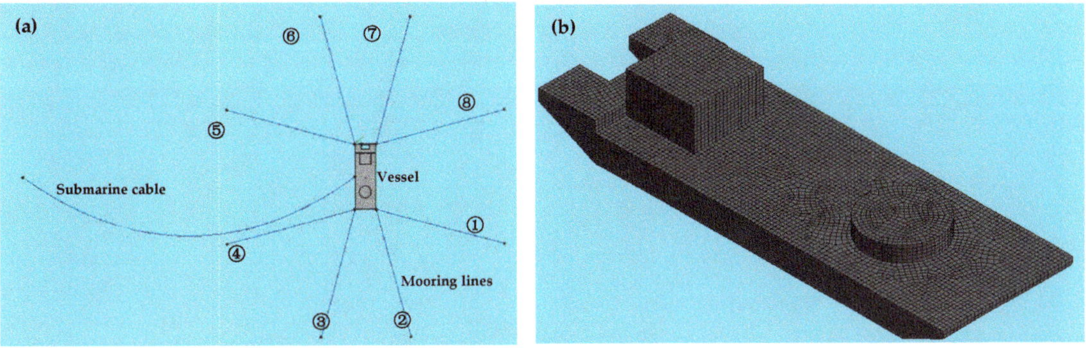

Figure 2. Numerical model for the moored vessel and landing submarine cable. (**a**) Model setup for the vessel, mooring lines ①–⑧, and submarine cable. (**b**) Mesh structures of the vessel.

The simulation modules in the software Ansys-AQWA can be accessed from the Hydrodynamic Responses and Hydrodynamic Diffraction modules in the Analysis Systems in Workbench. The module AQWA-LINE is responsible for the diffraction and radiation calculations, and AQWA-DRIFT is employed for the time-domain analysis of the second-order Morison forces in irregular waves. The AQWA-FER module can be used for the frequency-domain analysis with irregular waves. The nonlinear time-domain simulation of regular and irregular waves can be calculated for the survival conditions in the AQWA-NAUT. The module AQWA-Cable Dynamics can be integrated for simulations where the cables or mooring lines should be considered.

The cable-laying missions conducted by Qifan No. 9 shall employ the floaters to provide additional buoyancy to the landing cables in the future, as shown in Figure 3. Considering the primary function of the floaters, they are simplified as the uniform distributed buoyancy acting on the cable. During the time-domain calculations, the computational time step is 0.5 s with a total computation period of over 20,000 s.

Figure 3. Floaters for submarine cable landing. Adapted from [34].

3. Experimental Validation of the Numerical Model

3.1. Experiment Setup

An experimental test was conducted to validate the numerical model. The experiments were performed in the wave tank of the Shandong Provincial Key Laboratory of Ocean Engineering, Ocean University of China. Considering the vessel size, the tank size and capability, and the environmental conditions, a model scale ratio of 1:49 was used following the Froude similarity law. The model vessel was designed with the parameters listed in Table 1 and manufactured using wood, as shown in Figure 4a. In addition, the ballast of the vessel was carefully adjusted to satisfy the model weight, draught, and displacement. The truncation method was employed for the modeled mooring lines to adapt to the shallow testing water depth in the tank.

Table 1. Parameters for the prototype and model vessels.

Parameters	Prototype Vessel	Model Vessel
Vessel length (m)	99.0	2.02
Molded breadth (m)	32.0	0.65
Molded depth (m)	6.5	0.134
Maximum draft (m)	4.8	0.098
Maximum displacement (t)	14,300	0.122 *
Diameter of mooring lines (mm)	50.0	1.02

* Model weight: 0.069 t; ballast weight: 0.053 t.

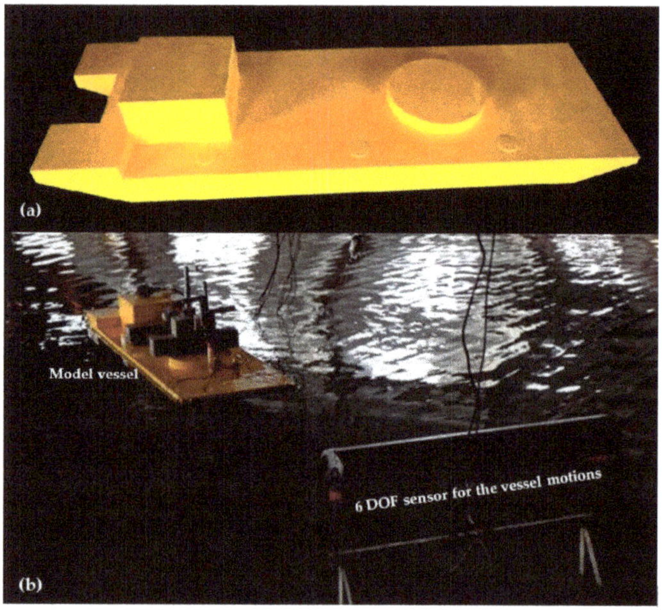

Figure 4. (**a**) Model vessel. (**b**) Experimental setup for validation.

The model vessel was moored following the same distribution pattern shown in Figure 2a. Only typical regular and irregular waves were used as the primary environmental conditions. The incident wave direction was perpendicular to the bow-to-stern of the vessel. The motions of the vessel in six DOFs were measured using an optical motion sensor, as shown in Figure 4b. The tensions in the mooring lines were recorded by the force transducers installed between the mooring lines and the vessel. The regular wave condition included a wave height of $H = 0.05$ m and a wave period of $T = 1.5$ s. Furthermore, the irregular wave condition included a significant wave height of $H_S = 0.05$ m and a significant wave period of $T_S = 1.5$ s. The JONSWAP spectrum was used to generate the irregular wave scenario with an enhanced peak factor of $\gamma = 3.3$.

The kinematic responses in six DOFs and tensions in the mooring lines and submarine cable are nondimensionalized. The nondimensional translational and rotational motions in six DOFs $\overline{\zeta}_i$ and $\overline{\zeta}_j$ can be expressed as follows [35]:

$$\overline{\zeta}_i = \frac{\zeta_i}{H_0}(i = 1,\ 2,\ 3), \overline{\zeta}_j = \frac{\zeta_j}{kH_0}\ (j = 4,\ 5,\ 6) \tag{18}$$

where k and H_0 represent the wave number and incident wave height, respectively. ζ_i and ζ_j are the corresponding dimensional motions. $i = 1, 2, 3$, ζ_i represent the heaving, swaying, and surging motions, respectively. $j = 4, 5, 6$, ζ_j represent the rolling, pitching, and yawing motions, respectively.

The nondimensional tension $\overline{F}$ can be expressed as follows:

$$\overline{F} = \frac{F}{T_M} \tag{19}$$

where F is the tension in the mooring lines or the submarine cable. T_M is the designed breaking loads for the mooring line and submarine cable with values of 950 kN and 190 kN, respectively.

3.2. Validation of Numerical Model

3.2.1. Experimental Case in Regular Waves

An experimental case tested in regular waves is employed for numerical model validation first. The testing conditions include a regular wave height of H_0 = 0.05 m, a regular wave period of T_0 = 1.5 s, and a vessel starboard that is perpendicular to the incident wave direction. The time histories of two typical motions of the vessel model are compared in Figure 5 between the experimental and numerical results. For the heaving amplitude in Figure 5a, the positive peaks in the numerical prediction are slightly larger than those in the experimental test, while the negative peaks are well-correlated. For the swaying motion in Figure 5b, the numerical amplitude also agrees well with experimental results. In addition, the numerical curve shape is more uniform, and the experimental negative peaks shift to the upstream of the time axis.

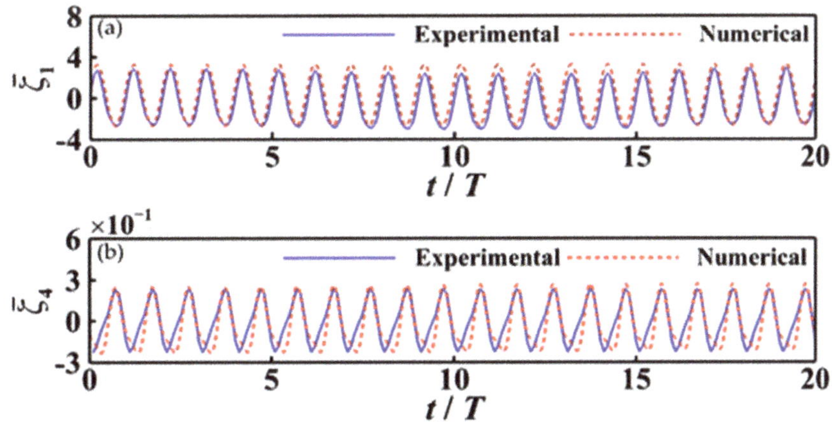

Figure 5. Time history comparisons between numerical and experimental results on two typical motions. (**a**) Heave. (**b**) Roll.

The comparison of nondimensional tensions in mooring lines between numerical and experimental results is shown in Figure 6. In most lines, the numerical model underestimated the tensions because more uncertainty factors influence the experimental testing process and results. The predicted results for four lines in the head waves have larger errors than the experimental data in the tension, while the differences between the four mooring lines on the other side are minor. Compared to the laboratory testing results, the numerical errors fall within a range from 5% to 12%.

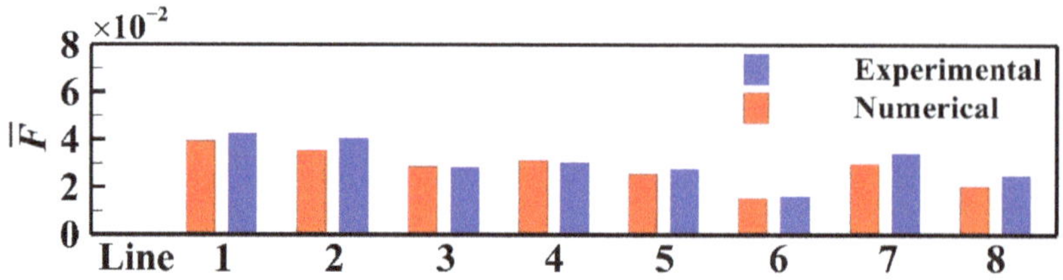

Figure 6. Comparison of average tensions in mooring lines between numerical and experimental results under the regular wave conditions.

3.2.2. Experimental Case in Irregular Waves

For the experimental case in irregular waves, the significant wave height and period are $H_S = 0.041$ m and $T_S = 1.43$ s with an incident direction perpendicular to the vessel hull. The comparison of peak amplitudes in the heaving and rolling motions between numerical and experimental results are shown in Figure 7. The numerical predictions overestimated the maximum amplitudes of the heaving and rolling motions because they were conducted in an idealized environment. Meanwhile, the experimental setup had several restrictions, such as the wave reflection by the end wall, bottom friction, and system errors from the measuring system for the model scale.

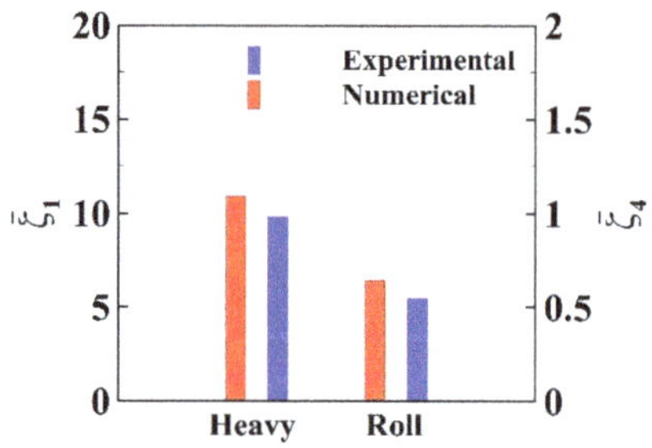

Figure 7. Comparison of peak amplitudes on two typical motions between numerical and experimental results.

The validation results of the numerical model on the tensions in mooring lines are shown in Figure 8. The numerical predictions on the average tensions in most mooring lines are overestimated, except for Line 1. In addition, the differences in the tensions of mooring lines in the head waves between numerical and experimental results are larger than those in the four lines on the other side. The average error of the numerical model for the irregular wave condition is approximately 10.5%, which is significantly larger than that in the regular wave condition because of the difficulties in force recording for the waves with smaller wave heights and periods. Regardless, the numerical model in this study shows reasonable accuracy in the prediction of critical performance indicators and will be employed in further investigations.

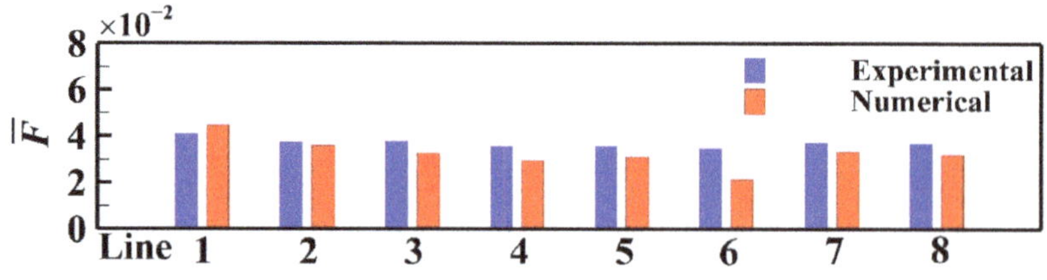

Figure 8. Comparison of average tensions in mooring lines between numerical and experimental results under the irregular wave condition.

4. Operating Performance in Actual Sea Conditions

4.1. Design of Testing Conditions

The numerical simulations for the vessel and submarine cable were all conducted at the prototype scale, which were set in Cartesian coordinates, as shown in Figure 9. The intersection angles of Lines 1 and 2 with the vessel hull are 22.5° and 112.5°, respectively. The other three series of mooring lines follow the same deployment format. The submarine cable is pulled from the center of the shipboard and the distance from the vessel to the anchor point on the coast is defined as L_c. The diameters and elasticity moduli of the mooring line and submarine cable are D_M = 0.05 m, D_C = 0.25 m, $E_M = 6.9 \times 10^3$ MPa, and $E_C = 4.5 \times 10^3$ MPa, respectively.

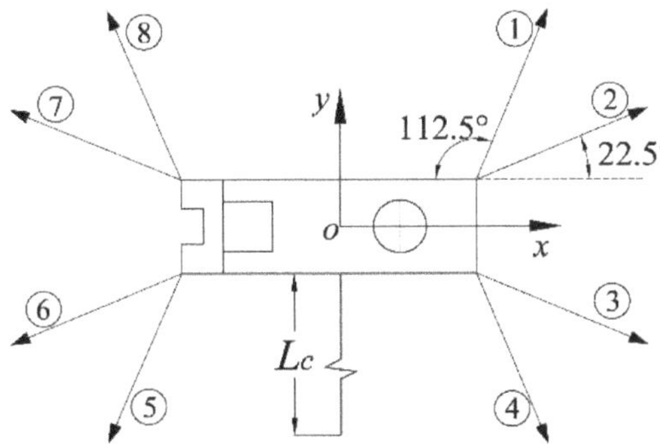

Figure 9. The vessel with mooring lines ①–⑧ and the submarine cable in Cartesian coordinates.

The sea around the Zhoushan Islands is considered the operating area, which has a great demand for submarine cable connections between islands. The water depth is between 5.0 m and 13.0 m with a maximum value of 50.0 m. The current velocity is approximately 4.0 knots, and the wind speed is around 6.0 m. The oceanic environmental factors are designed and listed in Table 2 as typically representative of sea conditions. As the vessel is close to the costal line during the cable-deploying operation, only the longshore current is considered with an incident direction of 180°. There are 16 testing cases in total required for further investigation.

Table 2. Designed oceanic environmental factors.

Factors	Designed Parameters	
Incident waves	H_S = 1.0 m T_S = 5.0 s	Direction [1] w: −45°, −90°, −135°
Current	Velocity (m/s): 1.5, 2.0, 2.5	Direction [1]: 180°
Wind	Wind speed: 6.0 m/s	Direction [1] f: 0, −45°, −90°, −135°, −180°
Length of submarine cable L_C	500.0 m, 520.0 m, 540.0 m, 560.0 m	

[1] Definition of direction: positive for counterclockwise rotation and negative for clockwise rotation from the positive x-axis in Figure 9.

4.2. Effects of the Submarine Cable Length

Based on typical operating conditions, four typical lengths of submarine cable are chosen in this section, as listed in Table 2. The incident current velocity is 2.0 m, and the incident wave and wind directions are w = −90° and f = −45°, respectively.

The effects of the submarine cable length on the peak amplitudes of the vessel motions are shown in Figure 10. As the cable length increases, the nondimensional values of the sway, surge, and roll increase. The peak values of $\overline{\zeta}_2$ and $\overline{\zeta}_3$ both exceed 20.0, while the peak value of $\overline{\zeta}_4$ is close to 1.0. The amplitude of yaw first increases and then converges as the value of L_C increases, while the heaving and pitching motions show little difference.

The decrease in buoyancy caused by the reduction in cable length results in an increase in the mooring function of the cable, and subsequently, the motions of the vessel all decrease. For a longer cable, the stability of the vessel is mainly provided by the mooring lines. The scattered dots on the right side represent motions without a submarine cable, which are larger than those with a cable.

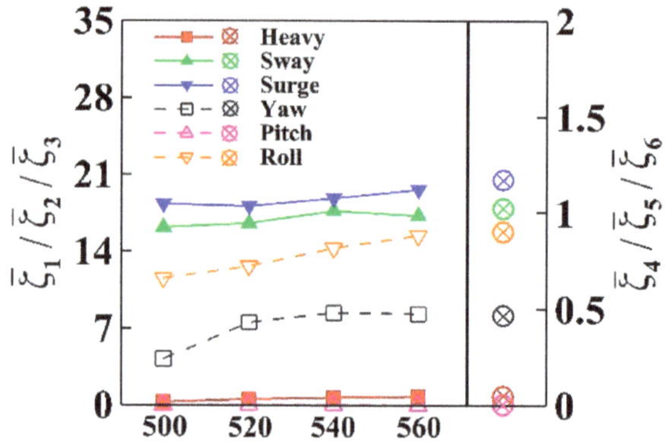

Figure 10. Effects of the submarine cable length on the vessel motions.

The effects of the submarine cable length on the tensions in the mooring lines and the cable are illustrated in Figure 11. As shown in Figure 11a, the tensions in Line 7 and 8 are larger than for the other cables because they are deployed in head waves to stabilize the vessel. As the cable length decreases, the peak values of the mooring line tensions all increase. For L_C = 500.0 m, the peak nondimensional tension in Line 8 of 0.071 is the maximum for all eight lines, which is much less than the safe threshold value of 1.0. For the average values of the tension in the mooring lines in Figure 11b, the distribution patterns are similar to those of the peak values. Furthermore, from Figures 10 and 11, the increase in the cable length results in a buoyancy increase due to the floaters and a decrease in the tension of the cable. Although the decrease in the cable length causes an increase in the vessel stability, the tension in the cable also significantly increases, and the insecurity of the cable also increases. Therefore, during deployment a floating length of the cable is maintained to keep it safe.

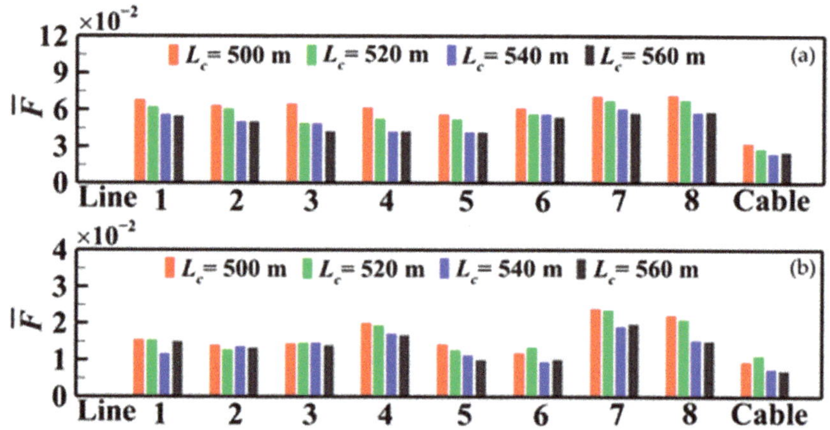

Figure 11. Effects of the submarine cable length on the tensions in the mooring lines and the cable. **(a)** Peak values. **(b)** Average values.

4.3. Effects of the Incident Current Velocity

The longshore current is the most critical type of current during cable landing from the vessel to the shore. Three values of the current velocity are employed in this study: 1.5 m/s, 2.0 m/s, and 2.5 m/s. Other typical operating conditions include: L_C = 500.0 m, $w = -90°$, and $f = -45°$.

The effects of the current velocity on the vessel stability are shown in Figure 12. The surging and swaying motions increase with the maximum nondimensional values of 20.6 and 17.7 as the current velocity increases due to the current acting on the vessel and the submarine cable. The dynamic positioning system is activated under large velocity conditions to stabilize the vessel. In addition, the rolling and yawing motions are restricted due to the tensioning cable as the current velocity increases. On the other hand, variations in the current velocity have little effect on the heaving and pitching motions, the values of which are minor.

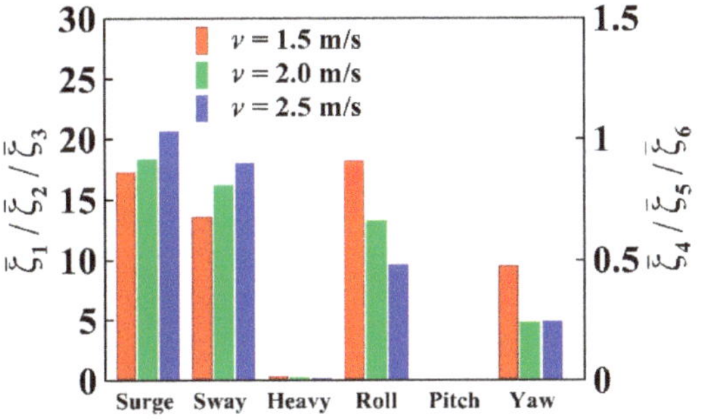

Figure 12. Effects of the incident current velocity on the vessel motions.

The effects of the incident current velocity on the tensions in the mooring lines and the cable are illustrated in Figure 13. The peak and average values of the tensions in the mooring lines and the cable all increase with an increase in the current velocity. Moreover, the mooring lines in head waves have larger tensions, which are focused on during the operation. During cable landing, the cable has a horizontal shift due to the current acting forces, and a significant increase in the cable tension can be observed. During cable deployment, if the current velocity is too great, the anchor boat is advised to push the floating cable to suitable positions to prevent over bending and over loading.

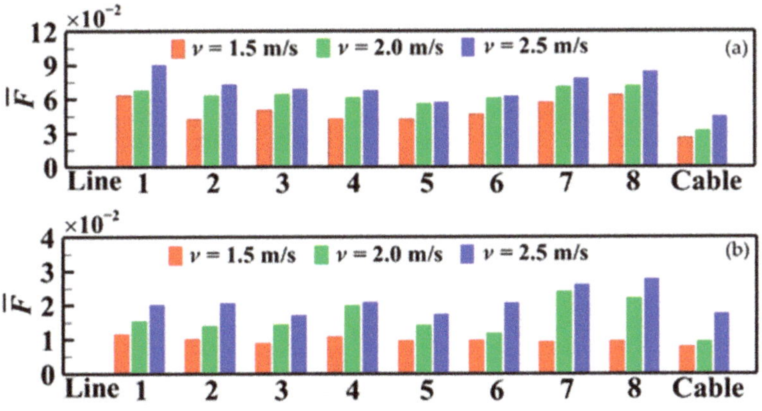

Figure 13. Effects of the incident current velocity on the tension in the mooring lines and the cable. (**a**) Peak values. (**b**) Average values.

4.4. Effects of the Incident Wave Direction

As the directions of the vessel and the current are determined by the cable-deploying operations, the incident wave direction is varied to identify the most negative method of offshore operation. Three incident wave directions were employed in this study: $-45°$, $-90°$, and $-135°$. Other typical operating conditions include L_C = 500.0 m, a current velocity of 2.0 m/s, and $f = -45°$.

The effects of the incident wave direction on the vessel stability are shown in Figure 14. As the wave direction changes from $-45°$ to $-135°$, the vessel bow is under the attack of the waves, resulting in a significant increase in surging motion with a maximum amplitude of 26.1. As $w = -90°$, the incident wave direction is perpendicular to the vessel hull, resulting in the maximum amplitudes of the swaying and rolling motions. In addition, for $w = -45°$ and $-135°$, the yawing amplitude is increased due to the oblique wave-acting forces.

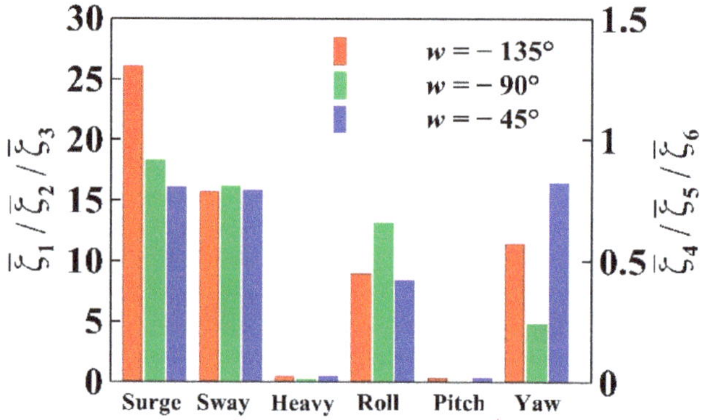

Figure 14. Effects of the incident wave direction on the vessel motions.

The effects of the incident wave direction on the tensions in the mooring lines and the cable are illustrated in Figure 15. For $w = -45°$, as shown in Figure 15a, the vessel stern is under the attack of the incident waves, resulting in larger tensions in Lines 5–8 with a maximum nondimensional tension in Line 7 of 0.075. On the other hand, for $w = -135°$, the vessel bow is under the wave attack, with the maximum tension on Line 4 of 0.074. If possible, a dynamic positioning system is recommended to provide additional power to the bow and stern of the vessel under these two directions of incident waves.

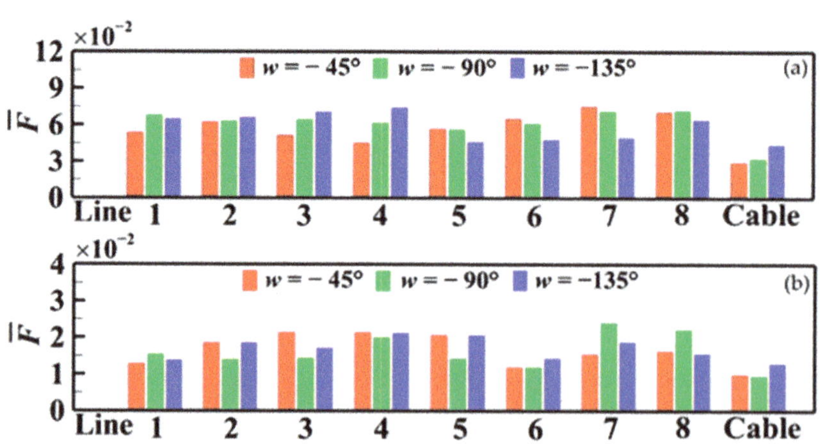

Figure 15. Effects of the incident wave direction on the tensions in the mooring lines and the cable. (**a**) Peak values. (**b**) Average values.

For the average tensions in Figure 15b, the values for different mooring lines have minor differences, and the distribution pattern is similar to that of the peak values. As $w = -135°$, the incident waves and current have joint acting forces on the submarine cable, and the tension in the cable is the largest with a value of 0.046. Therefore, under this circumstance, the anchor boat is used to push the floating cable to reduce the tension, which may lead to damage.

4.5. Effects of the Incident Wind Direction

During the cable deployment, the incident wind is another random environmental factor. As the wave direction of $w = -135°$ is identified as the most negative one, it is employed in the tests in this section. Five incident wind directions are employed in this study: 0, $-45°$, $-90°$, $-135°$, and $-180°$. Other typical operating conditions include $L_C = 500.0$ m and the current and wind velocities of 2.0 m/s and 6.0 m/s, respectively.

The effects of the incident wind direction on the vessel motions are shown in Figure 16. As the wind direction changes from 0 to $-180°$, the surging motion of the vessel is significantly enhanced, and a maximum nondimensional value of 22.5 is obtained. As $f = -90°$, the vessel is affected by the wind perpendicular to the hull, resulting in maximum swaying and rolling amplitudes with the same distribution pattern. Under these circumstances, the dynamic positioning system or the barge shall be employed to push the vessel and maintain the stability of the cable laying vessel. As the wind direction is parallel to the hull, the yaw motion is minor, while the maximum nondimensional yaw amplitude is obtained as 0.88. In addition, the heaving and pitching amplitudes are both minor.

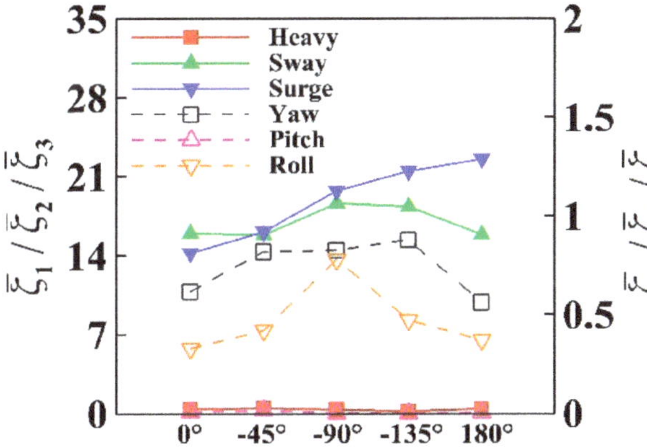

Figure 16. Effects of the incident wind direction on the vessel motions.

The effects of the incident wind direction on the tensions in the mooring lines and the cable are illustrated in Figure 17. For the peak values in Figure 17a, as the vessel bow is exposed to the incident wind ($f = -180°$), the peak tensions in Lines 1–4 are on average 23.4% larger than those in Lines 5–8; as the stern faces towards the wind ($f = 0$), the peak tensions in Lines 5–8 are on average 28.6% larger than those in Lines 1–4. The average tension values in Figure 17b have the same varying trend and distribution pattern as the peak values. For the submarine cable, as $f = -135°$ and $f = 180°$, the joint acting forces by the wave, wind, and current can be observed through the larger amplitude of the tension. Therefore, as the directions of the environmental forces tend to be the same, the tension in the cable is enhanced and the safety coefficient decreases, which should be avoided during cable landing operations.

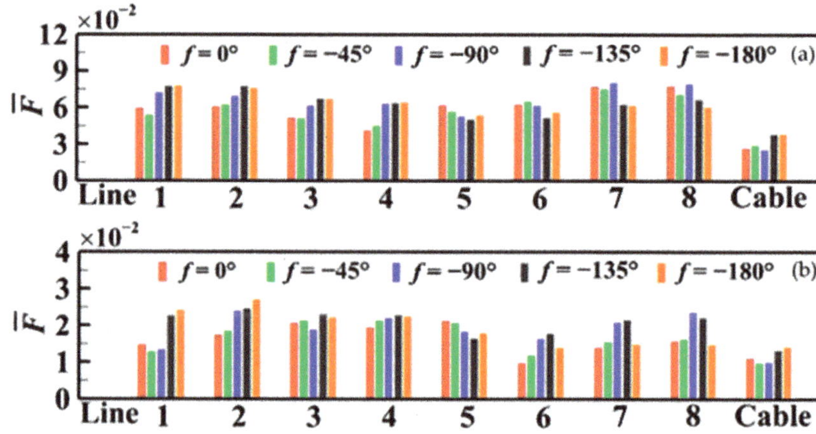

Figure 17. Effects of the incident wind direction on the tensions in the mooring lines and the cable. (**a**) Peak values. (**b**) Average values.

5. Conclusions

A numerical model was established based on Ansys-AQWA to simulate the dynamic interactions between a cable-laying vessel and a submarine cable during the cable landing process. The model was carefully validated by experimental data gleaned from a moored vessel model tested in a wave tank.

As the length decreases, the floating submarine cable with the floaters acts more like a mooring line for the vessel and enhances its stability with a significant increase in its tension. A stronger current increases the perpendicular forces acting on the cable, and the surging and swaying motions of the vessel are enhanced with increased tensions in both the mooring lines and the submarine cable. As the incident wave direction is −135°, the vessel is exposed to more positive impacts from the waves and the surging and rolling amplitudes are greater than those of the other DOFs. The joint effects from the waves and current cause the largest tension in the cable.

If the incident wind direction is parallel to the vessel hull, the wind has a minor influence on yawing motion, while its amplitude is enhanced at a wind direction of −135°. In addition, the tension in the cable is greatest under this wind direction due to the joint effects of all environmental forces, which should be avoided during the cable-laying process; otherwise, more safety steps should be implemented during operations.

Author Contributions: Conceptualization, Z.L.; methodology, J.K.; software, J.K., G.C. and Z.Y.; validation, G.C. and X.Q.; formal analysis, G.C., Z.Y. and X.Q.; writing—original draft preparation, Q.Y.; writing—review and editing, J.K. and Z.L.; visualization, J.K. All authors have read and agreed to the published version of the manuscript.

Funding: This research was funded by the Critical Technology Research Project on the Cable Landing Floaters, grant number 20190141.

Institutional Review Board Statement: Not applicable.

Informed Consent Statement: Not applicable.

Data Availability Statement: Not applicable.

Conflicts of Interest: The authors declare no conflict of interest.

References

1. Zhang, R.; Shen, G.Q.P.; Ni, M.; Wong, J.K.W. An overview on the status quo of onshore and offshore wind power development and wind power enterprise localization in China. *Int. J. Green Energy* **2019**, *16*, 1646–1664. [CrossRef]
2. Randolph, M.F.; Gaudin, C.; Gourvenec, S.M.; White, D.J.; Boylan, N.; Cassidy, M.J. Recent advances in offshore geotechnics for deep water oil and gas developments. *Ocean. Eng.* **2011**, *38*, 818–834. [CrossRef]
3. Zeng, Q.; Wang, Y.; Su, D. Safety Control of a Multiple Return 500 kV Submarine Cable Landing Project. *Soil Eng. Found.* **2018**, *32*, 10–14.
4. Evans, G.; Page, M. The planning and surveying of submarine cable routes. In *Submarine Cables*; Brill Nijhoff: Singapore, 2014; pp. 91–122.
5. Zhang, Z.; Dong, Y.; Yuan, Z.; Liu, Z. Innovative floating support technologies for submarine cable landing. In *Emerging Developments in the Power and Energy Industry*; CRC Press: Boca Raton, FL, USA, 2019; pp. 78–86.
6. Hu, Y.; Hu, Z.; Liu, Y.; Liu, J. Analysis of the large LNG ships moored against a quay based on AQWA. *Ship Sci. Technol.* **2012**, *34*, 70–73+110. [CrossRef]
7. Shigunov, V.; el Moctar, O.; Schellin, T.E.; Kaufmann, J.; Stute, R. Assessing the Dynamic Stability of an Offshore Supply Vessel. *J. Offshore Mech. Arct. Eng.* **2012**, *134*, 29–37. [CrossRef]
8. Liu, X.; Li, J. The Analysis of Some Deep-Sea Functional Ship's Structural Strength Based on AQWA. *Shipbuild. Vocat. Educ.* **2015**, *3*, 14–18. [CrossRef]
9. Putra, G.L.; Wibowo, H.T.; Agusta, F. Stability analysis of semi-trimaran flat hull ship for a sea transportation model. *Commun. Sci. Technol.* **2017**, *2*, 42–46. [CrossRef]
10. Roy, S.; Ghosh, V.; Dey, S.; Vimmadi, S.; Banik, A.K. A coupled analysis of motion and structural responses for an offshore spar platform in irregular waves. *Ships Offshore Struct.* **2017**, *12*, S296–S304. [CrossRef]
11. Banik, A.K.; Roy, S.; Saha, K.; Dey, S. Effect of wave spreading including multi-directional wave interaction on the responses of a spar platform. *Mar. Syst. Ocean. Technol.* **2019**, *14*, 153–165. [CrossRef]
12. Wang, L.; Tang, Y.; Zhang, X.; Zhang, J. Studies on parametric roll motion of ship under wave group by numerical simulation. *Ocean. Eng.* **2018**, *163*, 391–399. [CrossRef]
13. Su, Y.; Wang, J.; Zhuang, J.; Shen, H.; Bi, X. Experiments and CFD of a variable-structure boat with retractable twin side-hulls: Seakeeping in waves. *Ocean. Eng.* **2021**, *235*, 109358. [CrossRef]
14. Liu, Y. Dynamic Analysis of Mooring Cables and Their Damping Effect on the Low-Frequency Motion of Floating Platforms. Ph.D. Thesis, Chalmers University of Technology, Gothenburg, Sweden, 1997.
15. Sarkar, A.; Taylor, R.E. Effects of mooring line drag damping on response statistics of vessels excited by first-and second-order wave forces. *Ocean. Eng.* **2000**, *27*, 667–686. [CrossRef]
16. Sarkar, A.; Taylor, R.E. Dynamics of mooring cables in random seas. *J. Fluids Struct.* **2002**, *16*, 193–212. [CrossRef]
17. Gao, F.; Li, Y.; Shen, W.J. Test study on mooring characteristic for multi-body floating system under the action of wind, wave and flow. *J. Waterw. Harb.* **2017**, *38*, 555–560.
18. Mengxi, N.; Xusheng, W.; Xiaoming, W.; Lin, Z. Time domain approach for computing the mooring force of a mooring system subject to wind, waves and currents. *J. Tsinghua Univ.* **2004**, *9*, 1214–1217. [CrossRef]
19. Zheng, W.; Li, J.; Liu, S.; Sun, Y. Research on tensions of bow stern mooring of single ship under the combination impact of external environmental load factors. *J. Waterw. Harb.* **2016**, *37*, 490–496.
20. Pham, N.D.K.; Nguyen, X.P. Application of CFD for calculation and simulation of anchor-cable tensions in mooring ship. *J. Mech. Eng. Res. Dev.* **2019**, *42*, 182–186. [CrossRef]
21. Nguyen, X.P.; Pham Nguyen, D.K. Experimental Research on the Impact of Anchor-Cable Tensions in Mooring Ship at Vung Tau Anchorage Area. *Int. J. Adv. Sci. Eng. Inf. Technol.* **2019**, *9*, 1892–1899. [CrossRef]
22. Mingyu, F.; Aihua, Z.; Jinlong, X.; Jianfang, J. Adaptive dynamic surface tracking control for dynamic positioning cable laying vessel. In Proceedings of the 2012 IEEE International Conference on Automation and Logistics, Zhengzhou, China, 15–17 August 2012; pp. 266–271.
23. Yang, N.; Jeng, D.; Zhou, X. Tension analysis of submarine cables during laying operations. *Open Civ. Eng. J.* **2013**, *7*, 282–291. [CrossRef]
24. Tommasini, R.B.; de Oliveira Carvalho, L.; Pavanello, R. A dynamic model to evaluate the influence of the laying or retrieval speed on the installation and recovery of subsea equipment. *Appl. Ocean Res.* **2018**, *77*, 34–44. [CrossRef]
25. Cavefors, J.; Oscarsson, A. *Vessel Motions and Mooring Line Tensions in Very Shallow Water*; Chalmers University of Technology: Gothenburg, Sweden, 2016.
26. Zhang, D.; Bai, Y.; Zhao, W.; Zhu, K. Dynamic analysis of submarine cable during the process of laying back to the seabed. *Ships Offshore Struct.* **2020**, *15*, 153–161. [CrossRef]
27. Wang, Y.; Bian, X.; Zhang, X.; Xie, W. A study on the influence of cable tension on the movement of cable laying ship. In Proceedings of the OCEANS 2010 IEEE, Seattle, WA, USA, 20–23 September 2010.
28. Srikanth, N.; Rao, S.S. Subsea cable health monitoring system. In Proceedings of the 2017 Asian Conference on Energy, Power and Transportation Electrification (ACEPT) IEEE, Singapore, 24–26 October 2017; pp. 1–9.
29. Nexans. Available online: https://www.nexans.com/en/ (accessed on 23 May 2022).
30. Drissi-Habti, M.; Abhijit, N.; Sriharsha, M.; Carvelli, V.; Bonamy, P.-J. Concept of Placement of Fiber-Optic Sensor in Smart Energy Transport Cable under Tensile Loading. *Sensors* **2022**, *22*, 2444. [CrossRef] [PubMed]

31. Newman, J.N. *The Theory of Ship Motions: Advances in Applied Mechanics*; Elsevier: Amsterdam, The Netherlands, 1979; Volume 18, pp. 221–283.
32. Ghafari, H.; Dardel, M. Parametric study of catenary mooring system on the dynamic response of the semi-submersible platform. *Ocean. Eng.* **2018**, *153*, 319–332. [CrossRef]
33. Lu, Z.; Cao, C.; Ge, Y.; He, J.; Yu, Z.; Chen, J.; Zheng, X. Research on Improving the Working Efficiency of Hydraulic Jet Submarine Cable Laying Machine. *J. Mar. Sci. Eng.* **2021**, *9*, 745. [CrossRef]
34. Nexans. Lavrio Syros Interconnector Project. Available online: https://www.nexans.com/business/High-Voltage---Projects/Power-Networks-Subsea-Power-Cables/Lavrio---Syros.html,2017 (accessed on 15 October 2021).
35. Wang, W.; Wu, M.; Palm, J.; Eskilsson, C. Estimation of numerical uncertainty in computational fluid dynamics simulations of a passively controlled wave energy converter. *J. Eng. Marit. Environ.* **2018**, *232*, 71–84. [CrossRef]

Journal of
Marine Science and Engineering

MDPI

Article

Numerical Simulation of the Motion of a Large Scale Unmanned Surface Vessel in High Sea State Waves

Shuo Huang [1,2], Weiqi Liu [1], Wanzhen Luo [1,2,*] and Kai Wang [1,2]

[1] School of Marine Engineering and Technology, Sun Yat-sen University, Zhuhai 518000, China; huangsh97@mail.sysu.edu.cn (S.H.); liuwq27@mail2.sysu.edu.cn (W.L.); wangkai25@mail.sysu.edu.cn (K.W.)
[2] Southern Marine Science and Engineering Guangdong Laboratory (Zhuhai), Zhuhai 519082, China
* Correspondence: luowzh5@mail.sysu.edu.cn; Tel.: +86-188-4642-1436

Abstract: The motion stability of the Unmanned Surface Vessel (USV) is threatened by the action of waves under a rough sea state. In the present paper, the motion of a large-scale USV is numerically simulated under high sea state of level 5 and 7. The overset grid method and Reynolds Averaged Navier–Stokes (RANS) approach are employed to solve Navier–Stokes (*N-S*) equations. For the case of wave incident angle 0° and 30°, the heave, pitch and roll motion response of a large scale USV are investigated by using the six Degrees of Freedom (6-DOF) numerical model. The effects of different sea states, as well as different wave directions, on the motion of USV are compared. The comparative results indicate that the response of this USV in waves is the periodic free-motion according to the corresponding amplitude, which does not exceed the stable range, and there are no overturning and other situations that may affect the safety, in the case of level 5 and 7 sea states. The corresponding pressure at the bottom of this USV meets the range of material strength, and no structural damage or injury to the hull occurs, although the pressure varies at different wave periods. For the case of different wave directions, the analysis of the boundary layer thickness shows that the wave direction is of great importance to the boundary layer thickness distribution, both in the level 5 and level 7 sea states.

Keywords: high sea state; USV; large scale; motion in waves; safety of the trip

Citation: Huang, S.; Liu, W.; Luo, W.; Wang, K. Numerical Simulation of the Motion of a Large Scale Unmanned Surface Vessel in High Sea State Waves. *J. Mar. Sci. Eng.* **2021**, *9*, 982. https://doi.org/10.3390/jmse9090982

Academic Editor: Alessandro Ridolfi

Received: 24 July 2021
Accepted: 3 September 2021
Published: 8 September 2021

1. Introduction

In recent years, a variety of unmanned surface vehicles (USV) have emerged around the world for different operation requirements. As a new type of ocean carrying platform, USV is widely used in civil and military fields because of its small size, high mobility, intelligence, and all-weather operation. Spartan Scout of the United States, Silver Marlin of Israel, and Venus of Singapore have excellent performance in environmental monitoring, search and rescue activities, surveillance, and reconnaissance and patrol [1]. USV is becoming an important way to explore the ocean and an important tool for marine operations.

The motion response and resistance performance in a real sea state are very important for the safe and efficient completion of the task for USV. In real sea state, due to the interaction of waves, the resistance is larger than that in still water [2]. The maximum additional resistance of this part can reach 30% of the total resistance in still water [3]. At the same time, under the excitation of waves, the USV will produce obvious six degrees of freedom motion response [4], especially in high sea state, the excessive motion response will threaten its motion stability. The motion of ships in high sea states is related to the safety of navigation. Kim D. et al. [5] studied ship maneuverability under different wave conditions based on the CFD method. The main results of this study help to improve the understanding of ship maneuverability in waves, thus as to improve the safety of navigation. Jiri de Vos et al. [6] carried out statistical analysis on ship parameters of unmanned ship safely and obtained the length of an unmanned ship with maximum safety benefit. Chen et al. [7] did a detailed analysis of the relationship between ship responses and sea states under real rough sea conditions, which gives suggestions for ship safety under severe weather conditions. It is of great significance to study the hydrodynamic performance prediction of USV in high sea state for its shape design and hydrodynamic performance analysis.

Hydrodynamic prediction is an important basis for the performance research of USV. The common prediction methods include model experiments and numerical simulations. The model test has high reliability, but it needs a lot of time and money. Now it is mainly used as the final scheme verification and hydrodynamic mechanism research. Numerical simulation is widely used as the simplest and convenient method. The numerical methods for predicting ship motion response in waves include classical potential flow theory and unsteady Reynold's time-averaged equations (RANS). Tarafder and Suzuki analyzed the free surface flow problem of Wigley catamaran based on the potential flow theory [8]. Based on the potential flow theory, Fang and Chan predicted the motion responses (including heave, pitch, and roll) of the Wave Piercing Catamaran (WPC) in regular and irregular waves. The numerical results are in good agreement with the experimental data obtained in SSPA [9]. Based on the linear potential flow theory, Ma and Wan studied the nonlinear effects of heave and trim motion, wave resistance of four buoyant hull types (Wigley, S60, DTMB5415, KCS) sailing in still water. The results show that the numerical prediction is consistent with the experimental results, and the nonlinear free surface will increase wave making [10]. Min Guk Seo and others used the Rankine panel method based on potential flow theory to calculate the resistance of KVLCC2 ship in short wave. It was found that when the face element scale was small enough, the additional resistance consistent with the experimental measurement could be accurately predicted [11]. In most of the numerical calculation methods, potential flow theory has high efficiency and accuracy under certain conditions, but the method based on potential flow theory also has some shortcomings. This method is mainly for medium speed design and does not consider the fluid viscosity. In the prediction of ship motion response, the influence of the viscosity effect cannot be ignored. Simonsen et al. also emphasized that the effects of breaking waves, turbulence, and viscosity, which are ignored in potential flow theory, should be taken into account in the numerical calculation [12]. RANS method based on CFD (Computational Fluid Dynamics) breaks through the limitation of potential flow theory and is more accurate in motion response and additional resistance prediction. Tahsin found that in the case of high-speed navigation, the additional resistance and motion response obtained by the RANS equation are most different from those based on potential flow theory [13]. Further comparison with the model experiment shows that the prediction results of RANS equation are more accurate. Deng Rui et al. studied the additional resistance and motion response of the trimaran under different wave amplitude conditions by using the method of viscosity based on RANS. The results show that the peak value of the additional resistance and motion response of the trimaran is mainly affected by the wavelength [14].

In this paper, a catamaran USV is taken as the research object, and the mesh is completed based on overlapping grid technology. The RANS method and SST k-ω are applied as the turbulent model to solve the *N-S* equation and simulates the motion of the USV in the high sea states of 5 and 7 levels. The motion response and resistance performance of the USV under high sea states are analyzed. The differences of the motion response under different wave directions are compared, and the hull pressure and boundary layer distribution under different working conditions are studied.

2. Numerical Model

2.1. Governing Equations and Turbulence Models

The RANS method was applied to solve the *N-S* equation in this paper, which avoids the huge amount of calculation caused by the direct simulation of turbulent pulsations at various scales. The continuity and momentum equations can be written in Cartesian coordinates as follows [15]:

$$\frac{\partial u_i}{\partial x_i} = 0 \tag{1}$$

$$\frac{\partial u_i}{\partial t} + \frac{\partial u_j u_i}{\partial x_j} = \frac{\partial}{\partial x_j}\mu(\frac{\partial u_i}{\partial x_j}) - \frac{1}{\rho}\frac{\partial p}{\partial x_i} + S_j \quad (2)$$

where u_i and u_j (i, j = 1, 2, 3), p, ρ, and S_j represent the time-averaged values of velocity components, time-averaged pressure, fluid density, and source term, respectively. The fluid density was assumed constant for incompressible Newtonian fluids.

In this paper, The STAR-CCM+ commercial software package based on finite-volume computational method was used to analyze the interaction between USV and waves. The turbulence model used in this paper was the k-ω shear stress transport (SST) model [16]. The advantage of the SST k-ω turbulence model, compared with the standard k-ω model and k-ω model has higher accuracy and credibility. The free surface of the numerical wave tank was modeled using the two-phase volume-of-fluid (VOF) technique with the high-resolution interface capturing (HRIC) scheme [17].

2.2. USV Model

The calculation model is a full-scale model of a catamaran USV. The USV models from different perspectives are shown in Figure 1.

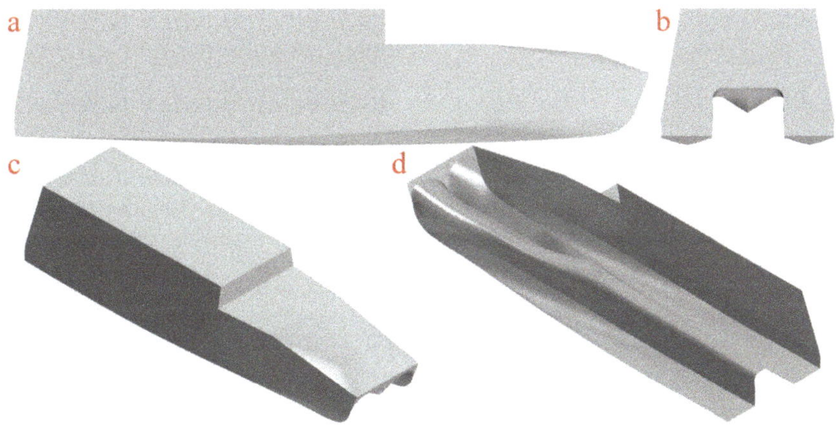

Figure 1. USV model (**a**) side view; (**b**) stern view; (**c**) side view from the bow; (**d**) side view from the stern).

The detailed main parameters of the model are shown in Table 1.

Table 1. Principal dimension of the USV.

Parameters	Symbol	Value
Length of overall (m)	L_{OA}	42.00
Length between perpendiculars (m)	L_{pp}	41.05
Breadth (m)	B	12.00
Design draft (m)	T	1.45
Depth (m)	D	4.50

2.3. Degree of Freedom Model

In the simulation, the dynamic fluid body interaction (DFBI) module [18] combined with overset mesh method simulates the heave, pitch, and roll motion of the USV in response to forces exerted by the wave. The 6-DoF (degree of freedom) solver computes fluid forces, moments, and gravitational forces on the USV, and pressure and shear forces were integrated over the surfaces. Two coordinate systems were applied to solve the

6 degrees of freedom equation. One is called the initial coordinate system (earth coordinate system $o'x'y'z'$), the other is called the non-initial coordinate system (hull coordinate system $oxyz$), as shown in Figure 2.

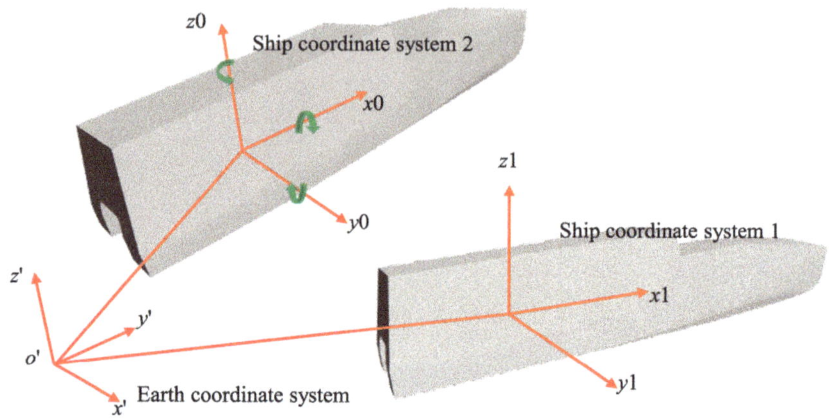

Figure 2. Hull coordinate system and earth coordinate system.

The forward direction of the USV is the positive direction of the x-axis, the positive direction of the y-axis points to the port side of the ship, and the z-axis points to the sky. According to the right-hand rule, in the heave motion of the ship, the positive value is the forward motion to the z-axis, and the negative value is the motion to the bottom. In the rolling motion of the ship, the positive value is about the starboard side of the USV in x-axis, and the negative value is about the port side of the ship in x-axis. In the pitch motion of the ship, the positive value turns around the y-axis with bow diving. The negative value is rotation around the y-axis with bow rising.

2.4. Computational Domain and Boundary Conditions

The computational domain of the USV in waves is shown in Figure 3. In order to realize the numerical simulation of the USV in different wave directions without changing the calculation grid and other conditions, the cylindrical calculation area is used.

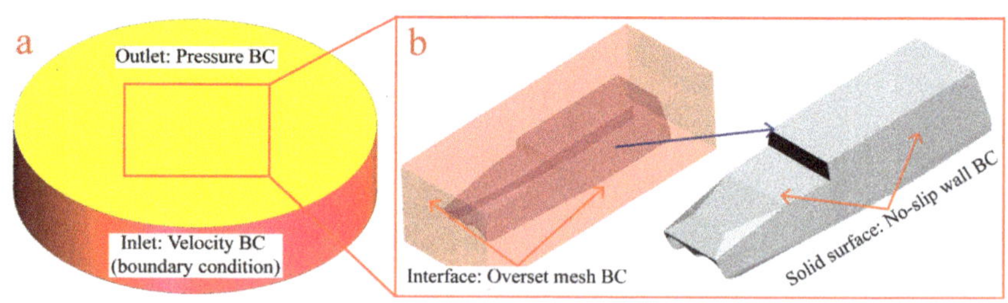

Figure 3. Computational domain: (**a**) background region, (**b**) overset region.

In order to ensure the full development of numerical wave and wake, there were enough interaction and fluid development areas between the USV and free surface in the domain. The axial range of the calculation area is $-5.5\ L_{oa} \leq x \leq 5.5\ L_{oa}$, the spanwise range is $-5.5\ L_{oa} \leq x \leq 5.5\ L_{oa}$, and the vertical range is $-1.25\ L_{oa} \leq z \leq 2.1\ _{Loa}$, where L_{oa} is the total length of the USV. The boundary conditions are shown in Table 2.

Table 2. Boundary conditions.

Boundary Name	Boundary Conditions
Inlet	1. Velocity inlet; 2. Fifth order Stokes waves based on VOF components of water vapor; 3. The relative velocity of the USV is 18 kn; 4. The turbulence intensity is 0.01.
Outlet	1. Pressure outlet; 2. Fifth order Stokes waves based on VOF components of water vapor; 3. Fifth order Stokes wave hydrodynamic pressure.
Bottom	Same as the inlet boundary condition
USV hull	No-slip wall
Boundary of Overset region	Overset mesh interface

2.5. Computational Mesh and Sensitivity Study

The mesh refinement blocks of the computational domain are shown in Figure 4, and Figure 4a is the mesh block of the background region, Figure 4b is the mesh refinement blocks of the overlapping zone and overset region, and Figure 4c is the mesh refinement blocks in the free surface wave zone. There are two mesh refinement blocks in the free surface wave zone, namely mesh refinement block of wave (Navier–Stokes Zone) and mesh refinement block of wave (Theoretical Zone), a mesh refinement block for the overlapping zone, a mesh refinement block for the overset region. For the wave zone, the Aspect Ratio (AR = length of the cell/height of the cell) of the mesh refinement in the free surface wave was 2. Moreover, about 40 cells per wave height (wave height = 9.0 m) and 300 cells per wavelength (wavelength = 150 m) according to the Aspect ratio. The mesh size of the overset boundary in the overset region was the same as the mesh size of the overlapping zone, thus as to facilitate data transmission between background region and overset region.

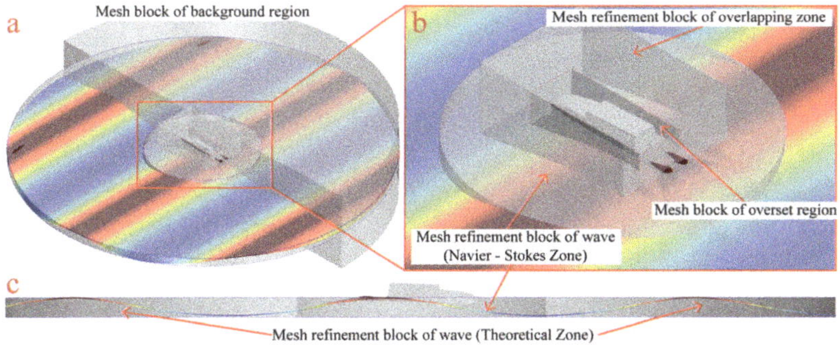

Figure 4. Mesh refinement blocks of computational domain: (**a**) mesh block of background region, (**b**) mesh refinement block of the overlapping zone and overset region, (**c**) mesh refinement block in the free surface wave zone.

The trimmed meshes and prismatic cell layers around the hull were generated using STAR-CCM+, and three sets of coarse, medium, and fine meshes of varying resolutions were generated and used for mesh sensitivity study. The level of the mesh resolution was changed by adjusting the base size of the meshes while keeping the other settings unchanged, and the total number of coarse, medium, and fine meshes meshes was 7.620, 12.207 m, and 22.220 million, respectively. In the present study, the Cartesian cut-cell method was used to generate the computational mesh. Roy et al. [19] presented methods for assessing the uniformity of the mesh refinement with Cartesian grid, and the mesh refinement ratio r_G was considered as follows:

$$r_G = \left(\frac{N_{fine}}{N_{coarse}}\right)^{1/d} \quad (3)$$

where the total number of meshes is noted as N, d is the dimensionality of the computing problem, and its value in the present study is 3. The mesh refinement factor for the fine/medium and medium/coarse meshes was 1.22 and 1.17, respectively. The mesh refinement ratio of fine to medium (medium to coarse) grids was approximately 1.2, which was consistent with references [20]. The heave and pitch motion plots of USV at 7 level sea state with different meshes are shown in Figure 5. The comparative analysis indicates that the three meshes with different resolutions were all able to accurately predict and calculate the motion of the USV in a high sea state. Compared to the fine grid, the use of a medium grid can reduce the computational cost considerably and still achieve a tolerable level of numerical accuracy. Thus, the meshes generation for the cases of this paper was dependent on the arrangement of a medium mesh. The time step in all cases was 0.005 s, and the free surface convection courant number meets the requirement, that is, Courant–Friedrichs–Lewy (CFL) number ($CFL = U \triangle t / \triangle x$, where CFL is the courant number, U is the reference velocity, $\triangle t$ is the time step, and $\triangle x$ is the grid size) needs to be less than 1.

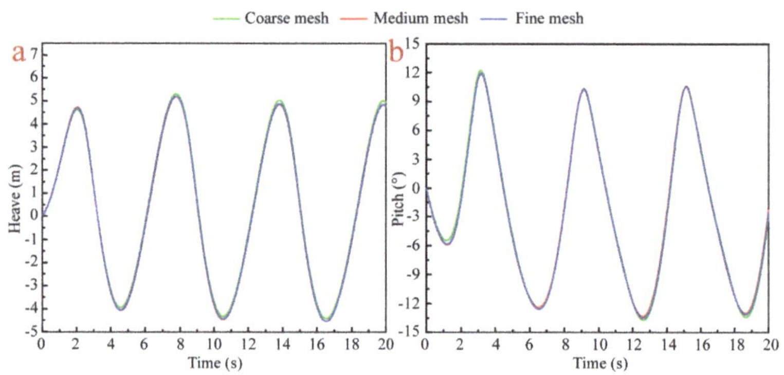

Figure 5. Heave and pitch motion plot of USV at 7 level sea state with different grids: (**a**) Heave motion, (**b**) Pitch.

The medium computational mesh of the USV in waves is shown in Figure 6, and Figure 6a is the overall meshes of the computational domain, Figure 6b is the overset mesh, and local mesh refinement of the free surface, overlap area, Figure 6c,d are the meshes around the hull and boundary layer mesh.

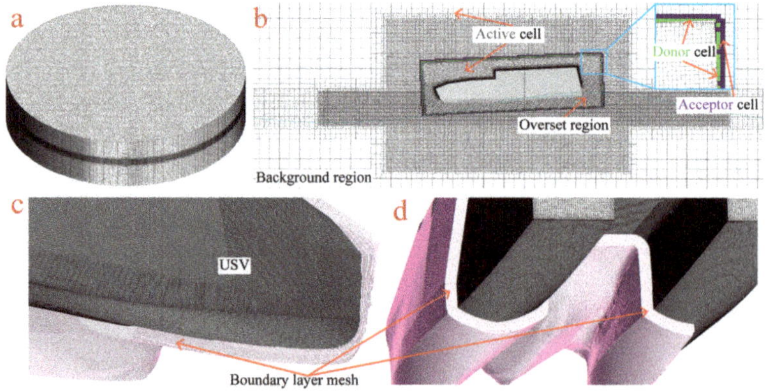

Figure 6. Computational mesh: (**a**) overall meshes of the computational domain, (**b**) overset mesh and local mesh refinement of the free surface, overlap area, (**c**,**d**) meshes around the hull and boundary layer mesh.

3. Numerical Wave Tank and Verification

3.1. Numerical Wave Tank

Wave generation of numerical wave tank was realized via the inlet and outlet boundaries, referred to as the boundary velocity input method. The velocity and pressure profile over the water depth of waves were set at the inlet and outlet boundaries. The wave generated by the fifth-order was more similar to a real wave than that generated by the first-order method, and a fifth-order wave modeled by approximating the fifth-order to the Stokes theory of waves was used as the numerical wave type in the study. The detailed description of wave profile, including the wave phase velocity, depends on the water depth, wave height, and current of fifth-order VOF waves can be found in the studies conducted by Fenton [21].

When numerical wave propagates to the physical boundary, the boundary wave reflection usually occurs, and the boundary wave reflection will interfere with the calculation and wave quality. The Euler-overlay method (EOM) [22] was adopted to address the reflections of surface waves at boundaries. Detailed numerical wave tank and illustration of the Euler overlay method are shown in Figure 7. There are three zones, namely the inner CFD (Navier-Stokes zone), overlay, and outer Euler wave (theoretical solution zone) zones. The overlay zone located between the outer Euler and inner CFD zones gradually blends the CFD and Euler solutions, applying source terms to VOF and momentum equations. The reflections of surface waves are addressed via the forcing solution of the discretized Navier–Stokes equations towards theoretical solution over a specified distance, which increases the calculation efficiency and reduces the calculation time cost owing to its use of reduced-size mesh refinement in the solution domain theoretical solution zone. A detailed explanation and application of the Euler-overlay method (EOM) can be seen in the references [23–26].

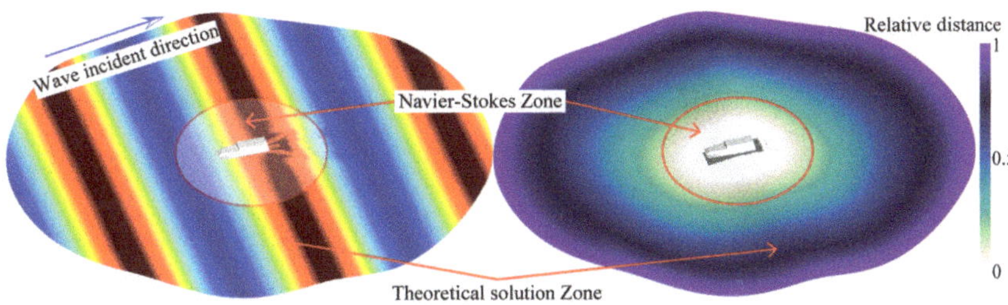

Figure 7. Illustration of the Euler overlay method in the cylindrical numerical wave tank.

The incident direction of waves and the sailing direction of the USV is shown in Figure 8.

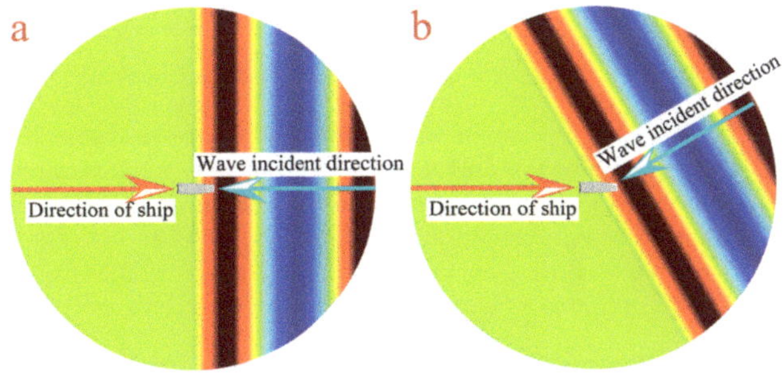

Figure 8. Sailing direction and wave direction of the USV: (**a**) 0° wave direction angle, (**b**) 30° wave direction angle.

3.2. Numerical Wave Tank Verification

In the verification study of the numerical wave, the fifth-order Stokes waves with level 7 sea state were simulated, and the obtained results were compared with the theoretical solutions (See Section 3.1) to verify the accuracy of the numerical wave tank. To record the change in water surface, a wave height detection point was set at the center of the tank. The wave height versus time is shown in Figure 9. The wave height position point distribution along the propagation direction is shown in Figure 10. The red curve is the numerical simulation result, and the green circle is the theoretical analytical solution. It can be seen from Figures 9 and 10 that the numerical simulation wave is in good agreement with the theoretical analytical solution of the wave.

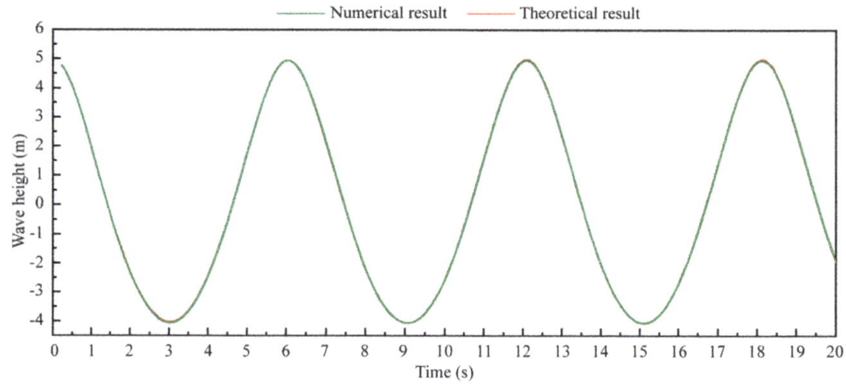

Figure 9. Numerical wave height with time history.

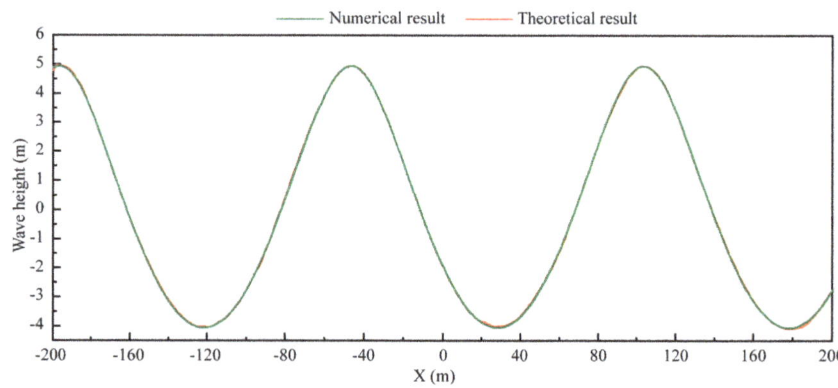

Figure 10. Distribution of wave height position points along the propagation direction.

The waveform of numerical wave propagation after 20 s is shown in Figure 11. The distribution of water and air composition after 20 s of numerical wave propagation is shown in Figure 12. The waveform and the distribution of water and air components are well preserved after propagation for 20 s in Figures 11 and 12. The verification of numerical results and theoretical analytical solutions can ensure that the calculation grid and calculation strategy of this study meet accuracy requirements, and the numerical wave tank is reliable.

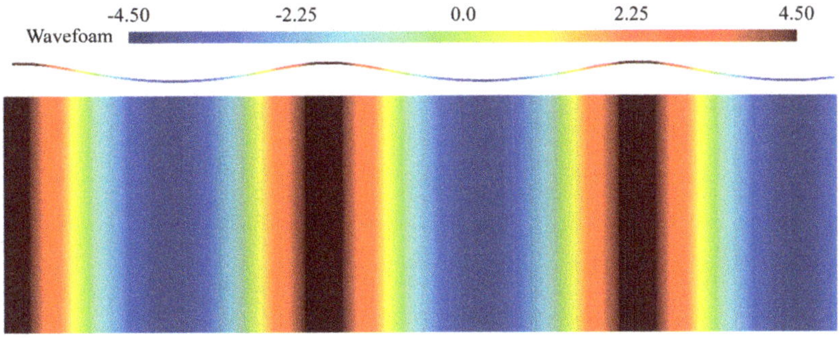

Figure 11. The waveform of numerical wave propagation after 20 s.

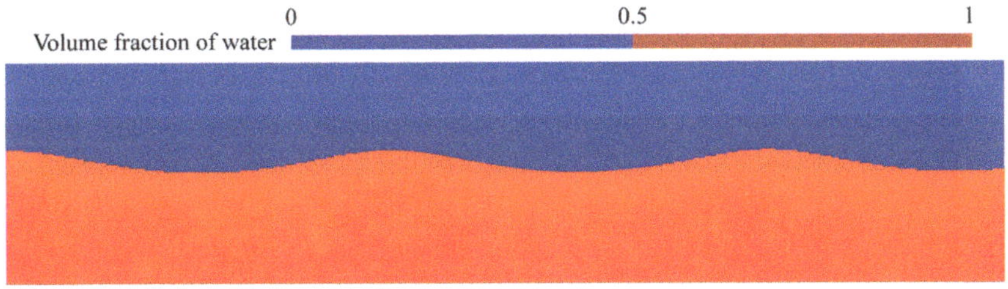

Figure 12. Distribution of water and air composition after 20 s of numerical wave propagation.

4. Results

In this paper, the motion of the USV in waves under 5 level and 7 level states is studied. The waves in the real sea state were all irregular waves. In this study, the waves in 5 level and 7 level sea states were simplified and analyzed by regular waves. The wavelength, height, and period of the simplified regular wave under 7 level states were 150 m, 9.0 m, and 9.629 s, respectively. The wavelength, height, and period of the simplified regular wave under 5 level states were 100 m, 3.25 m, and 7.961 s, respectively.

4.1. Six Degrees of Freedom Motion with Time History

The six degrees of freedom motion process was simulated at the 0° and 30° wave direction angles under 7 level sea state. Figures 13–15 show the heave motion, pitch motion, and roll motion with different wave direction angles, respectively, at 7 level sea state. At the 0° and 30° wave direction, the heave amplitude ranged from −5 m to 5 m, showing periodic variation. The pitch angle ranged from −12 to 12 degrees at 0° wave direction and from −12 to 10 degrees at 30° wave direction. The positive value for bow raising and the negative value for bow diving. The results show that the stable roll angle of the USV was very small in the 0° wave direction, which was close to 0 degrees and had no significant periodicity. The roll angle of the USV was obvious in the 30° wave direction, which ranged from −7 to 6 degrees.

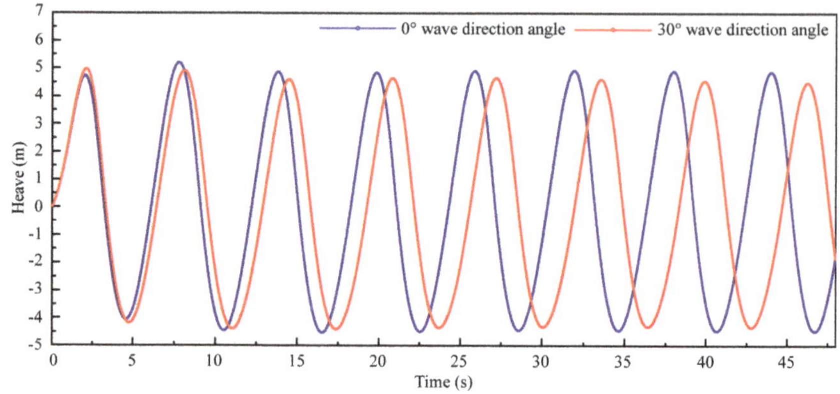

Figure 13. The heave motion with different wave direction angles at 7 level sea state.

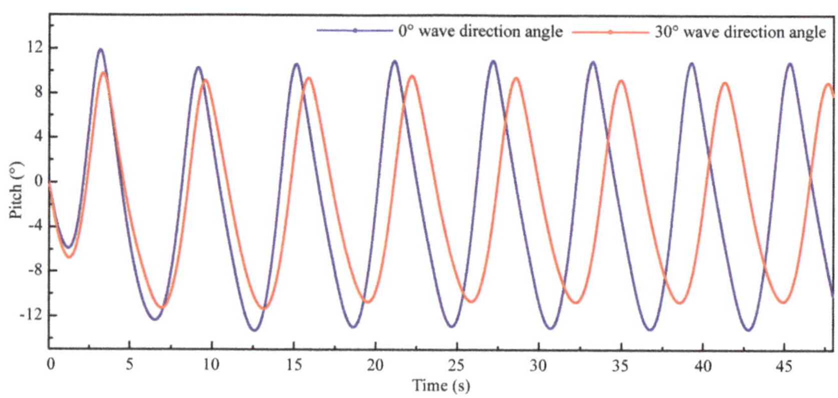

Figure 14. The pitch motion with different wave direction angles at 7 level sea state.

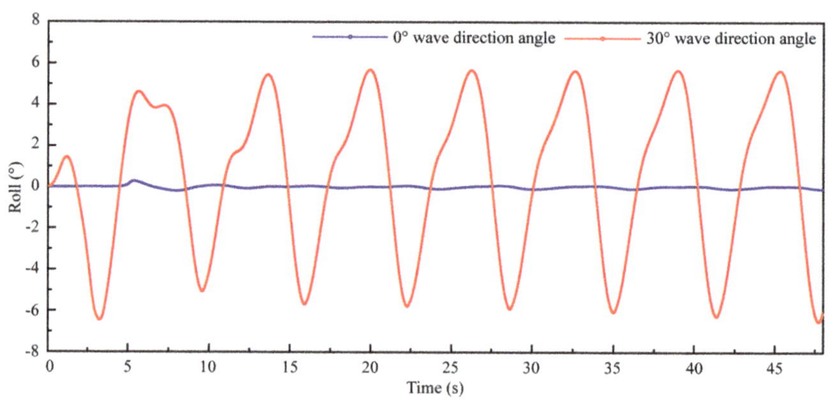

Figure 15. The roll motion with different wave direction angles at 7 level sea state.

Figures 16–18 show the heave motion, pitch motion, and roll motion with different wave direction angles, respectively, at 5 level sea state. At the 0° and 30° wave direction, the heave amplitude ranged from −1.75 m to 1.75 m, showing periodic variation. The amplitude of heave in the 0° wave direction was slightly larger than that in the 30° wave direction, and the period of heave change in the 0° wave direction was slightly smaller than that in the 30° wave direction. The pitch angle ranged from −7 to 5 degrees at the 0° and 30° wave direction. The amplitude of pitch in the 0° wave direction was slightly larger than that in the 30° wave direction. The positive value for bow raising and the negative value for bow diving. The results show that the stable roll angle of the USV was very small in the 0° wave direction, which was close to 0 degrees and had no significant periodicity. The roll angle of the USV was obvious in the 30° wave direction, which ranged from −4 to 4 degrees.

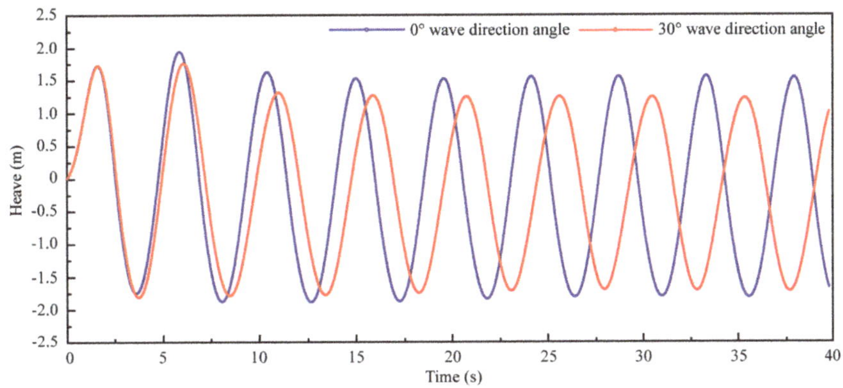

Figure 16. The heave motion with different wave direction angles at 5 level sea state.

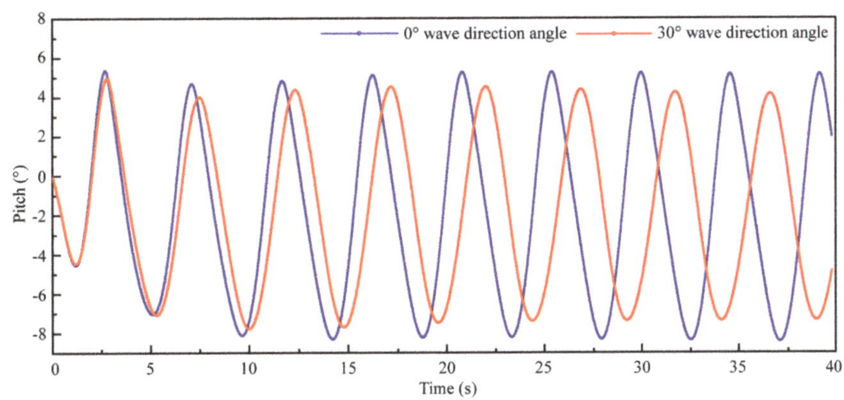

Figure 17. The pitch motion with different wave direction angles at 5 level sea state.

In summary, the six degrees of freedom motion of the USV under different sea states basically presents periodic changes, and the motion amplitude of the 0° wave direction was slightly greater than that of the 30° wave direction. The rolling motion was not obvious at 0°, and the motion of the USV under different sea states and does not exceed the stable range of motion, which meets the stability requirements, and there was no overturning phenomenon.

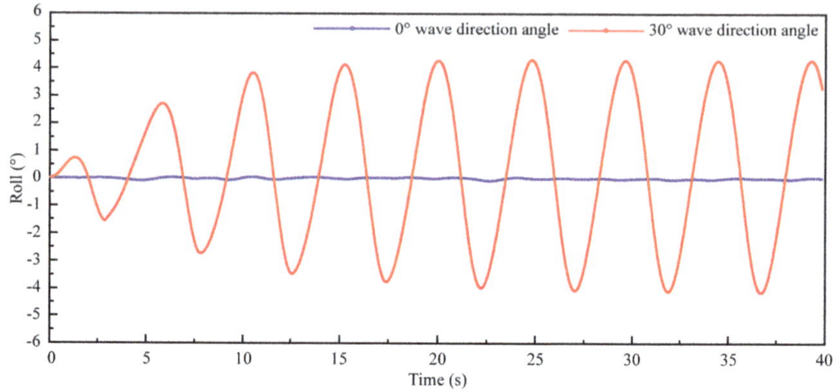

Figure 18. The roll motion with different wave direction angles at 5 level sea state.

4.2. Resistance

Figures 19 and 20 show the total resistance curves in different wave directions. The total resistance in the 0° wave direction was slightly greater than that in the 30° wave direction.

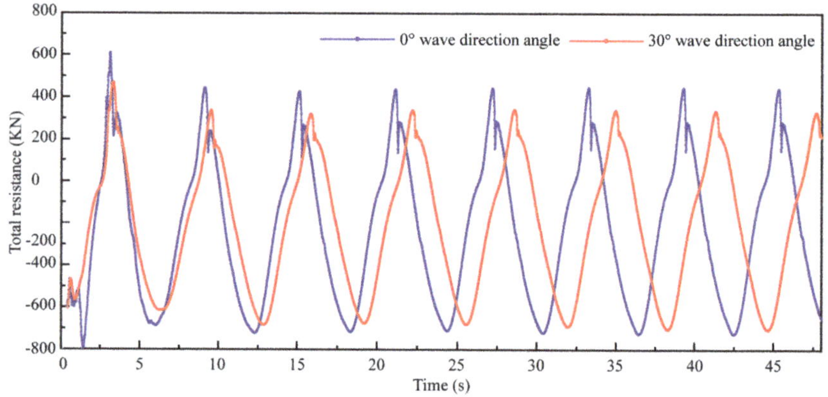

Figure 19. The total resistance in different wave directions with 5 level sea state.

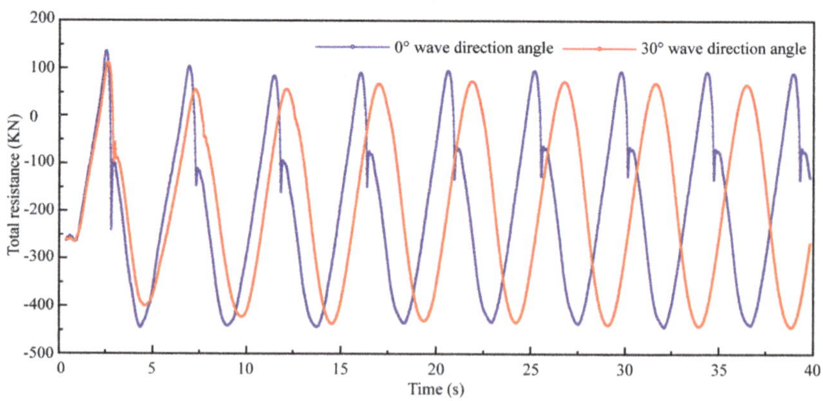

Figure 20. The total resistance in different wave directions with 7 level sea state.

4.3. The Attitude of USV in Waves at Typical Times under 7 Level Sea State

Figures 21 and 22 show the attitude of the USV in waves at a typical time when the wave direction was 0° and 30° under 7 level sea state, respectively. $T0$ is the period of the wave. When $T = 0$ to $1/4T0$, the incident wave peak propagates from the bow to the stern, and bow diving happens. The heave amplitude first increases and then decreases along the positive direction. When $T = 1/4T0$ to $2/4T0$, the wave trough gradually propagates to the stern, the heave amplitude gradually increases along the negative direction, the pitch amplitude gradually increases in the positive direction, and the bow rising happens. When $T = 2/4T0$ to $3/4T0$, the wave peak is transmitted to the bow, the heave amplitude reaches the maximum in the negative direction, and the pitch amplitude reaches the maximum in the positive direction, resulting in the maximum bow rising. When $T = 3/4T0$ to $4/4T0$, the wave trough is transmitted to the bow position, the heave amplitude reaches the maximum value in the positive direction, and the pitch amplitude reaches the maximum value in the negative direction. The maximum bow diving happens. When $T = 4/4T0$ to $5/4T0$, the wave trough gradually transmitted to the stern, and the bow rising happens. The motion law of the 0° wave direction and 30° wave direction was consistent, and the motion amplitude was slightly different. The motion amplitude of the 0° wave direction was larger than that of the 30° wave direction. In the whole process of USV movement around the waves, the stability of the USV meets the stability requirements, and no capsizing occurs.

4.4. Pressure Distribution at the Bottom of the USV at Typical Times under 7 Level Sea State

Figures 23 and 24 show pressure distribution at the bottom of the USV when the wave direction was 0° under 7 level sea state, respectively. The positive pressure at the bottom of the USV was mainly concentrated in the two side body areas, and the negative pressure was mainly concentrated in the area between the side bodies. In addition, when $T = 0$, the USV was in the initial wave state and had a large static pressure. The maximum pressure reached 29,570 Pa. When $T = 2/4T0$ and $T = 4/4T0$, the bottom pressure was large, including a large positive pressure and negative pressure, and the area was the largest. When $T = 4/4T0$, the negative pressure reached the maximum value −18,738 Pa.

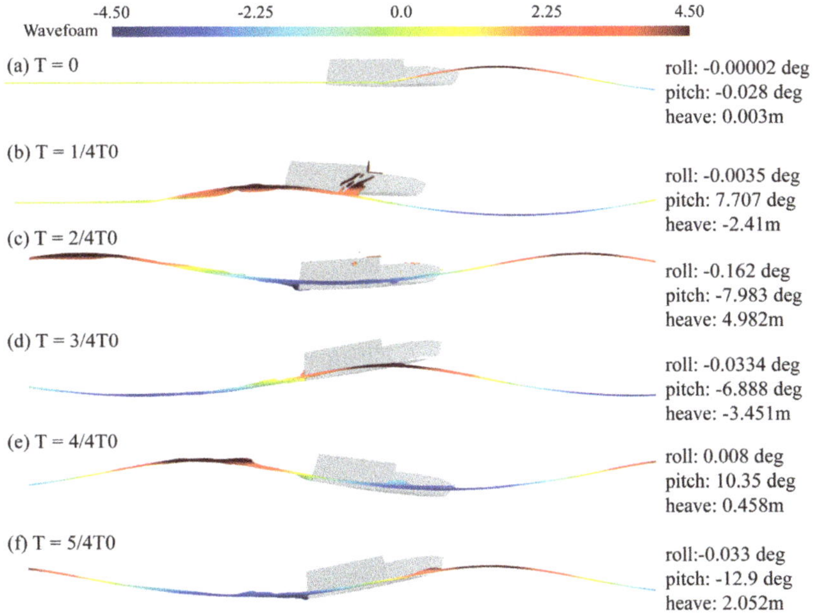

Figure 21. The attitude of USV in waves at typical times under 7 level sea state with the 0° wave.

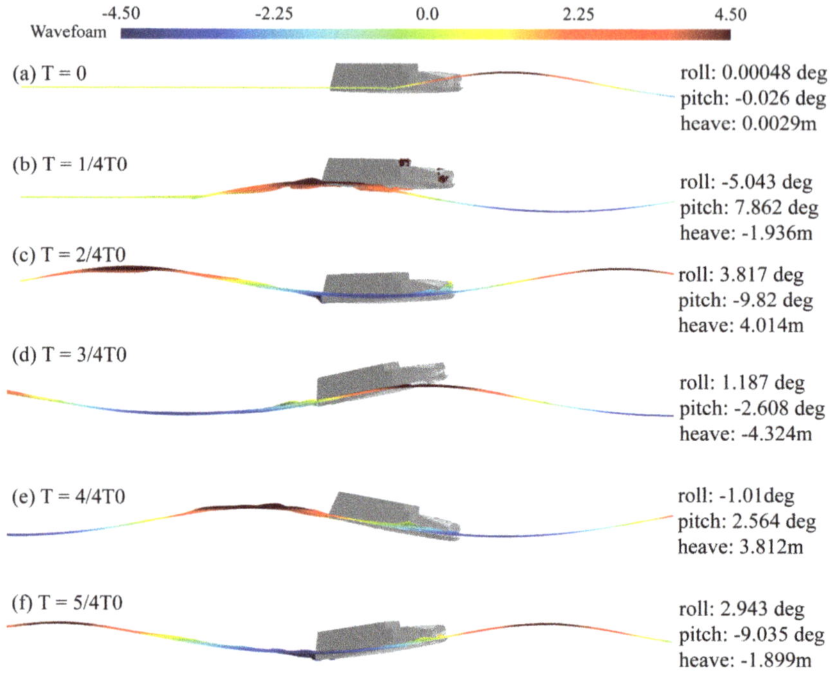

Figure 22. The attitude of USV in waves at typical times under 7 level sea state with the 30° wave direction.

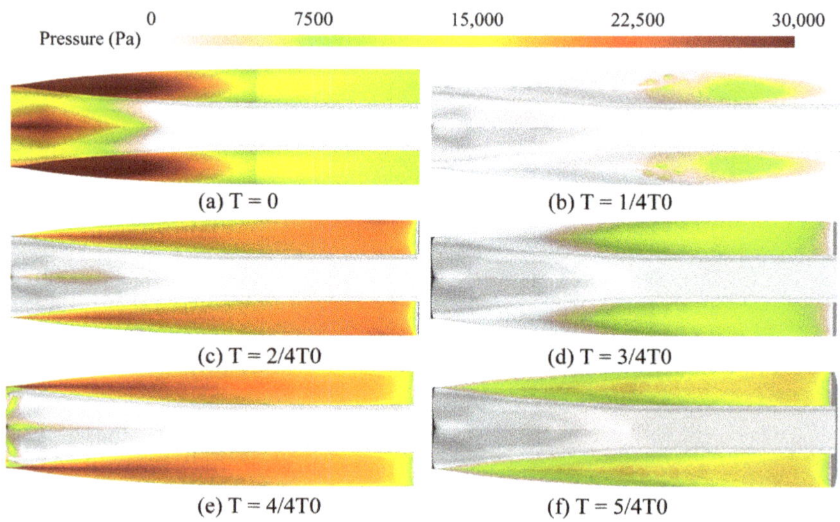

Figure 23. Pressure distribution at the bottom of the USV under 7 level sea state with the 0° wave direction (Positive pressure area).

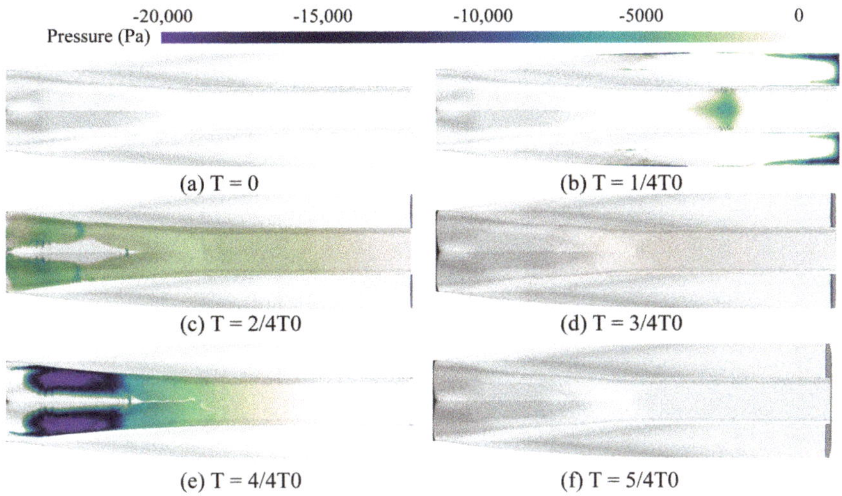

Figure 24. Pressure distribution at the bottom of the USV under 7 level sea state with the 0° wave direction (Negative pressure area).

Figures 25 and 26 show pressure distribution at the bottom of the USV when the wave direction is 30° under 7 level sea state, respectively. When $T = 0$ and $4/4T0$, the USV has a large static pressure. The maximum pressure reached 28,390 Pa. When $T = 2/4T0$ and $T = 5/4T0$, the bottom positive pressure was relatively large. When $T = 2/4T0$ and $4/4T0$, the bottom negative pressure was large. The maximum negative pressure reached −19,520 Pa.

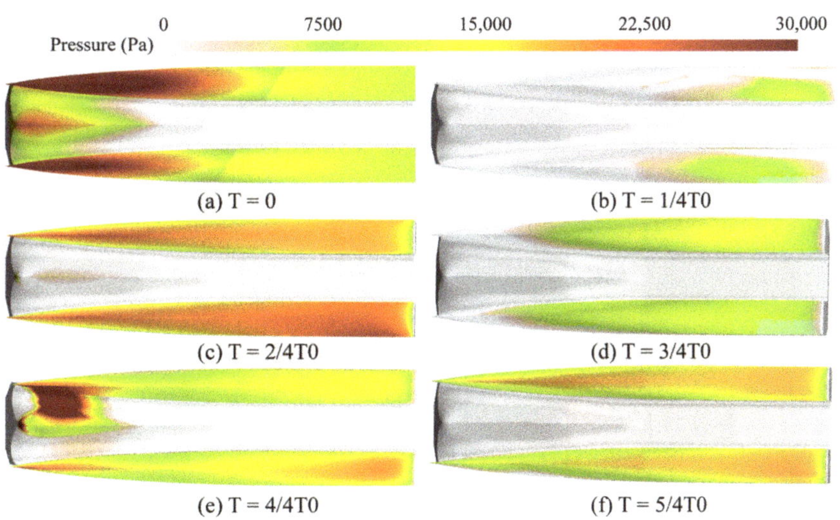

Figure 25. Pressure distribution at the bottom of the USV under 7 level sea state with the 30° wave direction (Positive pressure area).

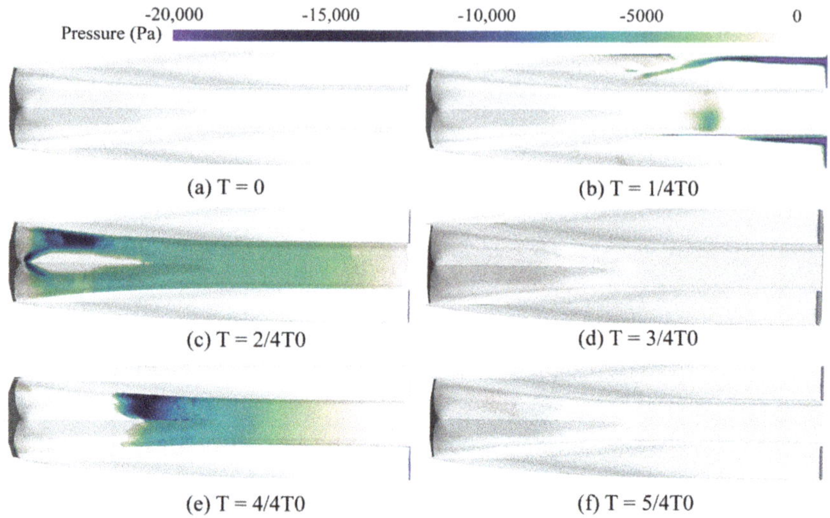

Figure 26. Pressure distribution at the bottom of the USV under 7 level sea state with the 30° wave direction (Negative pressure area).

4.5. Free Surface Waveform and USV Three-Dimensional Attitude

Figures 27 and 28 show free surface waveform and USV three-dimensional attitude at different times in different wave directions under 5 level and 7 level sea states, respectively. It can be seen from Figure 27 that when the wave direction was 0° and $T = 0$ to $1/4T0$, the bow was in the wave trough, and the stern passes through the wave peak position. The bow diving happens slightly. At this time, the stern causes wave and wake. When $T = 1/4T0$ to $2/4T0$, the whole USV hull is in the wave trough section, and the USV has large sinking and bow diving. When $T = 2/4T0$ to $3/4T0$, the whole USV is in the wave peak section, and the USV begins to rise, which changes from the trim by bow state to the trim by stern state, resulting in bow rising. When $T = 3/4T0$ to $4/4T0$, the ship moves from the wave peak to the wave trough. At $T = 4/4T0$, the bow is in the wave trough, and the USV sinks and bow diving happens. When $T = 4/4T0$ to $5/4T0$, the USV is between the wave trough and the wave peak and has experienced a change process from bow diving to bow rising. At $T = 5/4T0$, the stern is in the wave trough, and the bow is in the wave peak. In Figure 28, the process of the USV motion in 30° wave direction is similar, but the bow rising and bow diving amplitude in 30° wave direction are less than the 0° wave direction, and the wave making and wake intensity in stern area are less than the 0° wave direction. The USV rolling is obvious in 30° wave direction. The motion process of the USV at different times under the 7 level sea state follows the wave periodic variation, which is similar to the 5 level sea state. The difference is that the heave and pitch amplitude of the USV under the 7 level sea state is very severe, which is much larger than the 5 level sea state, but it still meets the stability requirements without capsizing.

4.6. Boundary Layer

Figure 29 is the boundary layer represented by slices colored with dimensionless axial velocity limited to $U_x/U_0 = 0.9$ at 5 level sea state with the 0° wave direction. Among them, Figure 29a–d are the underwater perspectives of $T = 1/4T0$, $T = 2/4T0$, $T = 3/4T0$, and $T = 4/4T0$, respectively. It can be found that the motion of wave and the USV has a different influence on the boundary layer at different times. When $T = 1/4T0$, the USV moves from wave trough to wave peak, and the bow just reaches the wave peak position. Most of the bow submerges into the water, and the midship to the bow area has a thick boundary layer. When $T = 2/4T0$, the whole USV is at the wave peak position (see

Figure 22), and the boundary layer is the thinnest. When $T = 3/4T0$, the whole USV is at the wave trough position (see Figure 22), and the bow is heading to the wave peak with a thick boundary layer at the bow area. When $T = 4/4T0$, the whole USV is at the wave peak position, the midship and the bow area have the thinnest boundary layer, and the stern area is relatively thick. Figure 30 is the boundary layer represented by slices colored with dimensionless axial velocity limited to $U_x/U_0 = 0.9$ at 5 level sea state with 30° wave direction. Among them, Figure 30a–d are the underwater perspectives of $T = 1/4T0$, $T = 2/4T0$, $T = 3/4T0$, and $T = 4/4T0$, respectively. It can be found that the thickness distribution of the boundary layer is very different with the 30° wave direction from that of the 0° wave direction. When $T = 1/4T0$ and $T = 3/4T0$, the USV is at the wave trough position, and the boundary layer is thin. When $T = 2/4T0$, the midship and the bow area are at the wave peak position with a thick boundary layer, and the stern area is relatively thin. When $T = 4/4T0$, the bow is heading from the wave trough to the wave peak. The bow area has a thick boundary layer.

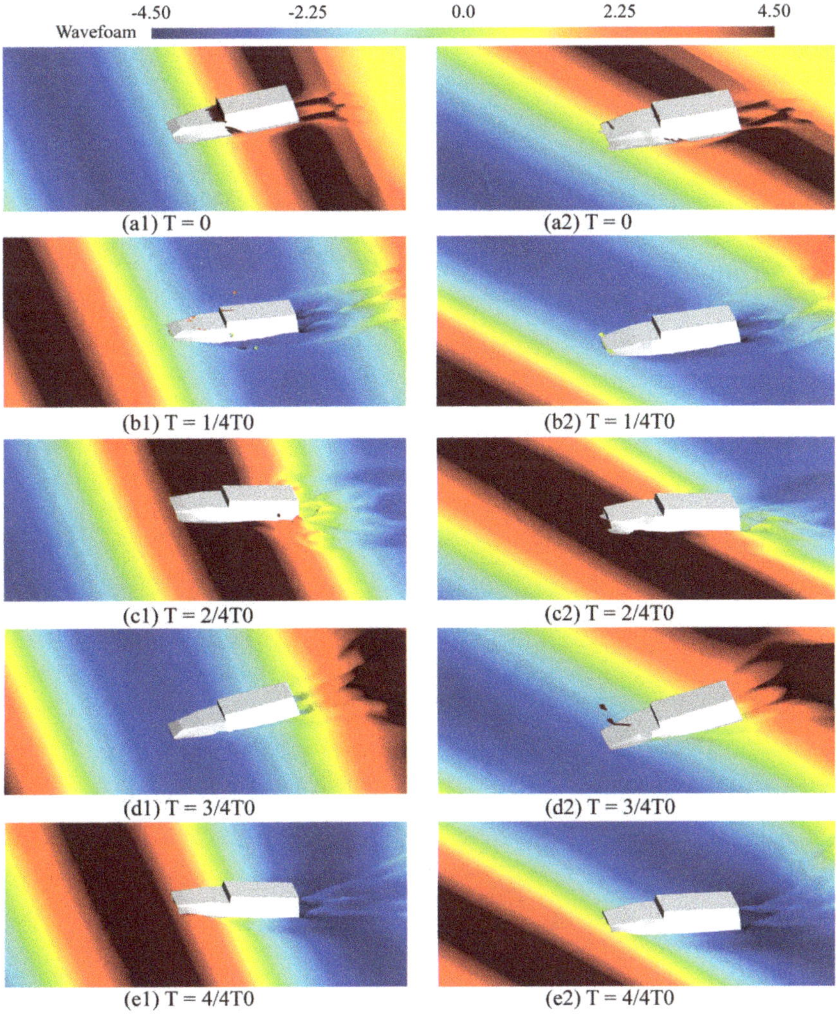

Figure 27. Free surface waveform and USV three-dimensional attitude at 7 level sea state ((**left**): 0° wave direction; (**right**): 30° wave direction).

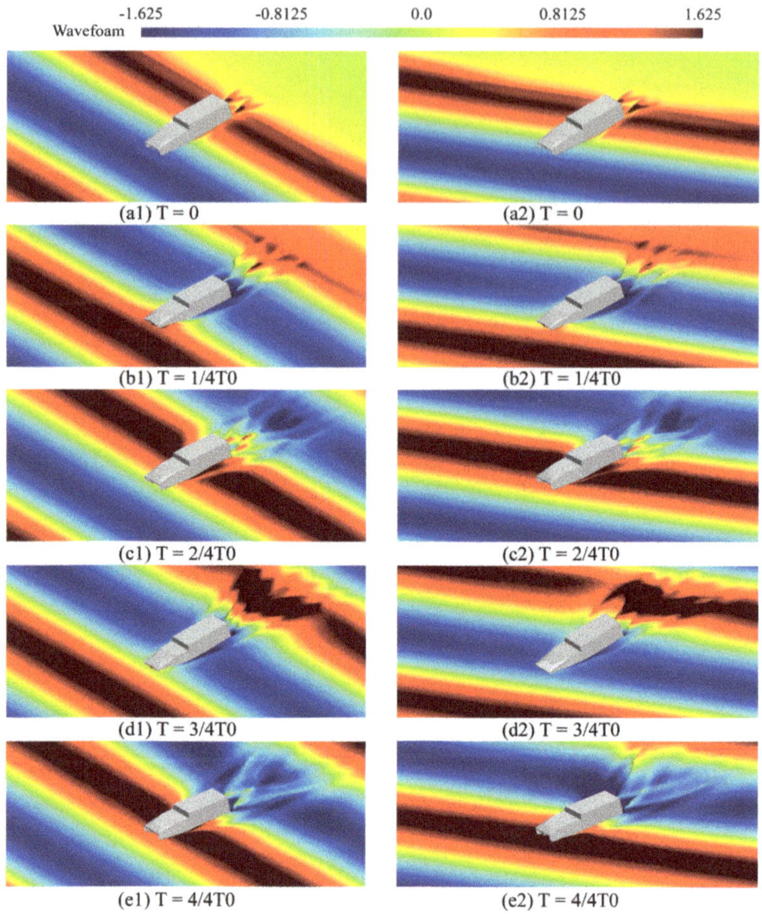

Figure 28. Free surface waveform and USV three-dimensional attitude at 5 level sea state ((**left**): 0° wave direction; (**right**): 30° wave direction).

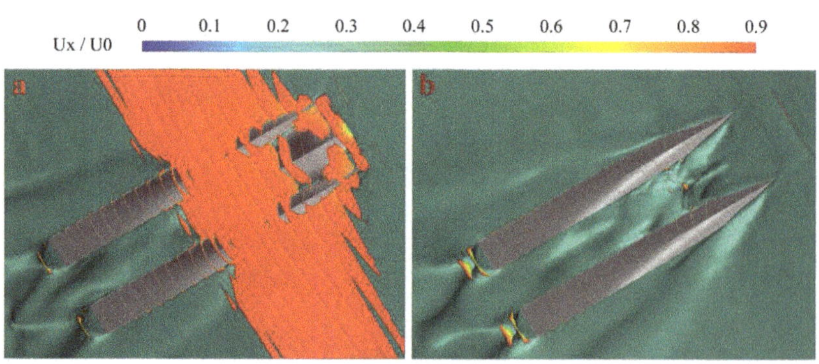

Figure 29. *Cont.*

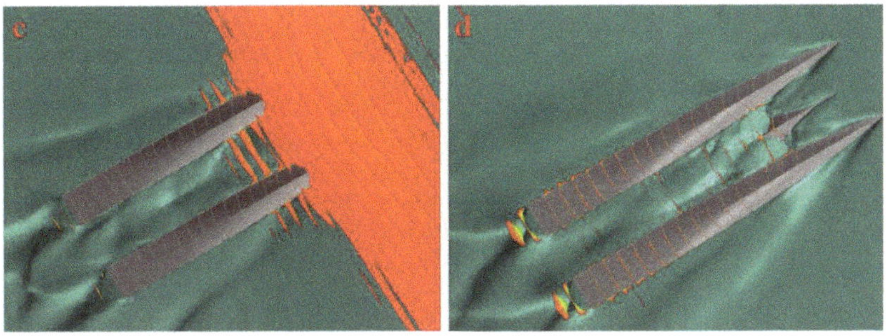

Figure 29. Boundary layer represented by slices colored with dimensionless axial velocity limited to U_x/U_0 = 0.9 at 5 level sea state with the 0° wave direction. (**a**) T = 1/4T0; (**b**) T = 2/4T0; (**c**) T = 3/4T0; (**d**) T = 4/4T0.

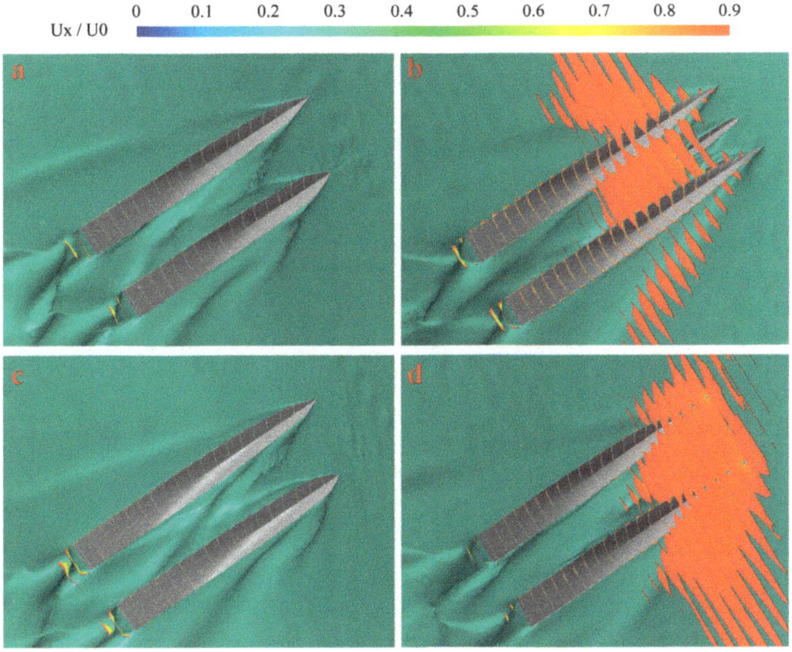

Figure 30. Boundary layer represented by slices colored with dimensionless axial velocity limited to U_x/U_0 = 0.9 at 5 level sea state with the 30° wave direction. (**a**) T = 1/4T0; (**b**) T = 2/4T0; (**c**) T = 3/4T0; (**d**) T = 4/4T0.

Figure 31 shows the boundary layer represented by slices colored with dimensionless axial velocity limited to U_x/U_0 = 0.9 at 7 level sea state with the 0° wave direction. Among them, Figure 31a–d are the underwater perspectives of T = 1/4T0, T = 2/4T0, T = 3/4T0, and T = 4/4T0, respectively. When T = 1/4T0, the USV moves from wave peak to wave trough, and the bow just reaches the wave trough position, the stern reaches the wave peak position (see Figure 21) with a thin boundary layer. When T = 2/4T0, the whole USV is at the wave trough position (see Figure 21), the midship to the bow area has a thick boundary layer. When T = 3/4T0, the whole USV is at the wave peak position (see Figure 21), and the bow is heading to the wave peak with a thick boundary layer at the bow area. The stern is at the transition region between the wave trough and wave peak, and the stern boundary layer is thick. When $T = 4/4T0$, the whole USV is at the wave trough position, the midship and the bow area have the thinnest boundary layer, and the stern area is relatively thick. Figure 32 shows the boundary layer represented by slices colored with dimensionless axial velocity limited to $U_x/U_0 = 0.9$ at 7 level sea state with the 30° wave direction. Among them, Figure 32a–d are the underwater perspectives of $T = 1/4T0$, $T = 2/4T0$, $T = 3/4T0$ and $T = 4/4T0$, respectively. It can be found that when $T = 1/4T0$, the whole USV is at the wave peak position, the bow is in the transition position from the peak to the trough, and the stern boundary layer is thick. When $T = 2/4T0$, the whole USV is at the wave trough position with a thin boundary layer. When $T = 3/4T0$, the bow area is at the wave peak position. The stern is at the transition region between wave trough and wave peak, and the bow area has a thick boundary layer. When $T = 4/4T0$, the whole USV is at the wave trough position with a thin boundary layer.

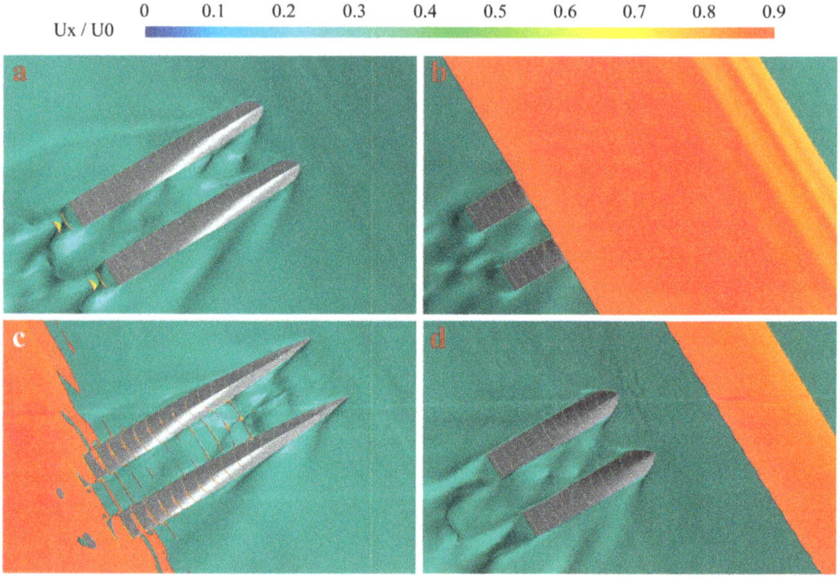

Figure 31. Boundary layer represented by slices colored with dimensionless axial velocity limited to $U_x/U_0 = 0.9$ at 7 level sea state with the 0° wave direction. (**a**) $T = 1/4T0$; (**b**) $T = 2/4T0$; (**c**) $T = 3/4T0$; (**d**) $T = 4/4T0$.

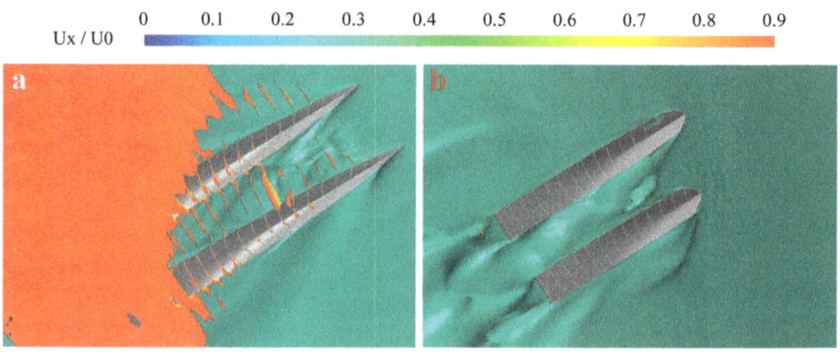

Figure 32. *Cont.*

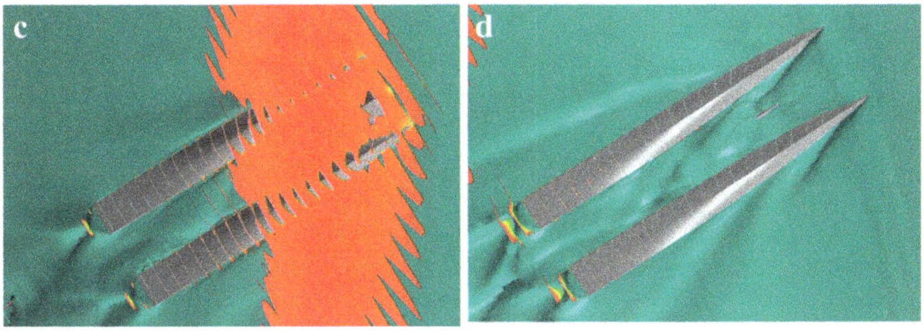

Figure 32. Boundary layer represented by slices colored with dimensionless axial velocity limited to U_x/U_0 = 0.9 at 7 level sea state with the 30° wave direction. (**a**) $T = 1/4T0$; (**b**) $T = 2/4T0$; (**c**) $T = 3/4T0$; (**d**) $T = 4/4T0$.

5. Conclusions

In this paper, the interaction between the wave and a catamaran USV is studied based on the CFD method. The simulation results are concluded as follows:

(1) The amplitude of the heave motion in the 0° and 30° wave directions is basically the same under the 7 level sea state, and it changes periodically in the range of −5 m to 5 m. The pitch angle ranged from −12 to 12 degrees at the 0° wave direction and from −12 to 10 degrees at the 30° wave direction. The positive value for bow raising and the negative value for bow diving. The stable roll angle of the USV is very small in the 0° wave direction. The roll angle of the USV is obvious in the 30° wave direction, which ranged from −7 to 6 degrees. The total resistance in the 0° wave direction is slightly greater than that in the 30° wave direction. The six degrees of freedom (6DOF) motion of the USV under 5 level sea state is similar to that of 7 level sea state, but the 6DOF motion amplitude of the 5 level sea state is smaller than that of 7 level sea state.

(2) Taking the 7 level sea state as an example, the attitude of the USV in waves at different times is analyzed. The results show that the USV moves with six degrees of freedom following the wave cycle, and in the whole process of USV movement around the waves, the stability of the USV meets the stability requirements, and no capsizing occurs. From the pressure distribution analysis of the USV, the positive pressure at the bottom of the USV is mainly concentrated in the two side body areas, and the negative pressure is mainly concentrated in the area between the side bodies.

(3) The thickness of the boundary layer in the 0° and 3° wave directions of 5 level and 7 level sea states is analyzed, respectively. The results show that the wave direction has a great influence on the boundary layer thickness. At the 0° wave direction, the distribution of the boundary layer is thicker when the USV is at the wave trough position. At the 30° wave direction, the distribution of the boundary layer is thicker when the USV is at the wave peak position. 5 level and 7 level sea states have the same conclusion.

Author Contributions: Conceptualization, W.L. (Wanzhen Luo); methodology, S.H.; software, W.L. (Wanzhen Luo); validation, K.W. and S.H.; formal analysis, W.L. (Weiqi Liu); investigation, W.L. (Weiqi Liu); resources, K.W.; data curation, S.H.; writing—original draft preparation, W.L. (Wanzhen Luo); writing—review and editing, W.L. (Wanzhen Luo); visualization, S.H.; supervision, S.H.; project administration, K.W.; funding acquisition, W.L. (Wanzhen Luo) and S.H. All authors have read and agreed to the published version of the manuscript.

Funding: This work was financially supported by the China Postdoctoral Science Foundation (Grant No. 2019M663243); National Natural Science Foundation of China (52101379, 52171289); Natural Science Foundation of Guangdong Province, China (2021A1515012134, 2021A1515011771); the Guangdong Basic and Applied Basic Research Foundation (2019A1515110721); the Fundamental Research Funds for the Central Universities (Grant NO. 2021qntd04).

Institutional Review Board Statement: Not applicable.

Informed Consent Statement: Not applicable.

Data Availability Statement: The data that support the findings of this study are available within the article.

Conflicts of Interest: The authors declare no conflict of interest.

References

1. Shetty, V.K.; Sudit, M.; Nagi, R. Priority-based assignment and routing of a fleet of unmanned combat aerial vehicles. *Comput. Oper. Res.* **2008**, *35*, 1813–1828. [CrossRef]
2. Arribas, F.P. Some methods to obtain the added resistance of a ship advancing in waves. *Ocean Eng.* **2007**, *34*, 946–955. [CrossRef]
3. Ozdemir, Y.H.; Barlas, B. Numerical study of ship motions and added resistance in regular incident waves of KVLCC2 model. *Int. J. Nav. Archit. Ocean Eng.* **2017**, *9*, 149–159. [CrossRef]
4. Huang, S.; Jiao, J.; Chen, C. CFD prediction of ship seakeeping behavior in bi-directional cross wave compared with in uni-directional regular wave. *Appl. Ocean Res.* **2021**, *107*, 102426. [CrossRef]
5. Kim, D.; Song, S.; Jeong, B.; Tezdogan, T. Numerical evaluation of a ship's manoeuvrability and course keeping control under various wave conditions using CFD. *Ocean Eng.* **2021**, *237*, 109615. [CrossRef]
6. Vos, J.D.; Hekkenberg, R.G.; Banda, O. The impact of autonomous ships on safety at sea—A statistical analysis. *Reliab. Eng. Syst. Saf.* **2021**, *210*, 107558. [CrossRef]
7. Chen, C.; Sasa, K.; Prpić-Oršić, J.; Mizojiri, T. Statistical analysis of waves' effects on ship navigation using high-resolution numerical wave simulation and shipboard measurements. *Ocean Eng.* **2021**, *229*, 108757. [CrossRef]
8. Tarafder, M.S.; Suzuki, K. Computation of wave-making resistance of a catamaran in deep water using a potential-based panel method. *Ocean Eng.* **2007**, *34*, 1892–1900. [CrossRef]
9. Fang, C.C.; Chan, H.S. Investigation of seakeeping characteristics of high-speed catamarans in waves. *J. Mar. Sci. Technol.* **2004**, *12*, 7–15. [CrossRef]
10. Ma, C.; Zhu, Y.; He, J.; Zhang, C.; Wan, D.; Yang, C.; Noblesse, F. Nonlinear corrections of linear potential-flow theory of ship waves. *Eur. J. Mech.-B/Fluids* **2018**, *67*, 1–14. [CrossRef]
11. Min-Guk, S.; Kyung-Kyu, Y.; Dong-Min, P.; Yonghwan, K. Numerical analysis of added resistance on ships in short waves. *Ocean Eng.* **2014**, *87*, 97–110.
12. Simonsen, C.D.; Otzen, J.F.; Joncquez, S.; Stern, F. EFD and CFD for KCS heaving and pitching in regular head waves. *Mar. Sci. Technol.* **2013**, *18*, 435–459. [CrossRef]
13. Tezdogan, T.; Demirel, Y.K.; Kellett, P.; Khorasanchi, M.; Incecik, A.; Turan, O. Full-scale unsteady RANS CFD simulations of ship behaviour and performance in head seas due to slow steaming. *Ocean Eng.* **2015**, *97*, 186–206. [CrossRef]
14. Deng, R.; Luo, F.; Wu, T.; Chen, S.; Li, Y. Time-domain numerical research of the hydrodynamic characteristics of a trimaran in calm water and regular waves. *Ocean Eng.* **2019**, *194*, 106669. [CrossRef]
15. Wilcox, D.C. *Turbulence Modeling for CFD*; DCW Industries Inc.: La Canada, CA, USA, 1994; pp. 15–19.
16. Menter, F.R. Two-equation eddy-viscosity turbulence models for engineering applications. *AIAA J.* **1994**, *32*, 1598–1605. [CrossRef]
17. Muzaferija, S. Computation of free surface flows using interface-tracking and interface-capturing methods. In *Nonlinear Water Wave Interaction*; Computational Mechanics: Southampton, UK, 1998.
18. *STAR-CCM+ User Guide*; Version 15.02; CAE Software: Melville, NY, USA, 2020.
19. Roy, C.J.; Heintzelman, C.; Roberts, S.J. Estimation of Numerical Error for 3D Inviscid Flows on Cartesian Grids. In Proceedings of the 45th AIAA Aerospace Sciences Meeting and Exhibit, Reno, NV, USA, 1–13 January 2007.
20. Guo, C.Y.; Wu, T.C.; Zhang, Q.; Lou, W.Z. Numerical simulation and experimental studies on aft hull local parameterized non-geosim deformation for correcting scale effects of nominal wake field. *Brodogradnja* **2017**, *68*, 77–96. [CrossRef]
21. Fenton, J.D. A Fifth-Order Stokes Theory for Steady Waves. *J. Waterw. Port Coast. Ocean Eng.* **1985**, *111*, 216–234. [CrossRef]
22. Kim, J.W.; Jang, H.; Baquet, A.; O'Sullivan, J.; Lee, S.; Kim, B.; Read, A.; Jasak, H. Technical and Economic Readiness Review of CFD-Based Numerical Wave Basin for Offshore Floater Design. In Proceedings of the Offshore Technology Conference, Houston, TX, USA, 2–5 May 2016.
23. Wu, T.C.; Luo, W.Z.; Jiang, D.P.; Deng, R.; Huang, S. Numerical Study on Wave-Ice Interaction in the Marginal Ice Zone. *J. Mar. Sci. Eng.* **2020**, *9*, 4. [CrossRef]
24. Jiao, J.L.; Huang, S.X.; Soares, C.G. Numerical investigation of ship motions in cross waves using CFD. *Ocean Eng.* **2021**, *223*, 108711. [CrossRef]
25. Kim, D.; Song, S.; Tezdogan, T. Free running CFD simulations to investigate ship manoeuvrability in waves. *Ocean Eng.* **2021**, *236*, 109567. [CrossRef]
26. Perić, R.; Abdel-Maksoud, M. Analytical prediction of reflection coefficients for wave absorbing layers in flow simulations of regular free-surface waves. *Ocean Eng.* **2018**, *147*, 132–147. [CrossRef]

Journal of
Marine Science and Engineering

MDPI

Article

Study on Sloshing Characteristics in a Liquid Cargo Tank under Combination Excitation

Qiong Zhang [1,2], Bo Shui [1] and Hanhua Zhu [3,*]

1 School of Transportation and Logistics Engineering, Wuhan University of Technology, Wuhan 430063, China
2 School of Materials Science and Engineering, Wuhan Institute of Technology, Wuhan 430205, China
3 School of Naval Architecture, Ocean and Energy Power Engineering, Wuhan University of Technology, Wuhan 430063, China
* Correspondence: hh.zhu@163.com; Tel.: +86-027-86551193

Abstract: Sloshing is a common flow phenomenon in liquid cargo tanks and has a great negative impact on the stability and safety of ship navigation. It is important to understand the sloshing process of tanks under the excitation of complex external conditions for the transportation of liquid cargo. In this paper, the sloshing characteristics of a liquid cargo tank are studied under the combination excitation conditions of roll and surge. The pressure distribution characteristics at different positions of the cargo tank are discussed, along with the influence of different excitation conditions on the pressure of the cargo tank. The results show that under the condition of combination excitation, the fluid sloshes along the diagonal direction of the tank, and the peak liquid height and peak pressure are located on the diagonal corner of the tank. The peak pressure at the lowest point on the diagonal of the tank is proportional to the amplitude of the roll angle and surge, and the change in roll angle amplitude has a significant impact on the pressure and liquid height at different positions.

Keywords: combination excitation; liquid cargo tank; sloshing

Citation: Zhang, Q.; Shui, B.; Zhu, H. Study on Sloshing Characteristics in a Liquid Cargo Tank under Combination Excitation. *J. Mar. Sci. Eng.* **2022**, *10*, 1100. https://doi.org/10.3390/jmse10081100

Academic Editor: Eva LOUKOGEORGAKI

Received: 28 July 2022
Accepted: 10 August 2022
Published: 11 August 2022

1. Introduction

With the continuous and in-depth implementation of international marine environmental protection, liquefied natural gas (LNG) fuel has become increasingly used due to its high efficiency and cleanliness. When an LNG carrier encounters adverse sea conditions, the excitation of waves causes the mass center and free surface of the LNG in the cargo hold to deviate from the original equilibrium state, resulting in sloshing. A strong impact on the bulkhead structure when the ship encounters intense sloshing results in property losses and casualties. However, LNG carriers in actual navigation encounter more complex excitations (roll and surge) than single excitations (roll or surge). Therefore, the study of the sloshing behavior and anti-sloshing design under the combination excitation is significant for the navigation safety of LNG carriers.

In terms of theoretical research, Moiseev [1] proposed a nonlinear sloshing theory based on potential flow theory using the asymptotic and modal methods. Faltinsen et al. [2] used the boundary element method to predict the nonlinear sloshing model of a rectangular tank under resonance excitation conditions. In another study, Faltinsen et al. [3] used the modal analysis method to study the two-dimensional liquid sloshing process in a circular tank. Budiansky [4] theoretically calculated the natural modes and frequencies of circular tubes and spherical tanks at different liquid loading heights.

In terms of experimental research, Kim et al. [5] visualized the sloshing process using a high-speed camera and discussed the causes of pressure measurement errors in the sloshing experiment from both macroscopic and microscopic aspects. Trimulyono et al. [6] evaluated the equivalent damping coefficient of an FPSO model storage tank under roll excitation at different loading rates. The results showed that ship damping is inversely proportional to the loading rate and the amplitude of the roll angle. Yu et al. [7] studied the sloshing mitigation characteristics of vertical grids under different frequency excitation conditions. Akyildiz et al. [8] designed an experimental device for sloshing to study the pressure distribution characteristics at different positions in the tank and the three-dimensional effect caused by sloshing. Zou et al. [9] conducted a sloshing model experiment of viscous liquid to study the influence of liquid viscosity on sloshing. The results show that viscous friction has a certain dissipation effect on the sloshing pressure. Doh et al. [10] used a panoramic particle image velocimetry (PIV) system to conduct experimental research on the sloshing characteristics in the LNG ship cargo tank model. The results showed that the oscillation amplitude has a large impact on the horizontal velocity of the liquid.

The numerical simulation of sloshing can be further divided into the finite difference method (FDM) [11], the finite element method (FEM) [12], the finite volume element method (FVM) [13], the boundary element method (BEM) [14] and the smooth particle dynamics method (SPH) [15]. Considering the influence of the turbulence model on the numerical simulation results of sloshing, Liu et al. [16] used the large eddy simulation method (LES) and the Smagorinsky subgrid scale (SGS) to simulate the turbulence effect and verified the accuracy of the model with experimental data. Tang et al. [17] used STAR-CCM+ software to numerically simulate the sloshing process in the tank. The results showed that adjusting the value of the gradient parameter k in the k-ε model can make the sloshing waveform in the tank more realistic. In terms of free surface capture, the volume of fluid (VOF) free surface capture algorithm has been used for sloshing numerical simulation research. For example, Borg et al. [18] studied the sloshing characteristics of fluid in a rectangular box with a chamfer and the sloshing reduction characteristics of the baffled tank under different loading rates. Hoch et al. [19] studied the sloshing behavior of liquid in a storage tank. The results showed that the VOF method better describes the change in the liquid level, but the predicted sloshing impact pressure is lower than the experimental measurement. Based on the OpenFOAM simulation platform, Xue et al. [20] numerically simulated the anti-sloshing characteristics of porous media materials in cylindrical tanks. In addition, some scholars have used the level-set method to capture sloshing free surfaces. For example, Xin et al. [21] used the three-dimensional gradient-augmented level set (GALS) two-phase flow model to capture the sloshing nonlinear free surface. The results showed that the numerical simulation is in good agreement with the experiment.

In summary, the relevant theoretical analysis and research on liquid sloshing characteristics in holding tanks have been relatively developed, but there are also some limitations. For example, when the sloshing in the tank is intense and there are obvious liquid level breaking and attacking phenomena, the theoretical analysis solution is no longer applicable. Although numerical simulation studies can obtain more comprehensive flow information of the flow field in the tank, the simulation results still need to be compared and verified with experiments. Furthermore, in the existing studies, for both numerical simulation and experimental research, scholars mostly focus on the sloshing characteristics of tanks under a single degree of freedom, while the ship navigation process is mostly multi-degree-of-freedom complex motion under the action of wind and waves. Therefore, it is necessary to study the sloshing characteristics of liquid tanks under combination excitation involving multiple degrees of freedom.

In this paper, the sloshing process of fluid in a cargo tank under the combination excitation of roll and surge is the research object. Through numerical simulation and experimental verification methods, the trend of the sloshing liquid level under combination excitation is clarified, and the impact pressure distribution of the bulkhead under combined excitation is explored. The research conclusions obtained in this paper provide theoretical support for tank structure design and anti-sloshing design under combination excitation.

2. Numerical Model and Experimental Platform

2.1. Geometric Model

Figure 1 shows the geometric model of the cargo tank studied in this paper. The model is a membrane-type cargo tank, commonly used in LNG carriers and proportionally reduced to laboratory scale. The size of the real cargo tank is 15.75 m × 15.55 m × 8.9 m (width × length × height) at a scale of 1:50. The model size is 0.315 m × 0.311 m × 0.178 m (width × length × height). The data monitoring points in the numerical simulation are selected as the intersection points of the vertical lines and the horizontal lines as shown in Figure 1. Among them, vertical lines P2 and S2 are located at the geometric centers of the two sidewalls, while S4 is located at the edge of the geometric model, and straight lines H89, H50 and H34 are horizontal lines with heights of 89 mm, 50 mm and 34 mm, respectively. Monitoring points were named, for example, P1H50, which represents the intersection of the vertical line P1 and the horizontal line at a height of 50 mm.

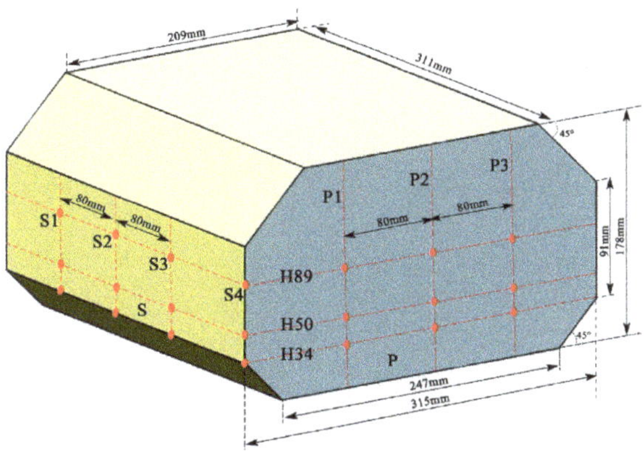

Figure 1. Three-dimensional model of cargo tank in experimental test and numerical simulation.

2.2. Numerical Model

The dynamic mesh method in the open source CFD tool OpenFOAM was adopted to solve the Navier-Stokes and other transport equations under the combination excitation. The surfaces of the tank are the wall boundaries. The Navier-Stokes equation was solved by the PISO method, and the turbulence model adopted the standard k-ε model. The Courant number was set to 0.2, and the time step was adaptively adjusted. The second-order schemes Gauss vanLeer and Crank Nicolson were used for the spatial and time discretization, respectively. Hydrostatic pressure and zero velocity initialized the flow field. In all the cases of sloshing, the values of α_l and α_g were both 0.5.

2.2.1. Volume Fraction Equation

The primary focus of this paper is on gas–liquid two-phase flow, in which the liquid phase is water and the gas phase is air. The interface between the two phases can be obtained by solving the continuity equation of the volume fraction. Equation (1) describes the volume fraction equation of the liquid phase as follows:

$$\frac{\partial}{\partial t}(\rho_l \alpha_l) + \frac{\partial}{\partial x_j}(\rho_l \alpha_l u_i) = 0 \tag{1}$$

where the subscripts l and g represent the liquid and gas phases, respectively; ρ is the density; α is the volume fraction; and the volume fraction of the liquid phase and the gas phase are represented by α_l and α_g, respectively. Their relationship is shown in Equation (2) as follows:

$$\alpha_l + \alpha_g = 1 \tag{2}$$

2.2.2. Momentum Equation

In the VOF model, the entire computational domain shares a set of momentum equations, and the velocity field obtained by solving is composed of the gas phase and the liquid phase. The momentum equations are shown in Equations (3)–(5) as follows:

$$\frac{\partial(\rho u_i)}{\partial t} + \frac{\partial(\rho u_i u_j)}{\partial x_j} = -\frac{\partial p}{\partial x_i} + \rho g_i + \frac{\partial}{\partial x_j}\left(\mu \frac{\partial u_i}{\partial x_j} - \rho \overline{u'_i u'_j}\right) + \vec{F}_{CSF} \tag{3}$$

$$\rho = \alpha_l \rho_l + (1 - \alpha_l)\rho_g \tag{4}$$

$$\mu = \alpha_l \mu_l + (1 - \alpha_l)\mu_g \tag{5}$$

where ρ is the average density of the fluid in the grid; μ is the average viscosity of the fluid in the grid; $\vec{u}$ is the vector velocity; and $\vec{F}_{CSF}$ is the surface tension. The continuum surface force (CSF) surface tension model [22] was selected. The CSF surface tension model is shown in Equation (6) as follows:

$$\vec{F}_{CSF} = \sigma_{l.g} \frac{\rho k_l \nabla \alpha_l}{\frac{1}{2}(\rho_l + \rho_g)} \tag{6}$$

where $\sigma_{l,g}$ is the surface tension coefficient between the gas and liquid phases where $\sigma_{l,g} = 0.071n/m$; k is the surface curvature at the gas–liquid interface; and n is the surface normal vector of the interface, which is defined in Equation (7) as follows:

$$n = \nabla \alpha_l \tag{7}$$

The surface curvature k at the gas–liquid interface is defined as the divergence of the unit vector $\hat{n}$ as follows:

$$\hat{n} = \frac{n}{|n|} \tag{8}$$

$$k = \nabla \cdot \hat{n} \tag{9}$$

2.2.3. Transport Equations for the Standard k-ε Model

The turbulence kinetic energy, k, and its rate of dissipation, ε, are obtained from the following transport equations:

$$\frac{\partial}{\partial t}(\rho k) + \frac{\partial}{\partial x_i}(\rho k u_i) = \frac{\partial}{\partial x_j}\left[\left(\mu + \frac{\mu_t}{\sigma_k}\right)\frac{\partial k}{\partial x_j}\right] + G_k + G_b - \rho\varepsilon - Y_M \tag{10}$$

$$\frac{\partial}{\partial t}(\rho \varepsilon) + \frac{\partial}{\partial x_i}(\rho \varepsilon u_i) = \frac{\partial}{\partial x_j}\left[\left(\mu + \frac{\mu_t}{\sigma_\varepsilon}\right)\frac{\partial \varepsilon}{\partial x_j}\right] + C_{1\varepsilon}\frac{\varepsilon}{k}(G_k + C_{3\varepsilon}G_b) - C_{2\varepsilon}\rho\frac{\varepsilon^2}{k} \tag{11}$$

$$\mu_t = \rho C_\mu \frac{k^2}{\varepsilon} \tag{12}$$

In these equations, G_k represents the generation of turbulence kinetic energy due to the mean velocity gradients, calculated as described in Modeling Turbulent Production in the k-ε Models; G_b is the generation of turbulence kinetic energy due to buoyancy, calculated as described in Effects of Buoyancy on Turbulence in the k-ε models; and Y_M represents the contribution of the fluctuating dilatation in compressible turbulence to the overall dissipation rate, calculated as described in Effects of Compressibility on Turbulence in the k-ε models.

2.3. Six-Degrees-of-Freedom Motion Experimental Platform

As shown in Figure 2, the six-degrees-of-freedom experimental motion platform built mainly includes a sloshing tank, hydraulic cylinder and a six-degrees-of-freedom motion platform. The sloshing tank was installed above the platform. The motion platform is driven by six hydraulic cylinders controlled by the host to move in six degrees of freedom and further drive the movement of the tank to realize the sloshing of the liquid in the tank. The motion of the experimental platform is shown in Equation (13) as follows [23]:

$$Z = A \cdot sin(2\pi f \cdot t) \tag{13}$$

where Z is the response output, A is the motion amplitude, f is the motion frequency, and t is the time. During the experiment, the motion platform was controlled by adjusting the values of A and f. The experimental platform can run continuously for more than 12 h, and the position drift of each hydraulic cylinder meets the requirements of less than 0.00025 m.

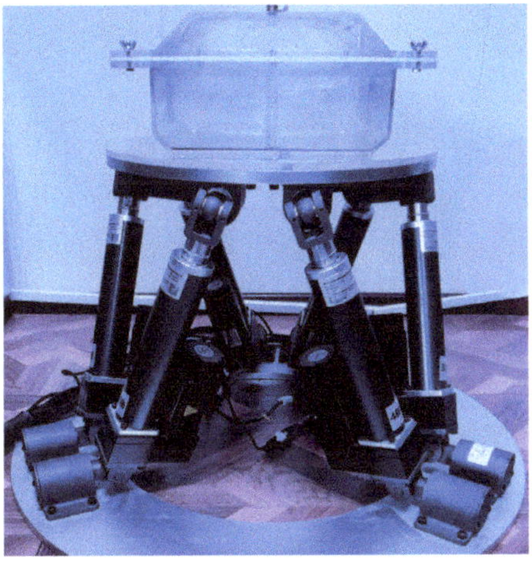

Figure 2. Six-degrees-of-freedom motion experimental platform.

3. Model Validation

To verify the accuracy of the numerical simulation method used in this paper, the experimental results and numerical simulation results obtained under the same conditions were compared and analyzed, including qualitative comparisons of the gas-liquid interface profile (as shown in Figure 3) and quantitative comparisons of dimensionless liquid level height variation characteristics (as shown in Figure 4). The selected conditions for validation are shown in Table 1.

Table 1. Model validation parameter settings.

Type	Roll Angle φ (°)	Surge Amplitude λ (mm)	Frequency f (Hz)	Filling Rate
Roll excitation	10	-	0.58	0.5
Surge excitation	-	50	0.58	0.5
Combination excitation	10	50	0.58	0.5

Figure 3 shows that the liquid surface topography obtained from the experiment and simulation under the rolling, surge and combination excitation are consistent. Compared with Figure 4, it can be seen that there is a slight deviation between the numerical simulation results and the experimental data at the position near the wall, which is due to the wall attachment effect of the fluid, but, in general, the numerical simulation is consistent with the experimental curve, which means that the numerical model adopted can describes the sloshing process of the tank exactly.

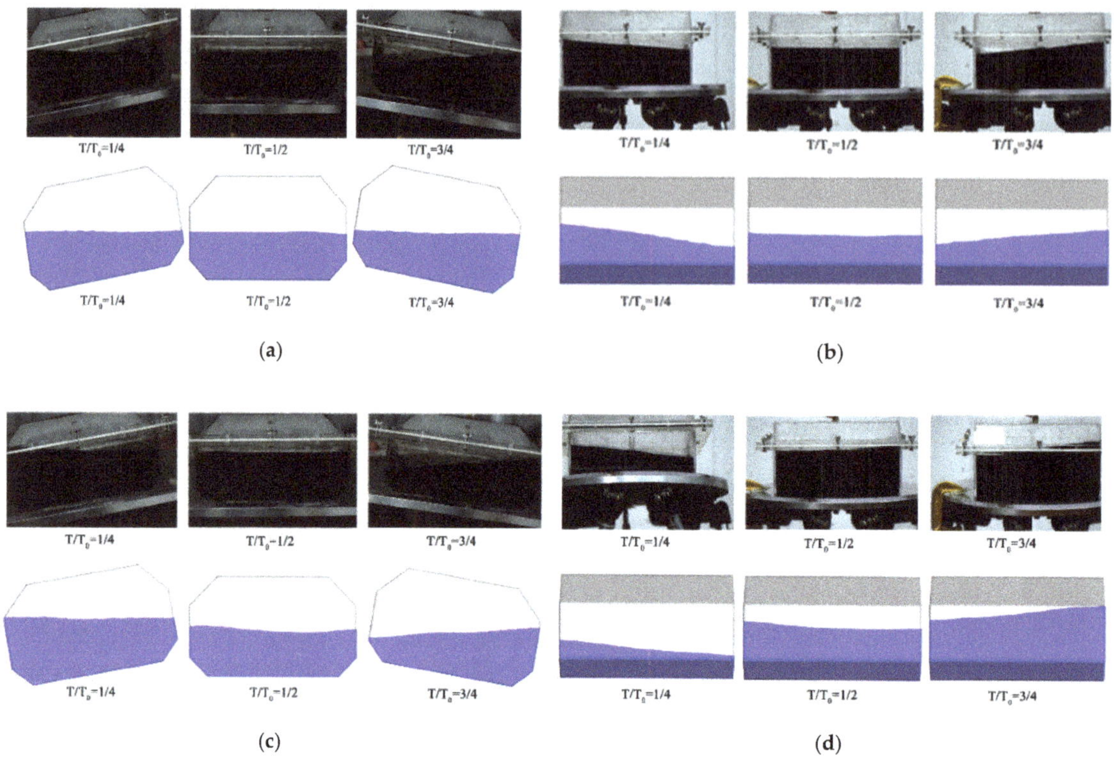

Figure 3. Comparison of the liquid-surface profile at different times: (**a**) roll, (**b**) surge, (**c**) front view under combination excitation, (**d**) side view under combination excitation.

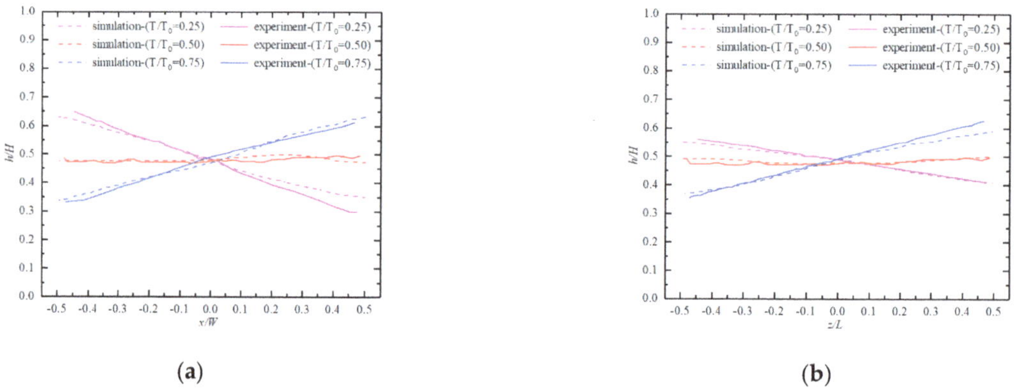

Figure 4. *Cont.*

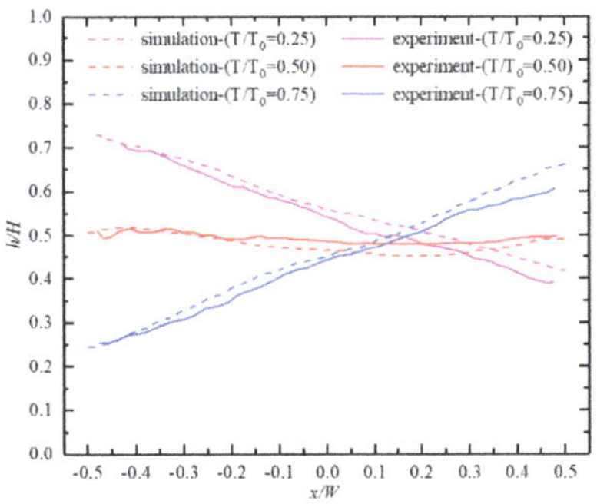

(c)

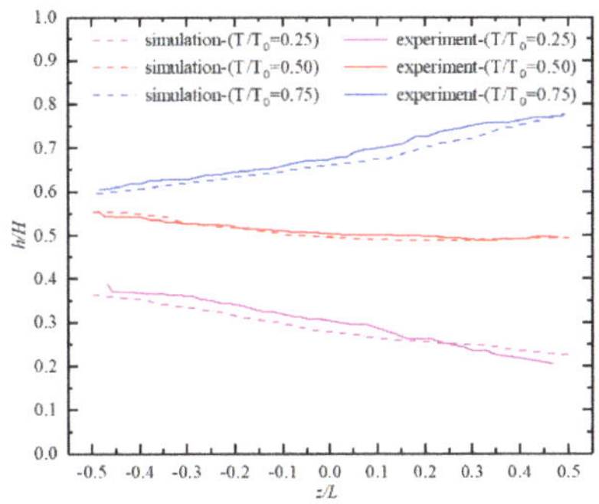

(d)

Figure 4. Liquid level distribution for (**a**) roll, (**b**) surge, (**c**) front view under combination excitation, (**d**) side view under combination excitation.

4. Results and Discussion

To study the sloshing behavior of the tank under different excitation conditions, numerical simulations of the tank motion under roll and surge conditions were carried out, and the variation in tank pressure at each monitoring point on the sidewall was analyzed. Then, the numerical simulation of the multiple-degrees-of-freedom sloshing process under the combination excitation of roll and surge was carried out. The pressure variations at different positions are mainly discussed, and the topographic variation characteristics of the liquid surface at different times are quantitatively analyzed. The numerical simulation conditions are as shown in Table 2.

Table 2. Numerical simulation case settings.

Type	Roll Angle φ (°)	Surge Amplitude λ (mm)	Frequency f (Hz)	Filling Rate
Roll excitation	10	-	0.58	0.5
Surge excitation	-	50	0.58	0.5
Combination excitation	3, 6, 10	30, 40, 50	0.58	0.5

4.1. Liquid Sloshing Characteristics under Roll Excitation

Figure 5 shows the variation of the liquid level on the vertical line at different positions of the tank under the excitation conditions of rolling (f = 0.58 Hz and φ = 10°). The liquid level on the S plane changes more than that on the P plane. The liquid level heights at lines P1 and P3 show periodic changes in the range of 0.25 H, and the two curves have opposite trends, while the liquid level height at the P2 position changes little. The liquid level at S1–S4 also changes periodically, with a similar trend.

Figure 6 shows the pressure distribution at different positions during the roll. Because of the impact of the periodic liquid sloshing on the tank wall, the pressure of each monitoring point changes periodically. At the same height, the pressure of the monitoring point on the S plane is much greater than that on the P plane. Figure 6a,b show that for the monitoring points on the same vertical line and at different levels, the more significant pressure fluctuations are closer to the bottom of the tank. Figure 6c,d show the pressures of monitoring points on the P plane and the S plane at the same height, respectively. Figure 6c shows that the trend of pressure at the monitoring point on the P plane is consistent with the trend of the liquid level in Figure 5a. This is because the pressure changes on the P plane during rolling mainly come from the static pressure change caused by the liquid surface change. Figure 6d shows that the pressures of the monitoring points at the same level on the S plane are the same, and the pressure value is higher than that of the monitoring points at the same level on the P plane. The reason is that under roll excitation, the pressure on the S plane includes not only the static pressure change caused by the change in the liquid level but also the impact dynamic pressure on the S plane.

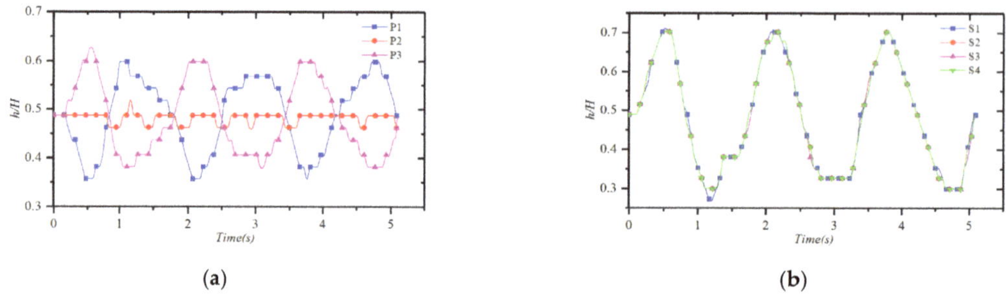

Figure 5. Variation of the liquid level at different positions: (**a**) variation of the liquid level on lines P1, P2, and P3 and (**b**) variation of the liquid level on lines S1, S2, S3, and S4.

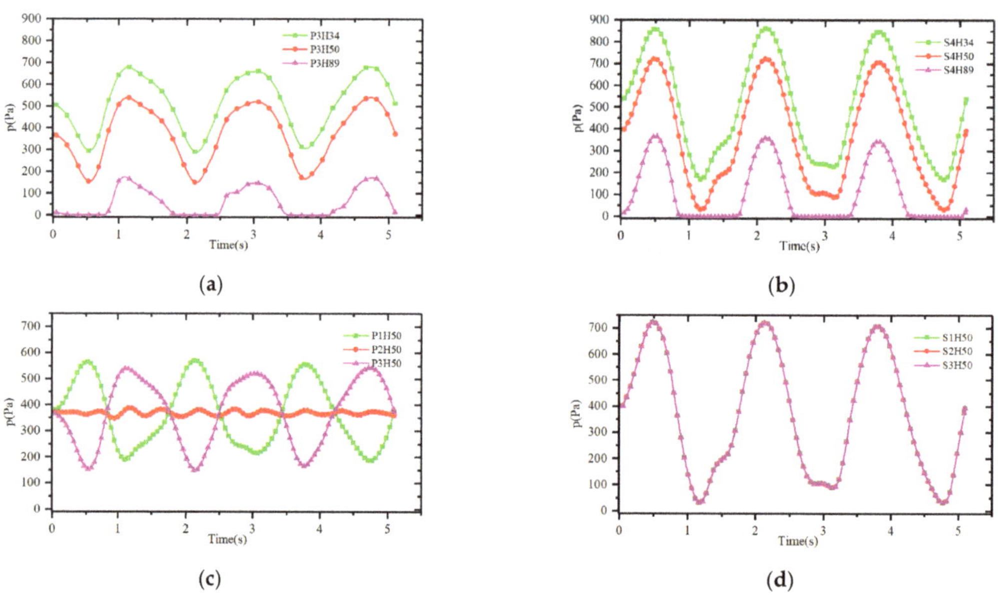

Figure 6. Pressure variation of different monitoring points under roll excitation; (**a**,**b**) are the pressures of monitoring points at different heights on lines P3 and S4, respectively; (**c**,**d**) are the pressures of monitoring points at 50 mm height on the P and S planes, respectively.

4.2. Liquid Sloshing Characteristics under Surge Excitation

Figure 7 shows the variation of the liquid level on lines P1–3 and S1–4 during the sloshing process of the tank under the condition of surge excitation. It is show that the liquid level on the P plane changes more dramatically than on the S plane, the liquid level changes on lines P1, P2 and P3 are similar, the liquid levels on S1, S2, S3 and S4 are different, the liquid level change of midline S2 is the smallest, and the liquid level change of S4 farther from the midline is the most severe. In addition, the liquid levels on S1 and S3 located at the same distance on both sides of S2 have the same degree of change, but in opposite directions.

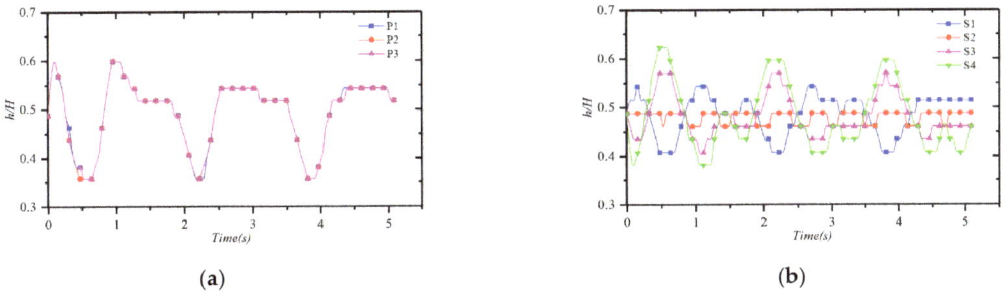

Figure 7. Variation of the liquid level under the conditions of surge excitation ($f = 0.58$ Hz, $\lambda = 50$ mm): (**a**) variation of the liquid level at P1–3; (**b**) variation of the liquid level at S1–4.

Figure 8 shows the pressure distribution of the monitoring points at different positions under the surge condition. Comparing Figure 8a,b, the pressure distribution trends of the monitoring points on lines P2 and S3 are similar, and at higher locations, the pressure of the monitoring points is smaller, and the range of variation is smaller. However, as the liquid surface mainly impacts the P plane during surge excitation, the pressure of monitoring points on P2 is larger than that of S3 in general. Figure 8c shows that the pressure distribution at the monitoring point at a height of 50 mm on the P plane is the same because the liquid level on the P plane is at the same level under surge excitation. Figure 8d shows that during surge excitation, the pressure change at the monitoring point on the P plane is caused by the change in the liquid level and the change in the static pressure at the monitoring point. Therefore, the trend of the pressure at the monitoring point on the P plane is the same as that shown in Figure 7b.

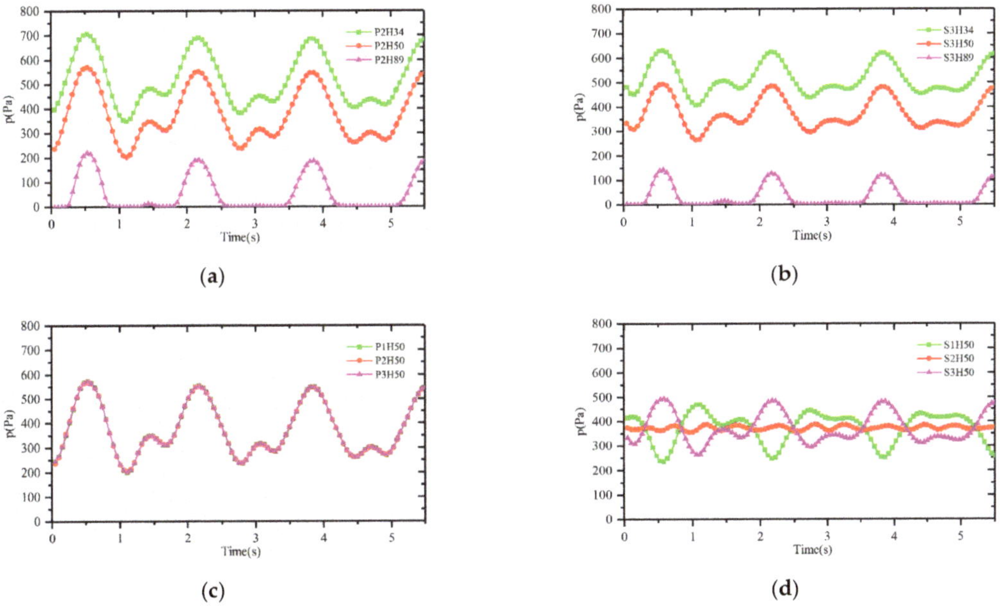

Figure 8. Pressure variation curves of different monitoring points under surge excitation; (**a**) and (**b**) are the pressure distribution curves of monitoring points on P2 and S3, respectively; (**c**,**d**) are the pressure of monitoring points at a height of 50 mm on the P and S planes, respectively.

4.3. Liquid Sloshing Characteristics under Combination Excitation

Considering that the excitations of carriers during navigation are mostly combination excitations with multiple degrees of freedom, combination excitation with two degrees of freedom, roll and surge, is studied. The morphological characteristics of the liquid level in the tank and the variation of the liquid level height at different positions are analyzed, and the sloshing behavior under combination excitation conditions is studied.

Figure 9a shows the change in the sloshing liquid surface profile under the combination excitation condition (f = 0.58 Hz, φ = 10°, λ = 50 mm). Compared with the liquid level variation under the single-degree-of-freedom excitation condition, the liquid level variation under the combination excitation condition is more complex and has significant asymmetry; overall, the liquid level changes periodically along the diagonal direction.

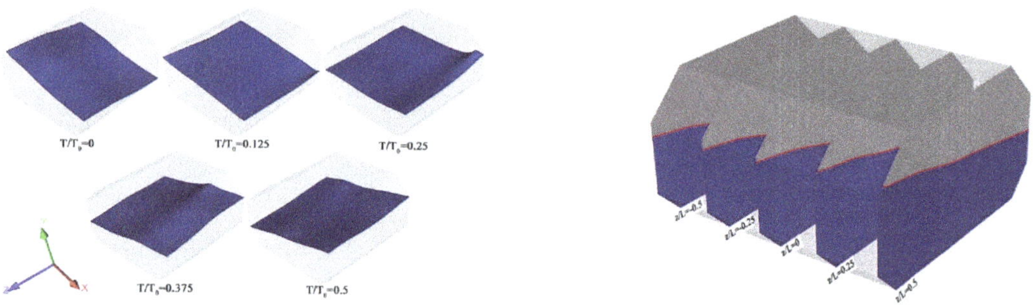

(a) (b)

Figure 9. Surface profile under combination excitation: (**a**) free surface change diagram and (**b**) different surface profiles.

To facilitate the observation and analysis of the liquid level change under combination excitation, the cross sections z/L = 0.5, z/L = 0.25, z/L = 0, z/L = −0.25 and z/L = −0.5 were selected as shown in Figure 9b, and the surface profile characteristics at different times were quantitatively analyzed. Figure 10 shows the distribution of the liquid level height on the section at different times. When T/T_0 = 0, the liquid level gradually decreased along the positive x/S and negative z/L directions. The highest point, h/H = 0.85, occurred on section z/L = 0.5, and the lowest point, h/H = 0.19, occurred on section z/L = −0.5. When T/T_0 = 0.125, compared with T/T_0 = 0, the liquid level increased along the positive x/S direction and the negative z/L direction, but the overall distribution is consistent with that at T/T_0 = 0. At the time of T/T_0 = 0.25, the liquid level had a positive movement trend along x/S, and the overall liquid level remained at h/H = 0.5. When T/T_0 = 0.375 and T/T_0 = 0.5, the liquid level gradually increased along the positive x/S direction and the negative z/L direction, and when T/T_0 = 0.5, the maximum positive x/S liquid level was h/H = 0.85.

Figure 11 shows the time-varying characteristic of the liquid level height at different positions. Compared with Figures 5 and 7, the liquid level variation curves at different positions under the combination excitation condition are different, and the peak value of the liquid level height increases. After the sloshing is stable, the closer to P1 on the P plane, the greater the liquid level fluctuation range. The maximum liquid level height at P1 can reach h/H = 0.65, and the closer to S1 on the S plane, the greater the liquid level fluctuation range. The maximum liquid level at S4 is h/H = 0.78.

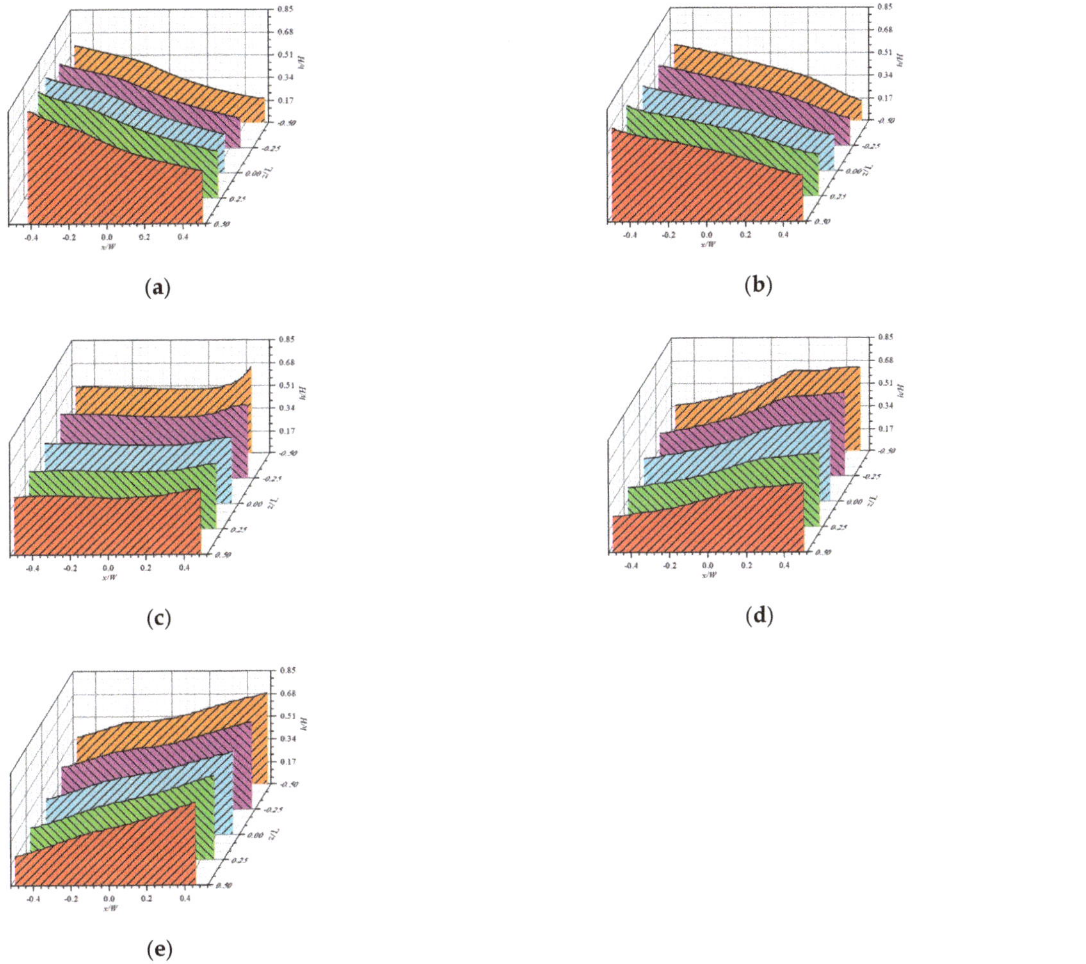

Figure 10. Changes in the liquid level at different times with combination excitation (f = 0.58 Hz, φ = 10°, λ = 50 mm): (**a**) T/T_0 = 0, (**b**) T/T_0 = 0.125, (**c**) T/T_0 = 0.25, (**d**) T/T_0 = 0.375, (**e**) T/T_0 = 0.5.

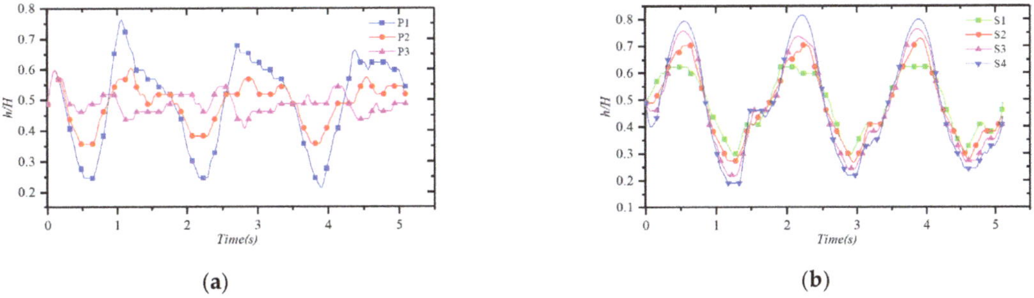

Figure 11. Liquid level variation curve under combination excitation: (**a**) P1–3 liquid level variation curve and (**b**) S1–4 liquid level change curve.

The time-varying characteristics of pressure at different monitoring points under combination excitation are further discussed below. Figure 12 shows the pressure time-varying curves of different monitoring points. The data in the figure show that under combination excitation, the degree of liquid sloshing is greater, and the sloshing form is more complex than that under the single degree of freedom excitation. The pressure peaks at all monitoring points are increased, and the pressure time-varying curves at some monitoring points are more complex. The data in Figure 12a,c show that due to the influence of hydrostatic pressure, the overall pressure and peak pressure at the monitoring points near the bottom of the tank are larger. Figure 12b,d show that under the condition of combination excitation, the liquid in the tank mainly sloshes along the diagonal direction of the tank. For the monitoring points at the same horizontal height, the closer to line S1 on the P plane, the greater the impact and the greater the pressure value. The pressure value at the monitoring point on centerline S2 on the S plane is the smallest, and the impact and pressure both increase with distance from the centerline. The maximum peak pressure at S4 can reach 900 Pa.

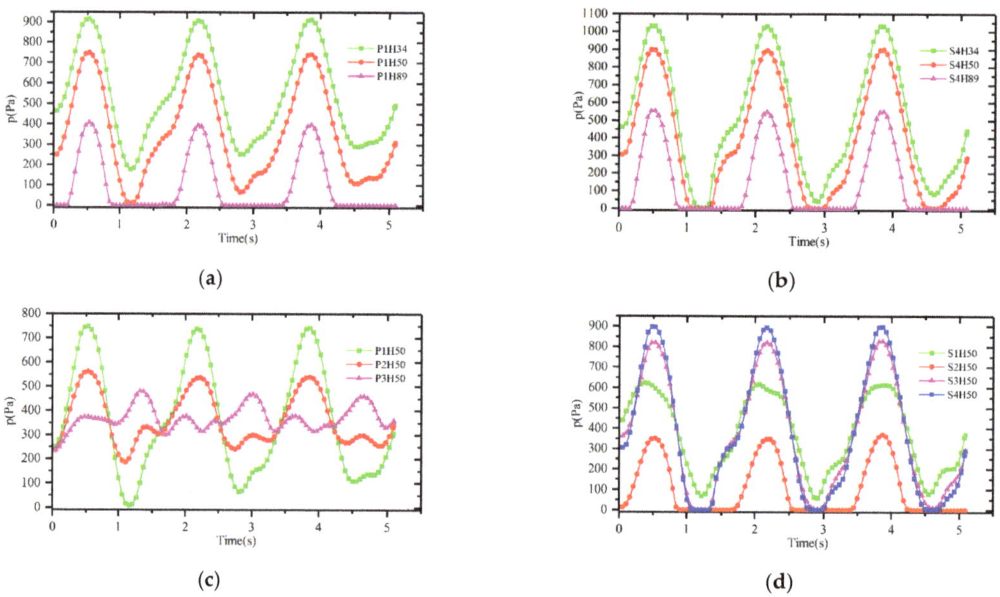

Figure 12. Liquid level variation curves under combination excitation (f = 0.58 Hz, φ = 10°, λ = 50 mm): (**a**,**b**) are the pressure distribution curves of monitoring points at different heights on P1 and S4, respectively; (**c**,**d**) are the pressure at the monitoring point with a height of 50 mm on the P and S planes, respectively.

4.4. Effect of Combination Excitation Intensity on Liquid Sloshing

To further explore the influence of combination excitation intensity on sloshing behavior, the variation characteristics of liquid surface height and the pressure time-varying characteristics at different positions under different combination excitation intensity conditions were compared and analyzed, and the influence of combination excitation intensity on peak pressure after sloshing stabilization was explored.

Figure 13 shows the variation curves of the liquid level on lines S1, S3, P1 and P3 with f = 0.58 Hz, λ = 40 mm at roll excitation angles φ of 3°, 6° and 10°. It is shows that with the increase in the amplitude of the roll angle, the peak height of the liquid level increases, and the liquid level at different positions changes periodically, with some differences. In addition, the change in the amplitude of the roll angle has a significant effect on the height change of the liquid level at lines P1 and S3. The reason is that the liquid level mainly sloshes along the diagonal direction of S4 under combination excitation, which makes the height change sharper at the position closer to S4.

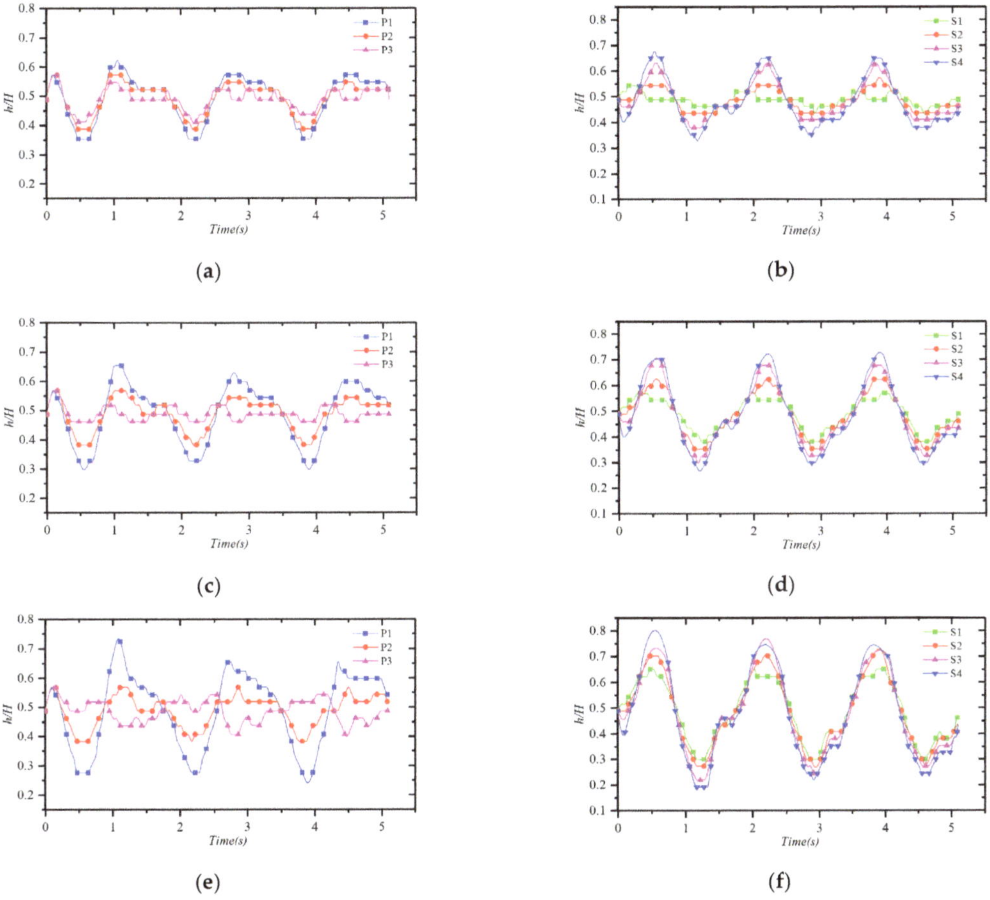

Figure 13. Liquid level variation curves under different roll angle amplitudes (f = 0.58 Hz, λ = 50 mm): (**a**) and (**b**), φ = 3°; (**c**) and (**d**), φ = 6°; (**e**) and (**f**), φ = 10°.

Figure 14 shows the variation of the liquid level when the sloshing frequency is 0.58 Hz, the roll angle amplitude is 10°, and the surge amplitudes are 30 and 50 mm. The data in Figure 14 show that the change in the surge amplitude has a more severe impact on the change in the liquid level on the P plane but has a lesser effect on the change in the liquid level on the S plane.

Figure 15 shows the pressure variation at monitoring points P3H50 and S4H50 under different combination excitation conditions. It can be seen that the change in roll angle amplitude has a greater impact on the pressure at the monitoring point, while the surge amplitude has a smaller impact. In addition, The data in Figure 15a,b show that when λ is 50 mm and the roll angle amplitude φ varies in the range of 3~10°, the impact pressure on the S plane increases with the increase in the amplitude of the roll angle, which makes the peak pressure at monitoring point S4H50 show an increasing trend, while the change in the amplitude of the roll angle has little impact on the impact pressure on the P plane. The pressure change at the monitoring point on the P plane mainly reflects the static pressure change caused by the change in the liquid height. Therefore, in the first half of the sloshing cycle, the peak pressure at monitoring point P3H50 increases with increasing amplitude of the rolling angle. In the second half of the sloshing period, the amplitude of the roll angle decreases with increasing roll angle amplitude. The data in Figure 15c,d show that for monitoring point S4H50, the peak pressure is in the first half cycle, during which the impact force of the fluid on the monitoring point is positively correlated with the surge amplitude; thus, the peak pressure of monitoring point S4H50 increases with the increase in the surge amplitude when the roll angle amplitude is certain. For monitoring point P3H50, the peak pressure occurs in the second half cycle, and the pressure at the monitoring point mainly reflects the static pressure change. Therefore, the larger the surge amplitude, the smaller the increase in liquid height. Hence, the surge amplitude increases, and the peak pressure at monitoring point P3H50 decreases.

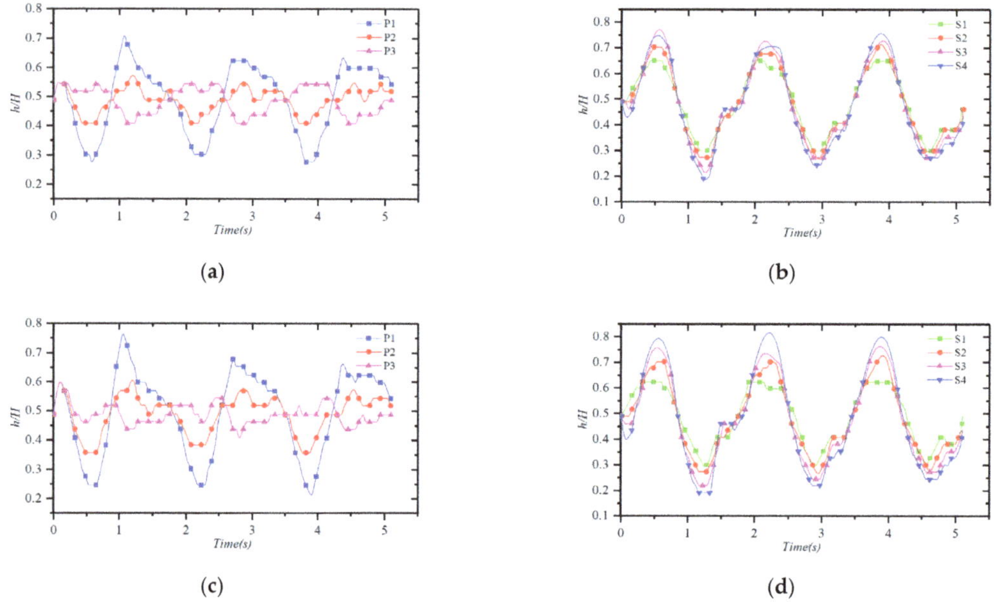

Figure 14. Liquid level variation curves under different surge amplitude conditions (f = 0.58 Hz, φ = 10°), (**a**,**b**) λ = 30 mm, (**c**,**d**) λ = 50 mm.

The above analysis shows that compared with single-degree-of-freedom excitation, the pressures of the monitoring points at different positions under the combination excitation condition are different, and the variation of the peak pressure with the excitation intensity is more complicated. The focus in engineering practice is on the maximum peak pressure generated by sloshing. The data in Figure 12 show that under the combination excitation condition, the monitoring point with the largest pressure peak (of the monitoring points shown in Figure 1) is S4H34. Thus, the following will take the peak pressure of monitoring point S4H34 as the research object to further explore the variation characteristics of the peak pressure at monitoring point S4H34 with changing combination excitation intensity.

As shown in Figure 16, the peak pressure of monitoring point S4H34 varies with the combination excitation intensity (φ and λ). It is shows that when the combination excitation intensity is φ = 3° and λ = 30 mm, the minimum peak pressure at the monitoring point is 731.5 Pa, while when the combination excitation intensity is φ = 10° and λ = 50 mm, the maximum peak pressure at the monitoring point is 1033.37 Pa. The peak pressure values under different combination excitation intensities are basically in the same plane in three-dimensional space. In addition, with the increase in roll angle amplitude and surge amplitude, the impact force of the fluid on the monitoring point will increase, and the peak pressure of monitoring point S4H34 will also increase, showing a linear positive relationship. Among these, the impact of roll angle amplitude on the peak pressure of monitoring point S4H34 is more significant.

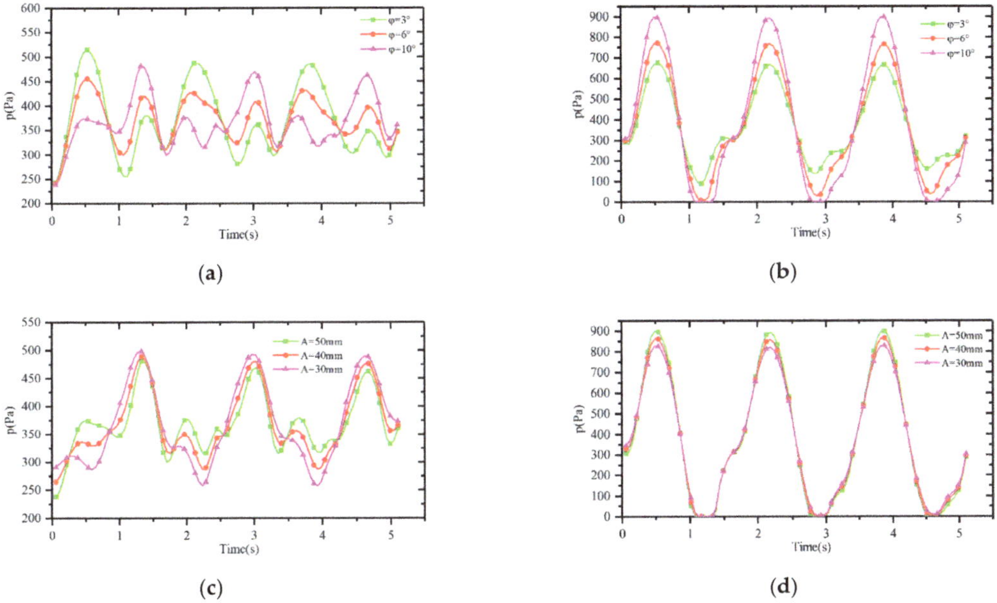

Figure 15. Pressure variations at different angles and amplitudes: (**a**,**b**) show the pressure curves of monitoring points P3H50 and S4H50 when the surge amplitude is 50 mm; (**c**,**d**) show the pressure variation curves of monitoring points P3H50 and S4H50 when the roll angle amplitude is 10°.

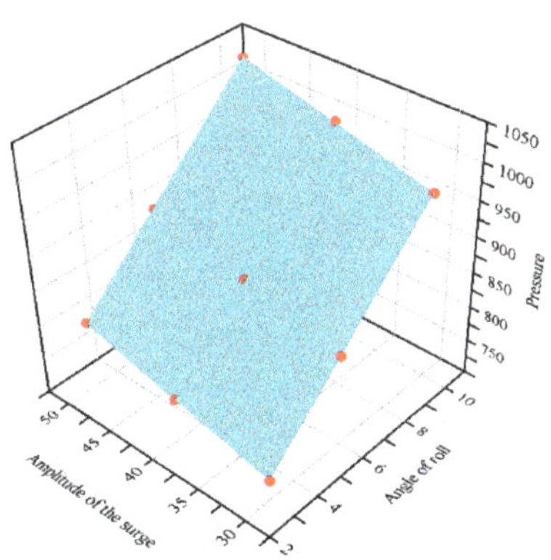

Figure 16. Maximum pressure fitting surface of S4H34 under combination excitation.

5. Conclusions

In this paper, the liquid sloshing characteristics of a tank under a single degree of freedom and combination excitation conditions are studied by numerical simulation combined with experimental verification. The characteristics of the sloshing liquid surface profile and pressure distribution of the liquid tank under typical combination excitation were compared and analyzed in detail. The influence of the combination excitation intensity on the liquid level and the time-varying characteristics of the monitoring point pressure and peak pressure were discussed. The main conclusions are as follows:

(1) The characteristics of the liquid level variations are obvious under excitation with a single degree of freedom. Meanwhile, the variation in the height of the liquid level noticeably increases with the intensity of single-degree-of-freedom excitation. Moreover, the pressure of the tank wall increases with sloshing. The height of the liquid level and the pressure of the wall have a linear increasing relationship with the sloshing intensity.
(2) The height of the liquid level varies periodically under the combination excitation. Furthermore, the liquid level in the cargo tank and the pressure of the wall vary more intensely compared with the single excitation. The peak pressure of the inner liquid, which is greater under the combination excitation, also changes significantly.
(3) When the liquid tank encounters the combination excitation (roll and surge), the angle of rolling has a great impact on the pressure of the inner wall of the liquid tank. There is a slight increase with the surge amplitude, and the change in the liquid level is also sensitive to the rolling angle, while the surge amplitude is not obvious. Under the combination excitation condition, the peak pressure at the lowest point of the tank diagonal increases linearly with increasing amplitude of the roll angle and the amplitude of the surge.

Author Contributions: Methodology, Q.Z.; software, Q.Z.; validation, Q.Z.; writing–original draft preparation, Q.Z.; resources, B.S.; writing–review and editing, B.S.; supervision, H.Z.; project administration, H.Z. All authors have read and agreed to the published version of the manuscript.

Funding: This research received no external funding.

Institutional Review Board Statement: Not applicable.

Informed Consent Statement: Not applicable.

Data Availability Statement: Data are contained within the article.

Acknowledgments: Our deepest gratitude goes to the anonymous reviewers and the editors for their careful work and thoughtful suggestions that have helped improve this paper substantially.

Conflicts of Interest: The authors declare no conflict of interest.

Nomenclature

Denomination (unit)

P	Internal pressure of an LNG ship (Pa)
H	Height of the cargo tank (m)
S	Width of the cargo tank (m)
L	Length of the cargo tank (m)
μ_l	Viscosity of water (Pa·s)
ρ_l	Density of water (kg/m^3)
μ_L	Viscosity of water (Pa·s)
α_l	Volume fraction of water (-)
u_l	Velocity of LNG (m/s)
VOF	Volume of fluid method (-)
Re	Reynolds number (-)
CSF	Continuum surface force (N)
u	Velocity of the flow field in the x-direction (m/s)
v	Velocity of the flow field in the y-direction (m/s)
w	Velocity of the flow field in the z-direction (m/s)
λ	Amplitude of surging (mm)
V_x	Velocity of the cargo tank in the x-direction (m/s)
V_y	Velocity of the cargo tank in the y-direction (m/s)
V_z	Velocity of the cargo tank in the z-direction (m/s)
T_0	Period of sloshing (s)
T	Time of cargo tank motion(s)
g	Acceleration of gravity (m/s^2)
Hi	Height of the surface P (-)
f	Frequency of sloshing (Hz)
φ	Angle of rolling (°)
Pi	Vertical lines of the surface P (-)
Si	Horizontal line of the sidewall (-)

References

1. Moiseev, N.N. On the theory of nonlinear vibrations of a liquid of finite volume. *J. Appl. Math. Mech.* **1958**, *22*, 860–872. [CrossRef]
2. Faltinsen, O.M. A nonlinear theory of sloshing in rectangular containers. *J. Ship Res.* **1974**, *18*, 224–241. [CrossRef]
3. Faltinsen, O.M.; Timokha, A.N. A multimodal method for liquid sloshing in a two-dimensional circular tank. *J. Fluid Mech.* **2010**, *665*, 457–479. [CrossRef]

4. Budiansky, B. Sloshing of Liquids in Circular Canals and Spherical Tanks. *J. Aerosp. Sci.* **1958**, *27*, 58. [CrossRef]
5. Kim, D.H.; Kim, E.S.; Shin, S.-C.; Kwon, S.H. Sources of the Measurement Error of the Impact Pressure in Sloshing Experiments. *J. Mar. Sci. Eng.* **2019**, *7*, 207. [CrossRef]
6. Trimulyono, A.; Hashimoto, H.; Matsuda, A. Experimental Validation of Single- and Two-Phase Smoothed Particle Hydrodynamics on Sloshing in a Prismatic Tank. *J. Mar. Sci. Eng.* **2019**, *7*, 247. [CrossRef]
7. Yu, L.; Xue, M.-A.; Zheng, J. Experimental study of vertical slat screens effects on reducing shallow water sloshing in a tank under horizontal excitation with a wide frequency range. *Ocean Eng.* **2019**, *173*, 131–141. [CrossRef]
8. Akyildiz, H.; Uenal, E. Experimental investigation of pressure distribution on a rectangular tank due to the liquid sloshing. *Ocean Eng.* **2005**, *32*, 1503–1516. [CrossRef]
9. Zou, C.-F.; Wang, D.-Y.; Cai, Z.-H.; Li, Z. The effect of liquid viscosity on sloshing characteristics. *J. Mar. Sci. Tech.-Jpn.* **2015**, *20*, 765–775. [CrossRef]
10. Doh, D.H.; Jo, H.J.; Shin, B.R.; Min, C.R.; Hwang, Y.S. Analysis of the Sloshing Flows of a LNG Cargo Tank. *J. Therm. Sci.* **2011**, *20*, 7. [CrossRef]
11. Unal, U.O.; Bilici, G.; Akyildiz, H. Liquid sloshing in a two-dimensional rectangular tank: A numerical investigation with a T-shaped baffle. *Ocean Eng.* **2019**, *187*, 106183. [CrossRef]
12. Chu, C.-R.; Wu, Y.-R.; Wu, T.-R.; Wang, C.-Y. Slosh-induced hydrodynamic force in a water tank with multiple baffles. *Ocean Eng.* **2018**, *167*, 282–292. [CrossRef]
13. Zhu, A.; Xue, M.-A.; Yuan, X.; Zhang, F.; Zhang, W. Effect of Double-Side Curved Baffle on Reducing Sloshing in Tanks under Surge and Pitch Excitations. *Shock Vib.* **2021**, *2021*, 6647604. [CrossRef]
14. Cho, I.H.; Choi, J.-S.; Kim, M.H. Sloshing reduction in a swaying rectangular tank by an horizontal porous baffle. *Ocean Eng.* **2017**, *138*, 23–34. [CrossRef]
15. Tao, K.; Zhou, X.; Ren, H. A Local Semi-Fixed Ghost Particles Boundary Method for WCSPH. *J. Mar. Sci. Eng.* **2021**, *9*, 416. [CrossRef]
16. Liu, D.; Lin, P. A numerical study of three-dimensional liquid sloshing in tanks. *J. Comput. Phys.* **2008**, *227*, 3921–3939. [CrossRef]
17. Tang, Y.-y.; Liu, Y.-d.; Chen, C.; Chen, Z.; He, Y.-p.; Zheng, M.-m. Numerical study of liquid sloshing in 3D LNG tanks with unequal baffle height allocation schemes. *Ocean Eng.* **2021**, *234*, 109181. [CrossRef]
18. Borg, M.G.; DeMarco Muscat-Fenech, C.; Tezdogan, T.; Sant, T.; Mizzi, S.; Demirel, Y.K. A Numerical Analysis of Dynamic Slosh Dampening Utilising Perforated Partitions in Partially-Filled Rectangular Tanks. *J. Mar. Sci. Eng.* **2022**, *10*, 254. [CrossRef]
19. Hoch, R.; Wurm, F.H. Numerical and experimental investigation of sloshing under large amplitude roll excitation. *J. Hydrodyn.* **2021**, *33*, 787–803. [CrossRef]
20. Xue, M.-A.; Jiang, Z.; Lin, P.; Zheng, J.; Yuan, X.; Qian, L. Sloshing dynamics in cylindrical tank with porous layer under harmonic and seismic excitations. *Ocean Eng.* **2021**, *235*, 109373. [CrossRef]
21. Xin, J.J.; Chen, Z.L.; Shi, F.; Shi, F.L.; Jin, Q. Numerical simulation of nonlinear sloshing in a prismatic tank by a Cartesian grid based three-dimensional multiphase flow model. *Ocean Eng.* **2020**, *213*, 107629. [CrossRef]
22. Brackbill, J.U.; Kothe, D.B.; Zemach, C. A continuum method for modeling surface tension. *J. Comput. Phys.* **1992**, *100*, 335–354. [CrossRef]
23. Jung, J.H.; Yoon, H.S.; Lee, C.Y.; Shin, S.C. Effect of the vertical baffle height on the liquid sloshing in a three-dimensional rectangular tank. *Ocean. Eng.* **2012**, *44*, 79–89. [CrossRef]

Journal of
Marine Science and Engineering

MDPI

Article

Investigation of the Hydrodynamic Characteristics of Two Manta Rays Tandem Gliding

Yunlong Ma [1,2], Qiaogao Huang [1,2,*], Guang Pan [1,2] and Pengcheng Gao [1,2]

[1] School of Marine Science and Technology, Northwestern Polytechnical University, Xi'an 710072, China
[2] Key Laboratory of Unmanned Underwater Vehicle, Northwestern Polytechnical University, Xi'an 710072, China
* Correspondence: huangqiaogao@nwpu.edu.cn

Abstract: Collective motion is a unique biological habit of manta rays. As the most basic unit, the hydrodynamic mechanism of tandem gliding deserves further study. In this paper, a numerical simulation method was used to explore the influence of the front-to-back distance and the angle of attack on the overall and individual hydrodynamic performance of a pair of manta rays gliding. Specifically, a numerical simulation of the hydrodynamic parameters and the distribution of pressure and velocity fields was carried out when the pair of manta rays were arranged at a distance of 0.25–1 times the body length and the angle of attack was -8° to 8°. The simulation results show that, when a pair of manta rays glide at close range, compared to a single manta ray gliding, the resistance of the leader is greatly reduced, and the lift changes little, while the resistance of the follower is greatly increased, and the lift is significantly reduced. For the average resistance of the system, in the specific scenario of a close range and a small negative degree angle of attack, the two manta rays' tandem gliding system can significantly reduce the resistance and play a role in reducing resistance and saving energy. The research content of this paper provides a theoretical basis for understanding the biological habits of manta rays and designing an underwater bionic robot group system.

Keywords: manta rays; fish schooling; tandem gliding; hydrodynamic characteristics

Citation: Ma, Y.; Huang, Q.; Pan, G.; Gao, P. Investigation of the Hydrodynamic Characteristics of Two Manta Rays Tandem Gliding. *J. Mar. Sci. Eng.* **2022**, *10*, 1186. https://doi.org/10.3390/jmse10091186

Academic Editor: Abdellatif Ouahsine, Alberto Ribotti, Peng Du, Haibao Hu, Xiaopeng Chen

Received: 8 July 2022
Accepted: 20 August 2022
Published: 25 August 2022

1. Introduction

Long-term natural evolution has endowed fish with extraordinary movement and survivability: long-distance cruise with low energy consumption, high-efficiency propulsion with large thrust, powerful explosive force and maneuverability, sensitive perception and positioning capabilities, excellent self-protection and precise attack capabilities. Manta rays are ray-like creatures with flat bodies, large aspect ratios, and cruising speeds of 0.25–0.47 m/s [1]. The Median and/or Paired Fin (MPF) mode is the steady-state cruising mode of manta rays [2], and the efficiency can be as high as 89%, which is higher than that of aquatic organisms such as eels and trevallies [3,4]. It has extraordinary advantages in mobility and concealment. Manta rays have obvious shape features, efficient propulsion performance, and energy-saving group swimming, which give them great bionic value.

Collective motion is a common phenomenon in fish biology [5], which is also present in manta ray species. Studies have shown that, for fish, swimming in groups is usually more efficient than swimming alone. In nature, more than 50% of fishes exhibit synchronized and coordinated group swimming at some point [6]. When migrating, many fish swim in groups in the same direction and maintain a near-constant separation from their neighbors [7]. In addition to the sociological advantages of avoiding natural enemies [8,9] and improving the success rate of predation [10–12], group swimming is also considered to be effective in reducing energy consumption [13–17].

In view of the high efficiency and energy-saving characteristics of fish group swimming, researchers have conducted a lot of research hoping to reveal the inherent mechanism of fluid mechanics. At present, the common research methods are as follows: 1. Fish colony observation method; 2. Computational fluid dynamics simulation method; 3. Experimental research method using bionic robotic fish.

The fish colony observation method used in earlier periods, due to the limitation of equipment and technology, mainly used the observation of the swimming characteristics of fish combined with the method of hydrodynamic theoretical analysis. Breder was the first to study the energy-saving mechanism of fish swimming. He believed that the complete wake vortex structure scale was an important factor affecting the distribution interval of individuals [18]. With the continuous development of observation technology, high-speed cameras and digital particle image velocimetry technology are used more and more by researchers to extract and observe the performance information, morphological structure, and eddy current characteristics of fish schools. Then, the energy consumption data can be obtained through theoretical calculation or DPIV experiment, so as to reveal the energy-saving mechanism of fish swimming. Marras et al., tested the collective motion of liza aurata at different flow rates and found that individuals in the best position in the flock reduced their tail-wagging frequency by up to 28.5% compared to swimming alone. It was also found by DPIV that the swing of an anterior individual helps its neighboring fishes to swim forward [19].

With the continuous development of computer technology and fluid simulation technology, the method of computational fluid dynamics is more and more widely used in the study of the mechanism of fish school energy saving. Unlike the observation method, which can only obtain some qualitative conclusions, the CFD simulation method can quantitatively measure the force of each individual in the fish group more accurately, making the analysis results more reliable and intuitive. Pan et al. [20], Shao et al. [21], and Xiao et al. [22,23] used the immersion boundary method to numerically simulate the passive motion of the flexible plate behind the D-shaped cylinder. The former can obtain thrust from the wake vortex of the former, which greatly improves the propulsion efficiency; Chao et al. [24,25] used a commercial CFD software package to numerically simulate the double-body system of a D-shaped cylinder and a NACA0012 flexible plate. It was found that the thrust of the downstream airfoil was improved due to the presence of the cylinder. Based on the theory of potential flow and viscosity, Fish et al., estimated the propulsion efficiency of manta rays and found that most of the thrust is generated by the distal end of the fin during its movement [26]. Zhang Dong et al., used the immersion boundary method to simulate the process of a manta ray flapping and advancing and explored the influence of motion parameters, such as flapping amplitude and frequency, on its propulsion effect [27].

On the basis of the first two methods, some researchers have begun to design experimental devices in recent years hoping to study the energy-saving mechanism of fish swimming through experimental methods. Dewey [28] and Boschitsch [29] experimentally studied the propulsion characteristics of two parallel and series flapping airfoils in a uniform incoming flow, respectively. The experimental system investigates the effects of wing-flap phase difference and separation distance. The study found that, for the tandem structure, the thrust and efficiency of the front wing were consistent with the single wing shape when the separation distance was large, while the dynamic characteristics of the rear wing were related to the separation distance and phase difference due to the influence of the wake of the front wing. For the parallel configuration, the propulsion characteristics of the two airfoils are the same: when flapping in-phase, efficiency increases and thrust decreases (relative to a single airfoil); when flapping in anti-phase, thrust increases while efficiency remains largely unchanged.

From the current research status, it can be seen that most of the work focuses on two-dimensional airfoils or flexible sheets, rather than on real three-dimensional creatures. Moreover, the research mainly focuses on the group travel over the two-dimensional plane and does not consider the effect of a real three-dimensional situation on the results. For the two-body tandem system, less research has been done on underwater targets, and more research has been focused on the water surface or in the air. Zhang Dong et al., used the CFD method to study the hydrodynamics of rotary underwater vehicles in series and gave the law of mutual influence between the front and rear individuals [30,31]. Yuan, Zhi-Ming et al., simplified the swimmer's model, used the CFD method to simulate the swimming process in an actual competition, and gave the law of influence of resistance on the back of the athlete with the distance [32]. Using the same method, they simplified the duck model to explain the hydrodynamic reasons why ducklings swim with their mother ducks in nature [33]. Blocken et al., used a combination of CFD and experiments to study the process of two cyclists following movement back and forth and gave the advantages of the two-body system and the wind resistance law of each single body [34]. None of the previous studies involved the simulation of manta rays group gliding, and this issue has not been systematically studied, so it is innovative and scientific to a certain extent. Research on the energy-saving mechanism of collective motion can, on the one hand, give us a deeper understanding of animal habits, and on the other hand, can provide inspiration and help for the design and control of underwater robots, especially underwater bionic robots, so as to improve operational efficiency and to develop and utilize marine resources more efficiently. Therefore, this paper focuses on a real, three-dimensional manta ray pair's tandem gliding system, explores the gliding hydrodynamic performance of manta rays at different distances, reveals the hydrodynamic mechanism of biological long-distance schooling fish migration, and provides a theoretical basis for the group design of underwater bionic robots.

The second chapter describes the goals and problems of the simulation and verifies the calculation method and grid independence. The third chapter records the simulation results, analyzes the hydrodynamic characteristics of the manta ray pair's gliding system., compares it with the single manta ray, and gives a formation scheme that can reduce resistance. The fourth chapter summarizes the full text and lists all the important conclusions.

2. Problem Definition and Methodology

2.1. Biological Model and Reference Dimensions

Through the real observation of the biological shape of the cow-nosed ray and three-dimensional software modeling, the final numerical simulation target model was obtained as shown in Figure 1. The dimensions of each model were defined as follows: body length (BL) = 1800 mm, spanwise length (SL) = 2900 mm, and maximum thickness length (TL) = 350 mm. This paper takes BL as the reference length and BL^2 as the reference area.

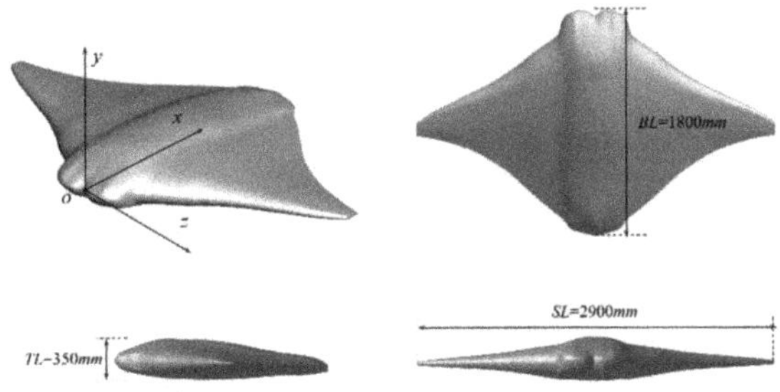

Figure 1. The 3D model of a manta ray.

2.2. Simulation Method Validation

In order to ensure the correctness of the numerical simulation results, the calculation method was first verified. At present, domestic and foreign scholars have not carried out numerical simulation work on manta ray gliding. In addition, the model has an irregular shape, a high proportion of curved surfaces, and is difficult to process, which can only be obtained by additive manufacturing. Machining the model and performing wind-tunnel or water-tunnel experiments on it would have disadvantages, such as a high cost and long time, which do not meet the requirements of previous mechanistic research. Therefore, the method of CFD simulation was used to study this problem. Observing the shape characteristics of the manta ray, it can be seen that its topology is similar to the three-dimensional hydrofoil. The scholar Zarruk [35] experimentally measured the hydrodynamic parameters of a graphical model 3D hydrofoil with different materials, Reynolds number (*Re*), and angle of attack (α). The root chord length of the model was 0.12 m, the wing chord length was 0.06 m, and the spread length was 0.3 m. The average chord length of 0.09 m was taken as the reference length. For this model, the grid was drawn as shown in Figure 2. The numerical simulation method and turbulence model of SST k-ω in this paper was used to calculate the variation, with the angle of attack law for the three-dimensional hydrofoil lift coefficient and resistance coefficient when the Reynolds number is 1×10^6. ν is the kinematic viscosity coefficient of water, which is 1.003×10^{-6}. The simulation results were compared with the experimental values, and the results are shown in Figure 3. It can be seen that the simulation results were in good agreement with the experimental data, indicating that the calculation method used in this paper has great reliability.

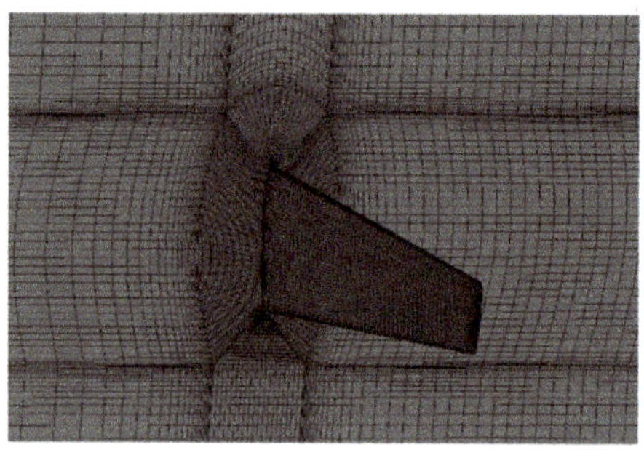

Figure 2. Grid of hydrofoil model for method validation.

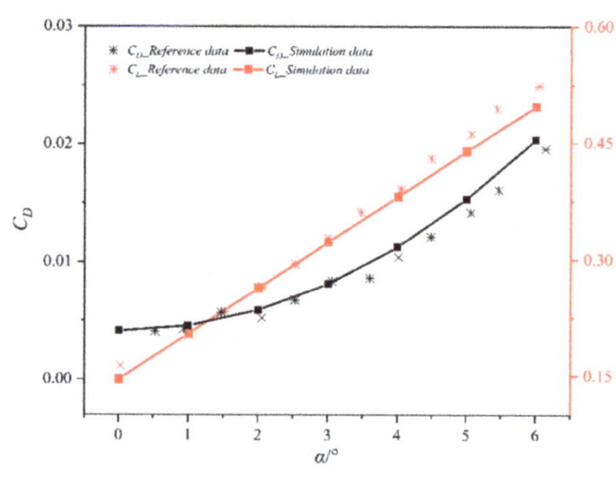

Figure 3. Method validation and comparison results.

2.3. Grid Independence Validation

The numerical simulation results are closely related to the number of cells. In order to ensure the correctness of the manta ray gliding simulation, it was necessary to carry out grid-independent verification. Draw the manta ray structure grid as shown in Figure 4. The typical y^+ value of the first grid point near the manta ray surface was set as 1, and the *Re* was 9×10^5. The overall grid distribution density was changed and the lift and resistance values of the manta ray model was recorded for different numbers of cells, as shown in Figure 5. It can be concluded from Figure 5 that, when the number of cells is small, the lift and resistance values under the same working conditions change drastically with the increase in the number of cells. When the number of cells reaches 4.5×10^6, the lift and resistance values no longer change with the grid number. On the premise of ensuring the reliability of the simulation results, in order to improve the calculation efficiency and save simulation time, the grids of all subsequent working conditions were drawn using the node distribution law with the number of cells being 4.5×10^6. The number of cells of the manta ray pair system increases to 6.1×10^6 due to the addition of one individual and will further grow as the distance between the two manta rays increases.

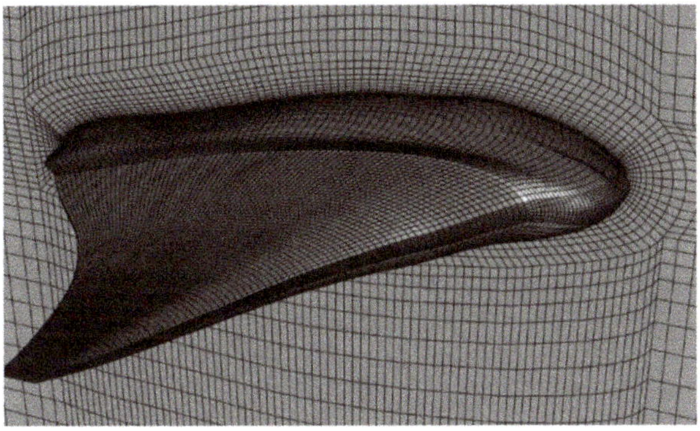

Figure 4. Structure grid of manta ray.

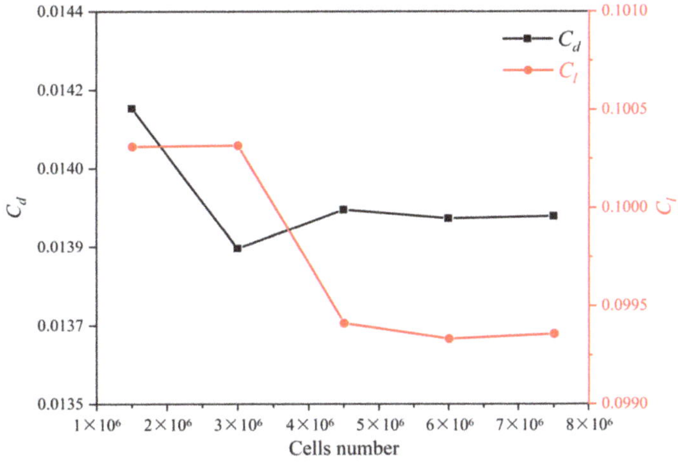

Figure 5. Simulation results under different numbers of cells.

In order to explore the hydrodynamic characteristics of the manta ray pair's tandem gliding system, the structure grid was drawn as shown in Figure 6. The two manta rays were distributed forward and backward along the x-axis, and the distance between the tail of the leader and the head of the follower was defined as Δs, which is measured as a multiple of the BL. Since the structure of the manta ray model was symmetrical, in order to reduce the number of cells and improve the calculation efficiency, half of the grid domain was drawn. The boundary condition on the left side was set as the symmetry plane, so the numerical simulation of the whole can be completed. In addition, the front side was set as the velocity inlet, the rear side was set as the pressure outlet, the surface of the manta ray model was set as a non-slip wall, and the other surfaces were set as free-slip walls. The overall computational domain was a cuboid. The height was set as 20 TL; the width was set as 2 SL; the leader's head was 4 BL from the velocity inlet; the follower's tail was 7 BL from the pressure outlet. The computational domain size and boundary condition settings are shown in Figure 6.

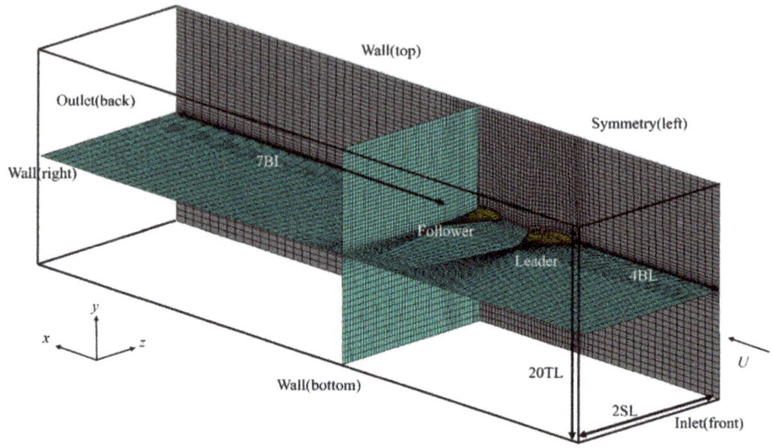

Figure 6. Computational domain size and boundary condition settings.

2.4. Simulation Domain and Boundary Condition Settings

After completing the grid-drawing work, we simulated the conditions by numerical simulation method. The finite volume method was used to discretize the Navier-Stokes equation, the second-order upwind style was used to discretize the momentum, the derivative was calculated based on the least-squares method, and the SIMPLEC algorithm was used to calculate the pressure in the continuous equation. As for the selection of the turbulence model, Lyu Da et al., used the SST k-ω model to accurately simulate the hydrodynamic parameters of the flying wing underwater glider [36]. The manta ray model in this study has certain similarities with the flying wing underwater glider. The related content regarding method verification in Section 2.2 also shows that the SST k-ω turbulence model can accurately simulate the underwater airfoil problem. In summary, we used the SST k-ω turbulence model for simulation calculations, and the convergence residual was set to 1×10^{-5}.

3. Result and Discussion

In this section, we compare and analyze the hydrodynamic performance differences between a single manta ray and a pair of manta rays when gliding at different angles of attack and consider the influence of distance on the system when two manta rays tandem glide. The resistance of the manta ray was defined as D and the lift as L. The above parameters were non-dimensionalized as:

$$C_D = \frac{D}{0.5\rho U^2 (BL)^2}, C_L = \frac{L}{0.5\rho U^2 (BL)^2} \tag{1}$$

where ρ is the density of the water, which was set to 1000 kg/m^3; U is the incoming velocity, which was set to 0.5 m/s; BL is the reference length. The Reynolds number at this time was calculated to be 9×10^5.

The average resistance and coefficient of the system was defined as:

$$D_{ave-system} = (D_{leader} + D_{follower})/2,\ C_{D_{ave-system}} = \frac{D_{ave-system}}{0.5\rho U^2 (BL)^2} \tag{2}$$

3.1. Numerical Simulation Results of a Single Manta Ray

In this part, we first study the hydrodynamic performance of a single manta ray gliding. Figure 7 shows the variation of the resistance coefficient and lift coefficient with the angle of attack when the manta ray is gliding. It can be seen from Figure 7 that in the range of −8 to 8° angle of attack, the lift received by the manta ray increases linearly with the change of the angle of attack, and the resistance shows a quadratic function with the angle of attack. It can be clearly seen from Figure 8 that the manta ray has a relatively flat lower surface and a raised upper surface in terms of biological structure. Therefore, when the angle of attack is 0°, the water velocity near the upper surface of the manta ray is fast, and the water velocity near the lower surface is slow. Therefore, the pressure difference in the vertical direction of the upper and lower surfaces causes the manta ray to experience a small lift at this time. When the angle of attack is −2°, the water velocity distribution near the upper and lower surfaces is relatively close, so the vertical pressure difference is small. Therefore, the lift force is approximately equal to 0, and the resistance on the manta ray reaches the minimum value at this time.

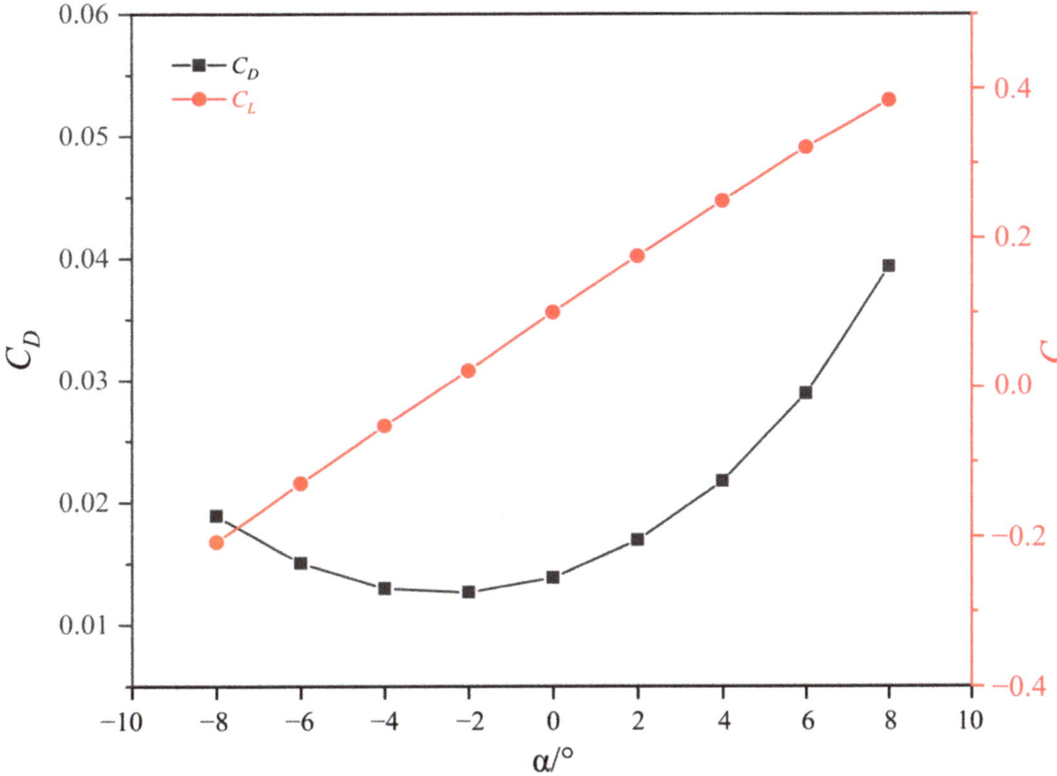

Figure 7. Variation of a single manta ray gliding: resistance coefficient and lift coefficient with angle of attack.

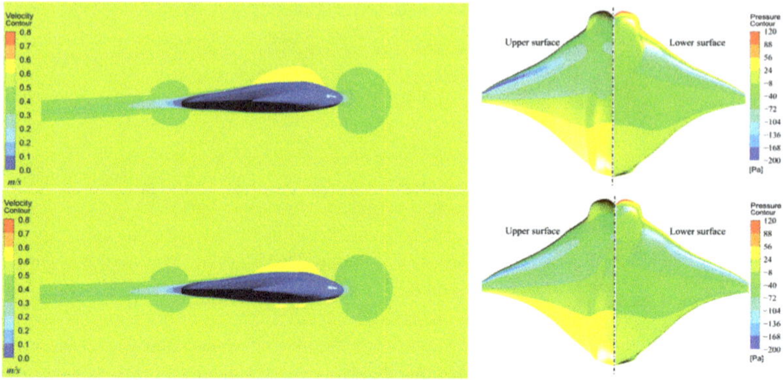

Figure 8. Longitudinal symmetry plane velocity distribution and manta ray surface pressure distribution during single gliding at $\alpha = 0°$ (**up**) and $\alpha = -2°$ (**down**).

3.2. Numerical Simulation Results of Two Manta Rays Tandem

This part studies the differences in the hydrodynamic performance of two manta rays gliding. As shown in Figure 6, the leader manta ray and the follower manta ray are arranged horizontally along the x-axis, and the angle of attack changes synchronously from −8 to 8°. The hydrodynamic performance at the distances Δs = 0.25 BL, 0.5 BL, 0.75 BL, 1 BL were simulated, respectively, and the results are shown in Figure 9.

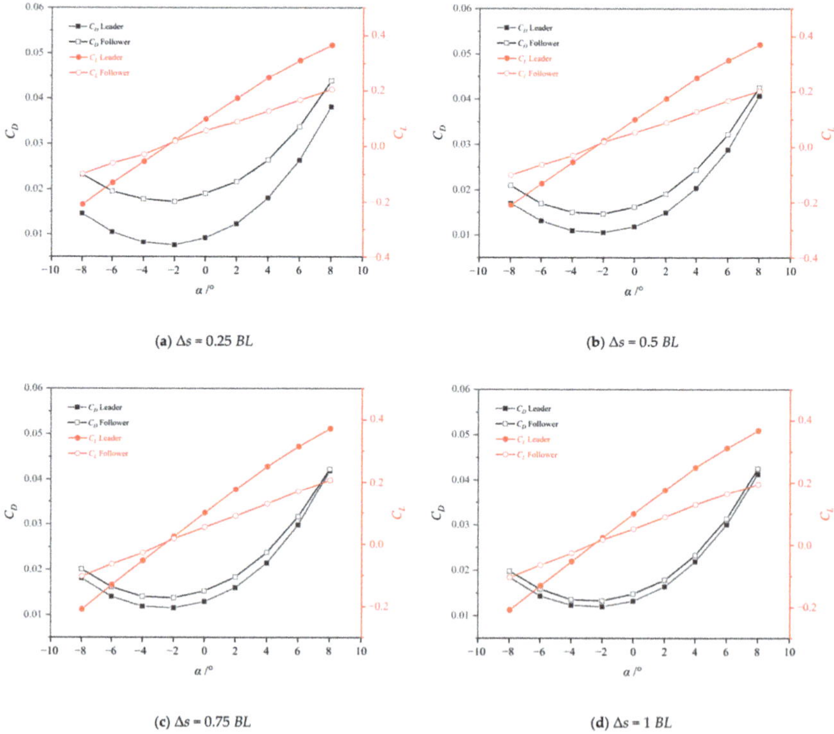

Figure 9. The variation of the resistance coefficient and lift coefficient with the angle of attack of two manta rays at different distances.

It can be seen from Figure 9 that, when the pair of manta rays are arranged horizontally along the x-axis, the hydrodynamic performance trend of each individual remains unchanged. The lift received by the manta ray increases linearly with the change of the angle of attack, and the resistance shows a quadratic function with the angle of attack. However, for the leader, compared with the single manta ray gliding, the resistance is greatly reduced, and the lift force is slightly increased, while for the follower, the resistance force is greatly increased, and the lift force is greatly reduced. This phenomenon—that the resistance of the leader decreases and the resistance of the follower increases—is close to the conclusion of Zhang et al. [30,31]. For targets on the water surface, such as swimmers or ducks, their followers can greatly reduce resistance with the help of the team [32,33], which is contrary to our conclusion. In addition, some studies have found that, when two cyclists ride in tandem, the resistance of each individual, whether the leader or the follower, is less than that of a single person. This conclusion is also different from our research results [34]. The main reason some results obtained from the above studies are different from our conclusions is that the research targets are located in different media and environments.

It is worth noting that, as the distance between the leader and the follower increases, the resistance of each individual quickly converges to the resistance data of a single manta ray, while the difference in lift decreases very slowly. Taking an angle of attack of 4° as an example, when the distance is 0.25 BL, the resistance of the follower is about 1.47 times that of the leader, and the lift is 51.5% that of the leader. When the distance increases to 1 BL, the resistance of the follower decreases to 1.06 times that of the leader, while the lift is 52.1% that of the leader, showing an increase of only 0.6%. In the process of increasing the distance, for the resistance value, the leader increases and the follower decreases to the single manta ray situation, respectively. For the lift value, the leader is not affected from the beginning, and it is always the same as the single manta ray situation; while the follower is strongly affected by the leader at close range and returns to the single manta ray level slowly, as shown in Figure 10. To summarize, for the manta ray tandem system, the resistance value of the manta ray is more sensitive to a change in the distance, and the lift value changes very slowly with the increase in the distance.

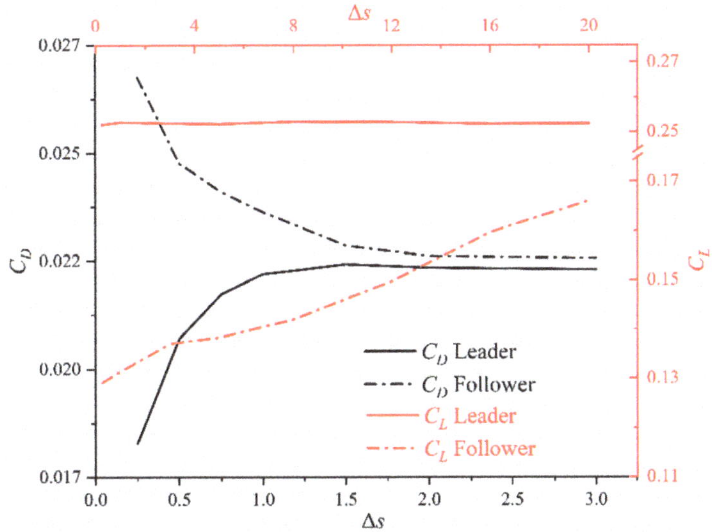

Figure 10. The variation of the gliding resistance coefficient and lift coefficient of two manta rays with distance at $\alpha = 4°$.

3.3. *Analysis of Resistance Reduction Effect and Flow Field Mechanism*

This part explores the hydrodynamic performance of the tandem gliding system under the influence of the two manta rays. Figure 11 shows the average resistance of the system $\overline{D}_{system}$ relative to the resistance of a single manta ray when gliding under different distances and angles of attack. It can be seen from Figure 11 that, for most working conditions, the average resistance of the system is higher than that of the single case. After 2° angle of attack, the relative value of average system resistance gradually increases with the increase in attack angle. For example, when the angle of attack is 8° and Δs is 1 BL, the average resistance of the system is about 6.6% higher than that of the single case. It shows that, in most cases, the average resistance of the pair of manta rays' tandem system is higher than the single manta ray's resistance, and the energy consumption of the system increases. Interestingly, however, we found that when the angle of attack is −6° to −2°, the average resistance of the system is lower than the single resistance at some certain distance cases, which can play a role in reducing resistance and saving energy. For example, when the angle of attack is −2° and Δs is 0.25 BL, the average resistance of the system is about 2.3% lower than that of the single manta case.

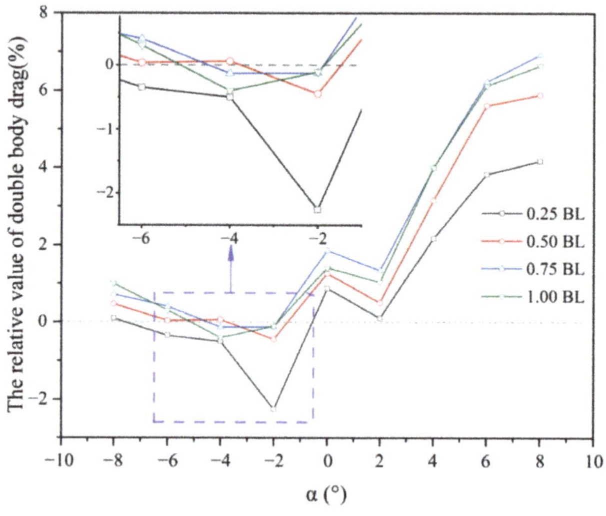

Figure 11. Relative value of the two manta rays gliding system average resistance coefficient.

In order to explore the fluid mechanism of the average resistance of the system, we give the pressure and velocity distribution of $\alpha = -2°$, $\Delta s = 0.25$ BL and $\alpha = 8°$, $\Delta s = 1$ BL, respectively, as shown in Figures 12 and 13. Figure 12 shows the surface pressure distribution of the tandem manta rays at $\alpha = -2°$, $\Delta s = 0.25$ BL, and the velocity distribution of some section planes are also given in the figure. Comparing Figures 8 and 12a, it can be seen that a continuous low-speed zone is formed between the tail of the leader and the head of the follower. Due to the presence of this area, the pressure on the leader's tail and the follower's head increases. Under the premise of little change in the rest, the front-to-back pressure difference of the leader decreases, resulting in decreased resistance. The front-to-back pressure difference of the follower increases, resulting in increased resistance. As for the overall system, the low-speed area at the tail of the leader has not closed and disappeared, and it is affected by the follower and continues to develop backward. Therefore, a coherent high-pressure area is formed between the leader and the follower, as shown in Figures 12 and 13, which makes the average resistance of the system decrease slightly.

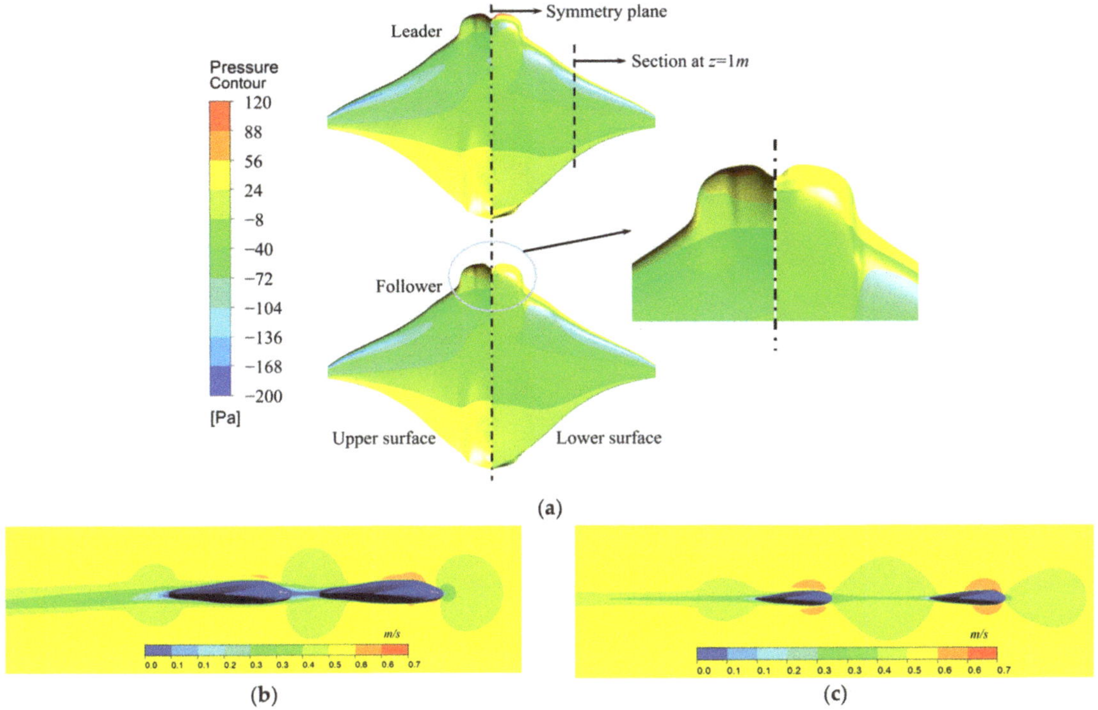

Figure 12. Surface pressure distribution and typical cross-sectional velocity distribution of two manta rays tandem system at $\alpha = -2^{\circ}$, $\Delta s = 0.25$ BL. (**a**): Surface pressure distribution. (**b**): Velocity distribution of symmetry plane. (**c**): Velocity distribution of section at $z = 1$ m.

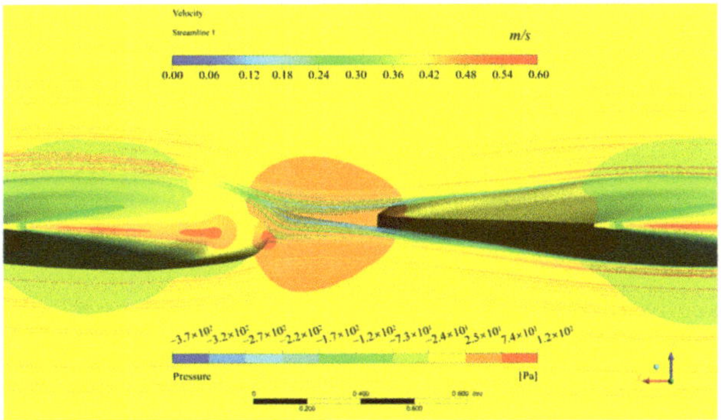

Figure 13. Surface pressure distribution and streamline diagram of the symmetry plane of two manta rays tandem system at $\alpha = -2^{\circ}$, $\Delta s = 0.25$ BL.

Figure 14a shows the surface pressure distribution of the tandem manta ray at $\alpha = 8^{\circ}$, $\Delta s = 1$ BL, and Figure 14b,c shows the velocity distribution of some section planes. Compared with the single manta ray gliding at 8°, the tail flow field of the leader and the lower flow field of the follower have obvious mutual influence. For the leader, superimposing the large distance factor, its surface pressure distribution is almost unaffected, so its lift and resistance are not much different from those of the single gliding; for the follower, the low-pressure area on the upper surface is significantly reduced, and the high-pressure area on the lower surface expands. The pressure difference between the front and rear leads to an increase in the resistance, and the pressure difference between the upper and lower sides decreases, so that the lift force is greatly reduced.

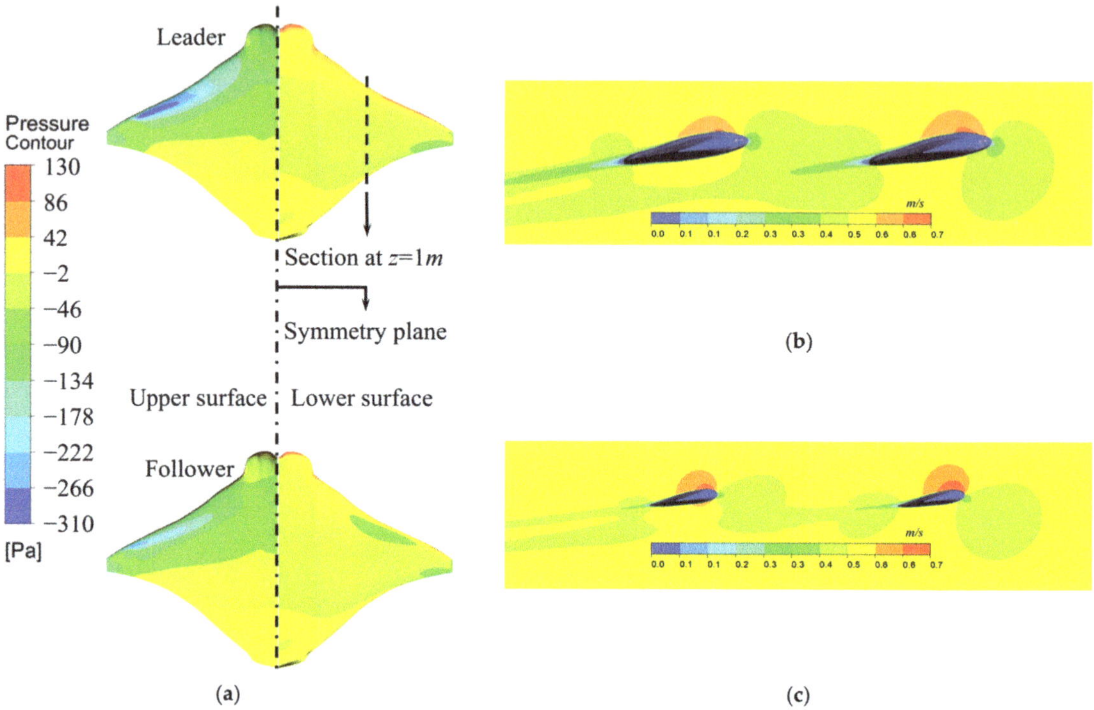

Figure 14. Surface pressure distribution and typical cross-sectional velocity distribution of two manta rays tandem system at $\alpha = 8^\circ$, $\Delta s = 1$ BL. (**a**): Surface pressure distribution. (**b**): Velocity distribution of symmetry plane. (**c**): Velocity distribution of section at $z = 1$ m.

4. Conclusions

In order to explore the hydrodynamic mechanism behind the group swimming phenomenon during the biological migration of manta rays in nature, we used a numerical simulation method to systematically analyze the basic group swimming component of a two-body tandem. The changes in hydrodynamic parameters and surrounding pressure field and velocity field when two manta rays glide at different distances and angles of attack were mainly discussed. The variation law of each individual and group in the two manta rays system under different working conditions was obtained, which provides a strong theoretical support for understanding the collective motion habits of manta rays and the group formation design of underwater bionic vehicles. The main simulation conclusions are as follows:

1. When a single manta ray glides, the lift coefficient is a linear function of the angle of attack, and the resistance coefficient is a quadratic function of the angle of attack. This law does not change when two manta rays are gliding in series. Whether it is the leader or the follower, the lift coefficient and resistance coefficient change with the same angle of attack, and the difference lies in the magnitude of the value.
2. In the two manta rays tandem system, compared with the single-body gliding, the resistance of the leader is greatly reduced, and the lift of the leader is almost unchanged, while for the follower, the resistance is greatly increased, and the lift is greatly reduced.

3. With the increase in the distance between the front and rear of the two manta rays, the resistance of the leader increases rapidly, and the resistance of the follower decreases rapidly, and the two tend to be the same; in terms of lift, the leader maintains the lift regardless of the distance. The lift is stable, not much different from the single-body gliding situation. The lift experienced by the follower is greatly reduced when the distance is small, and the growth recovers slowly with the increase in the distance.
4. For the average resistance of the system, in the specific scenario of a close-range, small negative degree angle of attack, the double manta rays series system can effectively reduce the resistance and play a role in reducing resistance and saving energy. For example, when the distance between the front and rear of the double manta rays is $\alpha = -2°$, $\Delta s = 0.25$ BL, and the two attitudes maintain a $-2°$ angle of attack, the average resistance of the system is reduced by about 2.3% compared with the single resistance. This is of great significance for understanding biological habits and designing underwater bionic vehicle group formations.

According to the above research results of a manta ray pair's tandem gliding, perhaps in the future, when unmanned underwater vehicles travel in groups, due to the increase in the resistance of the leader, we need to pay more attention to the energy consumption of the leader. Change the group formation in time, or let the leader carry more energy. It should be pointed out that the hydrodynamic characteristics of the two manta rays gliding are only the most basic and smallest unit in the study of manta ray group swimming. In our future series of research, we will further consider the problem of manta rays group swimming with more individuals, more complex movement modes, and multiple formations so as to reveal the hydrodynamic characteristics behind the group swimming of this species.

Author Contributions: Writing-original draft, Y.M.; study design, Y.M. and Q.H.; data collection, Y.M. and P.G.; data analysis, Y.M. and P.G.; review & editing, Q.H. and G.P.; supervision, G.P. All authors have read and agreed to the published version of the manuscript.

Funding: This work was supported by the National Natural Science Foundation of China (Grant No. 51879220 and 52001260), the National Key Research and Development Program of China (Grant No. 2020YFB1313201), Fundamental Research Funds for the Central Universities (Grant No. 3102019HHZY030019 and 3102020HHZY030018), and Innovation Foundation for Doctor Dissertation of Northwestern Polytechnical University (CX2022025).

Institutional Review Board Statement: Not applicable.

Informed Consent Statement: Not applicable.

Data Availability Statement: Not applicable.

Conflicts of Interest: The authors declare no conflict of interest.

References

1. Rosenberger, L.J. Pectoral fin locomotion in batoid fishes: Undulation versus oscillation. *J. Exp. Biol.* **2001**, *204*, 379–394. [CrossRef] [PubMed]
2. Graham, R.T.; Witt, M.; Castetllanos, D.W.; Remolina, F.; Maxwell, S.; Godley, B.; Hawkes, L. Satellite tracking of manta rays highlights challenges to their conservation. *PLoS ONE* **2012**, *7*, e36834. [CrossRef] [PubMed]
3. Borazjani, I.; Fotis, S. Numerical investigation of the hydrodynamics of carangiform swimming in the transitional and inertial flow regimes. *J. Exp. Biol.* **2008**, *211*, 1541–1558. [CrossRef] [PubMed]
4. Breder, J.; Charles, M. The locomotion of fishes. *Zoologica* **1926**, *4*, 159–291. [CrossRef]
5. Weihs, D. Hydromechanics of fish schooling. *Nature* **1973**, *241*, 290–291. [CrossRef]
6. Shaw, E. Schooling fishes: The school, a truly egalitarian form of organization in which all members of the group are alike in influence, offers substantial benefits to its participants. *Am. Sci.* **1978**, *66*, 166–175.
7. Pitcher, T.J. Functions of shoaling behaviour in teleosts. In *The Behaviour of Teleost Fishes*; Springer: Boston, MA, USA, 1986; pp. 294–337.
8. Godin, J.J.; Joanne, M. Predator avoidance and school size in a cyprinodontid fish, the banded killifish (*Fundulus diaphanus* Lesueur). *Behav. Ecol. Sociobiol.* **1985**, *16*, 105–110. [CrossRef]
9. Magurran, A.E.; Anthony, H. Information transfer across fish shoals under predator threat. *Ethology* **1988**, *78*, 153–158. [CrossRef]
10. Pitcher, T.J.; Magurran, A.E.; Winfield, I.J. Fish in larger shoals find food faster. *Behav. Ecol. Sociobiol.* **1982**, *10*, 149–151. [CrossRef]

11. Wolf, N.G. Schooling tendency and foraging benefit in the ocean surgeonfish. *Behav. Ecol. Sociobiol.* **1987**, *21*, 59–63. [CrossRef]
12. Ranta, E.; Kai, L. Assortative schooling in three-spined sticklebacks? *Ann. Zool. Fenn.* **1990**, *27*, 67–75.
13. Fish, F.E. Energetics of swimming and flying in formation. *Comments Theor. Biol.* **1999**, *5*, 283–304.
14. Herskin, J.; Steffensen, J.F. Energy savings in sea bass swimming in a school: Measurements of tail beat frequency and oxygen consumption at different swimming speeds. *J. Fish Biol.* **1998**, *53*, 366–376. [CrossRef]
15. Johansen, J.L.; Vaknin, R.; Stetffensen, J.F.; Domenici, P. Kinematics and energetic benefits of schooling in the labriform fish, striped surfperch *Embiotoca lateralis*. *Mar. Ecol. Prog. Ser.* **2010**, *420*, 221–229. [CrossRef]
16. Killen, S.S.; Maunrras, S.; Stetffensen, J.F.; McKenzie, D.J. Aerobic capacity influences the spatial position of individuals within fish schools. *Proc. R. Soc. B Biol. Sci.* **2012**, *279*, 357–364. [CrossRef] [PubMed]
17. Ross, R.M.; Thomas, W.H.B.; Karin, E.L. Group-size-mediated metabolic rate reduction in American shad. *Trans. Am. Fish. Soc.* **1992**, *121*, 385–390. [CrossRef]
18. Breder, C.M. Vortices and fish schools. *Zool. Sci. Contrib. N. Y. Zool. Soc.* **1965**, *50*, 97–114. [CrossRef]
19. Marras, S.; Killen, S.S.; Lindström, J.; McKenzie, D.J.; Steffensen, J.F.; Domenici, P. Fish swimming in schools save energy regardless of their spatial position. *Behav. Ecol. Soc.* **2014**, *69*, 219–226. [CrossRef]
20. Pan, D.Y.; Hua, L.; Shao, X.M. Studies on the oscillation behavior of a flexible plate in the wake of a D-cylinder. *J. Hydrodyn.* **2010**, *22*, 132–137. [CrossRef]
21. Shao, X.; Pan, D.; Deng, J.; Yu, Z. Hydrodynamic performance of a fishlike undulating foil in the wake of a cylinder. *Phys. Fluids* **2010**, *22*, 111903.
22. Shao, X.; Pan, D.; Deming, J.; Yu, Z. Computational study on near wake interaction between undulation body and a D-section cylinder. *Ocean. Eng.* **2011**, *38*, 673–683.
23. Xiao, Q.; Wendi, L.; Hu, J. Parametric study on a cylinder drag reduction using downstream undulating foil. *Eur. J. Mech. B/Fluids* **2012**, *36*, 48–62. [CrossRef]
24. Chao, L.M.; Zhang, D.; Cao, Y.-H.; Pan, G. Numerical studies on the interaction between two parallel D-cylinder and oscillated foil. *Mod. Phys. Lett. B* **2018**, *32*, 1850034. [CrossRef]
25. Chao, L.M.; Pan, G.; Zhang, D.; Yan, G.X. On the thrust generation and wake structures of two travelling-wavy foils. *Ocean. Engin.* **2019**, *183*, 167–174. [CrossRef]
26. Fish, F.E.; Schreiber, C.M.; Moored, K.W.; Liu, G.; Dong, H.; Bart-Smith, H. Hydrodynamic performance of aquatic flapping: Efficiency of underwater flight in the manta. *Aerospace* **2016**, *3*, 20. [CrossRef]
27. Zhang, D.; Huang, Q.-G.; Pan, G.; Yang, L.-M.; Huang, W.-X. Vortex dynamics and hydrodynamic performance enhancement mechanism in batoid fish oscillatory swimming. *J. Fluid Mech.* **2022**, *930*, A28. [CrossRef]
28. Dewey, P.A.; Quinn, D.B.; Boschitsch, B.M.; Smits, J. Propulsive performance of unsteady tandem hydrofoils in a side-by-side configuration. *Phys. Fluids* **2014**, *26*, 041903. [CrossRef]
29. Boschitsch, B.M.; Peter, A.; Alexander, J.S. Propulsive performance of unsteady tandem hydrofoils in an in-line configuration. *Phys. Fluids* **2014**, *26*, 051901. [CrossRef]
30. Zhang, D.; Liming, C.; Guang, P. Analysis of hydrodynamic interaction impacts on a two-AUV system. *Ships Offshore Struct.* **2019**, *14*, 23–34. [CrossRef]
31. Zhang, D.; Pan, G.; Shi, Y.; Wang, P.; Chao, L. Investigation of the resistance characteristics of a multi-AUV system. *Appl. Ocean. Res.* **2019**, *89*, 59–70. [CrossRef]
32. Yuan, Z.-M.; Pan, G.; Shi, Y.; Wang, P.; Chao, L. Stseady hydrodynamic interaction between human swimmers. *J. R. Soc. Interface* **2019**, *16*, 59–70. [CrossRef]
33. Yuan, Z.-M.; Li, M.; Ji, C.-Y.; Li, L.; Jia, L.; Cetcik, L. Wave-riding and wave-passing by ducklings in formation swimming. *J. Fluid Mech.* **2021**, *928*, 820. [CrossRef]
34. Blocken, B.; Defraeye, T.; Koninckx, E.; Carmeliet, J.; Hespel, P. CFD simulations of the aerodynamic drag of two drafting cyclists. *Comput. Fluids* **2013**, *71*, 435–445. [CrossRef]
35. Zarruk, G.A.; Brandner, P.A.; Pearce, B.W.; Phillips, A.W. Experimental study of the steady fluid-structure interaction of flexible hydrofoils. *J. Fluids Struct.* **2014**, *51*, 326–343. [CrossRef]
36. Lyu, D.; Song, B.; Pan, G.; Yuan, Z.; Li, J. Winglet effect on hydrodynamic performance and trajectory of a blended-wing-body underwater glider. *Ocean Eng.* **2019**, *188*, 106303. [CrossRef]

Journal of
Marine Science and Engineering

MDPI

Article

Numerical Research of the Pressure Fluctuation of the Bow of the Submarine at Different Velocities

Xing He [1,2], Qiaogao Huang [1,2,*], Guocang Sun [3] and Xihui Wang [1,2]

1 School of Marine Science and Technology, Northwestern Polytechnical University, Xi'an 710072, China
2 Key Laboratory for Unmanned Underwater Vehicle, Northwestern Polytechnical University, Xi'an 710072, China
3 Wuhan Second Ship Design and Research Institute, Wuhan 430205, China
* Correspondence: huangqiaogao@nwpu.edu.cn

Abstract: A numerical analysis based on stress-blended eddy simulation was conducted to investigate the pressure fluctuation of the bow of a submarine at various velocities (5.93 kn, 10 kn, and 12 kn). The simulation results were compared with the experimental data to demonstrate the validity of the numerical method. Self-power spectrum and wave-number frequency spectrum were discussed from the perspective of energy. The results show that with increasing submarine velocity, the pressure fluctuation in the axial direction increases, the transition point moves forward, and the frequency of the Tollmien–Schlichting wave raises.

Keywords: the bow of the submarine; turbulent boundary layer; pressure fluctuation; stress-blended eddy simulation

Citation: He, X.; Huang, Q.; Sun, G.; Wang, X. Numerical Research of the Pressure Fluctuation of the Bow of the Submarine at Different Velocities. *J. Mar. Sci. Eng.* **2022**, *10*, 1188. https://doi.org/10.3390/jmse10091188

Academic Editors: Md Jahir Rizvi and Unai Fernandez-Gamiz

Received: 20 July 2022
Accepted: 22 August 2022
Published: 25 August 2022

1. Introduction

Turbulent wall pressure fluctuation is an important hydrodynamic noise source, and it is necessary to carry out the corresponding calculation and experimental research. In addition, the sonar is installed on the bow of the submarine, and it is easily affected by self-noise. Therefore, it is important to understand the change in the flow field on the bow at different velocities to improve the performance of the submarine itself.

Benefiting from the development of computer technology, computational fluid dynamics (CFD) is increasingly applied to submarine hydrodynamic research. Huang et al. [1] analyzed the wake of the SUBOFF model with various appendages through experiments and provided the velocity information, which was used to develop the CFD capability for the prediction of flow fields around underwater bodies. Abedi et al. [2] used the Reynolds-averaged Navier–Stokes (RANS) to calculate the three-dimensional axisymmetric flow of the SUBOFF without any appendages and carried out numerical simulation and comparison of the cross-flow vortex structure of the submarine with the SUBOFF and DRDC STR precursors. The calculated data agreed with the experimental. Manoha et al. [3] used the large eddy simulation (LES) to calculate the pressure fluctuation of the unsteady flow field at the blunt trailing edge of the thick plate, then analyzed the pressure fluctuation on the wall of the trailing edge, and the evolution of its amplitude, frequency, and flow direction. Both were in good agreement with the measurements of blunt trailing-edge airfoils. Posa et al. [4] simulated the submarine wake flow field based on LES and it was shown that the wake of the body was affected mainly by the shear layer from the trailing edge of the fins and the turbulent boundary layer growing along the stern, while the influence of the wake of the sail was minimal. Broglia et al. [5] analyzed both the global loads and the vortices originating from the sail of submarine, but focused on the coherent structures shed from the tip and the root of the appendages, where the results showed that the pattern of the coherent structures produced by the stern appendages was substantially dependent on the maneuvering conditions. Ashok et al. [6] used PIV to investigate the turbulent wake of the submarine, that in yaw was expected to generate wakes that were inherently more persistent than one in pitch, and the strong asymmetries in yaw were expected to produce a net rolling moment on the body. In recent years, hybrid formulations such as the detached eddy simulation (DES) [7] and stress-blended eddy simulation (SBES) have been widely used. The starting point for the hybrid formulation was designed to cover all attached boundary layers in the RANS model and to only switch to the LES model in detached zones. Alin et al. [8,9] used LES, DES, and RANS models to investigate the effect of the different simulation methods and to demonstrate the feasibility of using DES and LES on the relatively coarse grids for submarine flows. Liu et al. [10] adopted the DES model to investigate the horseshoe vortex generated around the appendage-body junction of submarines and analyzed the flow characteristics around the appended submarine body, where a new method on the vortex control baffle was presented. Wang et al. [11] studied the effect of the enclosure shape on the hydrodynamic noise of the submarine by the SST-SBES turbulence model and analyzed the influence of the leading and trailing edges with different shapes on the flow field and noise of the underwater vehicle, where the results demonstrated that the front edge of the enclosure could effectively suppress the noise. Magionesi et al. [12] investigated the physical properties of the pressure fluctuation under the turbulent boundary layer on the ship wall through a self-power spectrum and cross-spectrum to reduce the vibration source that may cause noise. Bhushan et al. [13] applied the SST-URANS, SA-DDES, k-ω-DDES, and RANS/LES hybrid methods to simulate the flow field and vortex structure around the SUBOFF enclosure and wake, respectively. Magionesi et al. [14] used Farabee's semi-empirical formula to compare with measurements of the mean square value of the pressure fluctuation in the turbulent boundary layer of the bow sonobuoy, and found that the experimental values of the zero pressure gradient test section agreed better with the empirical values, and the experimental values of the inverse pressure gradient test section were significantly higher than the empirical values. Li et al. [15] used the IDDES model to capture the variation characteristic of the eddy and showed that the pump-jet propulsor had significant force fluctuation and more complex vortices evolution in drifts. Dietiker et al. [16] used the DES method to calculate the pressure fluctuation of the backstage flow, and the calculated main frequency was in good agreement with the test, and the power spectrum was consistent with the empirical model. Meng et al. [17,18] applied the LES method to calculate the pressure fluctuation in the low-speed case of a thin-plate wing with arching and combined it with the FW–H equation for radiated noise. The calculated frequency spectrum and spreading correlation of the pressure field in the edge region of the wing surface were in good agreement with the experiment.

At present, there are few researchers exploring the analysis of pressure fluctuation from the perspective of energy. Based on the analysis of the law of axial pressure change, this paper analyzed the variation of pressure fluctuation at the bow of the submarine from the perspective of energy change under different velocities by calculating the self-power spectrum and wave-number frequency spectrum.

2. Numerical Simulation Methods

The flow of the submarine was simulated based on the RANS equations. In this paper, the SST-SBES (stress-blended eddy simulation, SBES) model was used. The SBES model shielding properties were much improved relative to the current DES/DDES model formulations and SBES allowed for a much more rapid RANS-LES transition than DES/DDES [19]. SBES is expressed as

$$\frac{\partial(\rho k)}{\partial t} + \frac{\partial(\rho \overline{U}_j k)}{\partial x_j} = P_k - \rho \frac{k^{3/2}}{\min(L_t, C_{DES}\Delta_{\max})} + \frac{\partial}{\partial x_j}\left(\left(\mu + \frac{\mu_t}{\sigma_k}\right)\frac{\partial k}{\partial x_j}\right) \tag{1}$$

$$L_t = \frac{k^{3/2}}{\varepsilon} = \frac{\sqrt{k}}{\beta^* \omega} \tag{2}$$

$$\varepsilon_{SDES} = -\beta^* \rho w k F_{SDES} \tag{3}$$

$$F_{SDES} = \left[\max\left(\frac{L_t}{C_{SDES}\Delta_{SDES}}(1 - f_{SDES}), 1 \right) - 1 \right] \tag{4}$$

where ρ is the fluid density; p is the fluid pressure; μ is the fluid dynamic viscosity; $\overline{U}$ is the velocity vector; t is the time; k is the turbulent kinetic energy; and ω is the turbulent dissipation rate.

The SBES model takes the dissipative term ε replaced with Equation (3). Compared with DES, the shielding function of SBES can realize fast conversion between RANS and LES. The stress tensor is obtained by

$$\tau_{ij}^{SBES} = f_{SDES}\tau_{ij}^{RANS} + (1 - f_{SDES})\tau_{ij}^{LES} \tag{5}$$

where τ_{ij}^{RANS} and τ_{ij}^{LES} are the RANS and LES parts of the model stress tensor, respectively, f_{SDES} is the shielding function. If the two models are based on the eddy-viscosity, the formula can be simplified as

$$\nu_t^{SBES} = f_{SDES}\nu_t^{RANS} + (1 - f_{SDES})\nu_t^{LES} \tag{6}$$

where ν_t^{SBES} is the eddy viscosity and ν_t^{RANS} and ν_t^{LES} are the RANS and the LES eddy viscosity, respectively. In addition, it should be noted that the equations were solved based on the ANSYS Fluent V2020. The detailed information of other parameters can be found in [20].

After selecting the SBES model for the hybrid model, ANSYS Fluent provides four subgrid-scale (LES) models as options: the Smagorinsky–Lilly model, dynamic Smagorinsky–Lilly model, wall-adapting Local eddy-viscosity (WALE) model, and algebraic wall-modeled LES $S - \Omega$ (WMLES S-Omega) model. In this paper, WALE was selected as it provides the lowest eddy viscosity in 2D flow regions and thereby allows the flow to quickly develop 3D turbulence structures in separating the shear layers. The combination of SBES with the WMLES S-Omega formulation is likely not a useful combination, as the SBES-WALE model combination can by itself act in a WMLES mode [20].

In the simulation, the PISO (pressure-implicit with splitting of operators) algorithm was used. The Eulerian terms and viscous terms were discretized by a standard second-order scheme, and a bounded second-order scheme was used for the physical time discretization. Pressure-velocity coupling was SIMPLIC and the time-step was 5×10^{-5}.

3. Numerical Simulation of Submarine

3.1. Model Geometry

In this paper, the 1:24 scaling SUBOFF model was used. The geometric model was based on the SUBOFF geometric model published by the David Taylor Research Center [21]. The total length of the model was 4356 mm, the head was 1016 mm, the middle part was 2229 mm, and the tail was 1111 mm. The leading edge of the sail was located at 924 mm of the submarine, the length was 368 mm, and the height was 206 mm. The trailing edge of the stern appendage was 349 mm away from the tail end of the submarine, which was arranged in a cross shape. The SUBOFF model is shown in Figure 1.

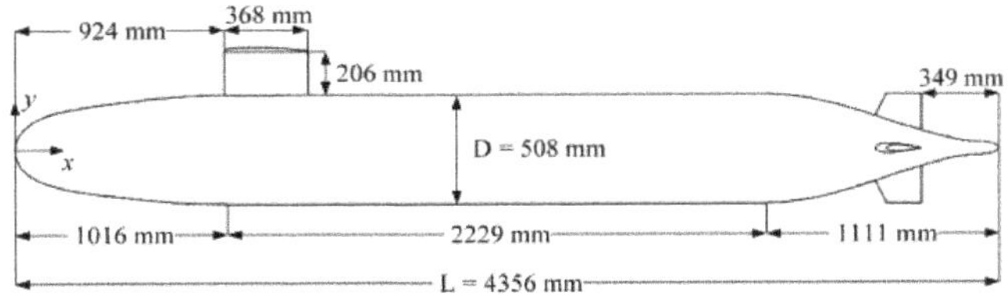

Figure 1. The shape of the SUBOFF.

3.2. Mesh and Numerical Setup

The entire computational domain was divided into two subdomains: the inner domain and outer domain. The outer domain is shown in Figure 2. The submarine worked in a cylinder tunnel with a length of 25 m, and a radius of 10 m. The blockage ratio of the water domain was 0.0025, which was small enough that the blockage effect of the SUBOFF model could be ignored. The inlet was located 8 m upstream of the submarine. For the outlet boundary, it was defined as the pressure outlet boundary. Smooth wall conditions (zero shear) were defined as the side of the cylinder boundary, and the SUBOFF surface was considered to be the no-slip wall.

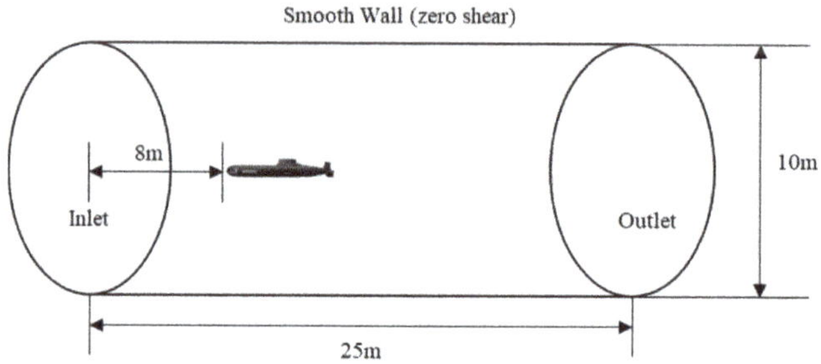

Figure 2. Computational domain and boundary conditions.

In order to obtain better turbulent structural characteristics of the submarines, mesh densification was carried out on the sail, the tail rudder, and the tail area. Meanwhile, considering the calculation power, the entire computing domain was divided into two parts: the inner domain and the outer domain. By refining the mesh, the outer domain can appropriately reduce the number of elements, so that the mesh can be effectively concentrated around the submarine. The submarine mesh is shown in Figure 3. In the numerical simulation of the submarine, high-quality results can be obtained only when all boundary layer solutions are sufficient. In this paper, the $y+$ was set to 1, the width of the first cell to the wall y was set to 1×10^{-5} m, and it increased exponentially according to 1.1.

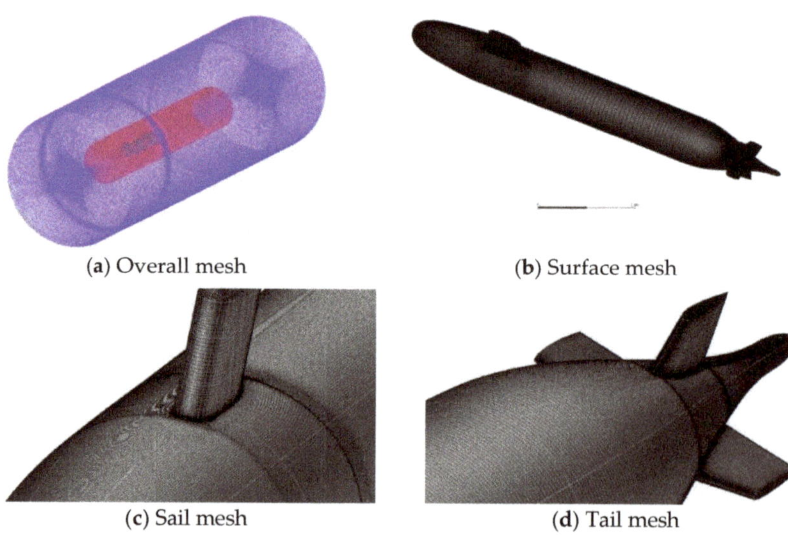

Figure 3. Mesh of SUBOFF.

3.3. Verification of Mesh

The mesh density and quality have an important influence on the calculation results, and CFD uncertainty analysis was carried out with the numerical results of the submarine drag. Three groups of meshes were generated for verification, having approximately 23 million (coarse mesh), 44.3 million (medium mesh), and 82 million (fine mesh) cells. The verification simulations are presented at an operating velocity of 5.98 kn. Table 1 lists the number of meshes with different scales and resolutions.

Table 1. The calculation results of the total drag coefficient.

Mesh Groups	ID	Number (M)	Drag Force (N)
Coarse	1	23	97.3
Medium	2	44.3	99.6
Fine	3	82	100.7

Next, two steps were performed to complete the mesh-independent verification.

Step 1: Calculate the convergence ration R_k according to the following formula:

$$R_k = \varepsilon_{k32}/\varepsilon_{k21} \tag{7}$$

$$\varepsilon_{k32} = S_{k3} - S_{k2} \tag{8}$$

$$\varepsilon_{k21} = S_{k2} - S_{k1} \tag{9}$$

where ε_{k21} is defined as the changes between ID1 and ID2, in the same way as ε_{k32}.

The R_k = 0.4783. Because $0 < R_k < 1$, the three different meshes presented monotone convergence, and the convergence was satisfied.

Step 2: Calculate the order of accuracy and one-term estimates:

$$p_{k1} = \frac{\ln(\varepsilon_{k32}/\varepsilon_{k21})}{\ln(r_k)} \tag{10}$$

$$\delta^*_{RE_{k1}} = \frac{\varepsilon_{k21}}{r_k^{p_{k1}} - 1} \tag{11}$$

The correction factor C_k is

$$C_{k1} = \frac{r_k^{p_{k1}} - 1}{r_k^{p_{kest}} - 1} \tag{12}$$

Because C_k = 1.8572 >1, according to the ITTC (2017) recommendation, when the C_k >1, the mesh uncertainty U_k is

$$U_k = \left|C_{k1}\delta^*_{RE_{k1}}\right| + \left|(1 - C_{Tk1})\delta^*_{RE_{k1}}\right| \tag{13}$$

The comparison error e_k is

$$E_k = D - S \tag{14}$$

The validation uncertainly U_v is

$$U_v = \sqrt{U_{SN}^2 + U_D^2} \tag{15}$$

After calculation, E_k = 1.60, U_v = 3.42. For validation of the uncorrected simulation, $|E_k| < U_v$, so that the validation of U_v was achieved. Therefore, the medium met the requirement of mesh independence. Meanwhile, the result of the medium was 99.6 N, the experimental [22] result was 102.3 N, and the error was 2.64%, which met the requirements. Therefore, the mesh with 44.3 million was chosen for further simulation.

3.4. Verification of Numerical Method

In order to verify the validity of the numerical calculation method adopted in this paper, the numerical simulation under the working condition of 5.93 kn was carried out, and the results were compared with the experimental results [23], which were used to verify the pressure coefficient C_p and the surface friction coefficient C_f along the upper meridian line of the submarine longitudinal section obtained by numerical simulation. The formulas of C_p and C_f are shown as follows:

$$C_p = \frac{p - p_{ref}}{0.5\rho_{ref} v_{ref}^2} \tag{16}$$

$$C_f = \frac{\tau_w}{0.5\rho_{ref} v_{ref}^2} \tag{17}$$

where p is the static pressure; τ_ω is the wall shear stress; p_{ref} is the reference pressure; ρ_{ref} is the reference density; and v_{ref} is the reference velocity, in this paper, v_{ref} refers to the flow velocity.

It can be seen from Figures 4 and 5 that C_p and C_f are in good agreement with the experimental value. The pressure coefficient C_p values of the bow, the sail, and the tail of the SUBOFF were larger, while the C_p values of other parts of the submarine were smaller, which is consistent with the actual flow field. It was noted that the lift force was the integral of the pressure on the surface of the submarine, so it can be represented with the integral of the pressure coefficient. The experimental [23] value was −0.03615 and the calculated value was −0.03721. The error was 2.93%, which met the requirement of less than 5%.

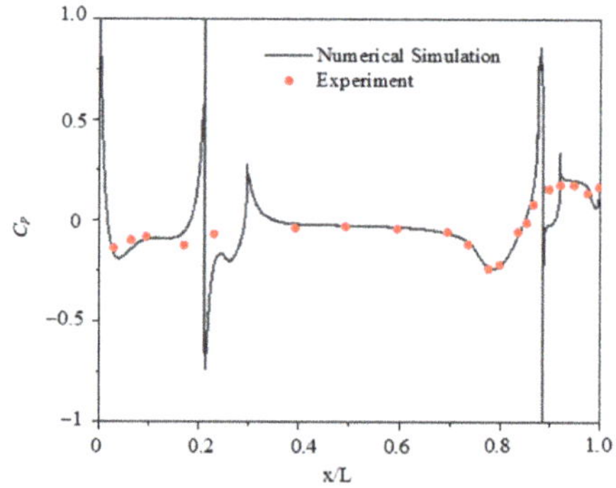

Figure 4. C_p along the upper meridian line.

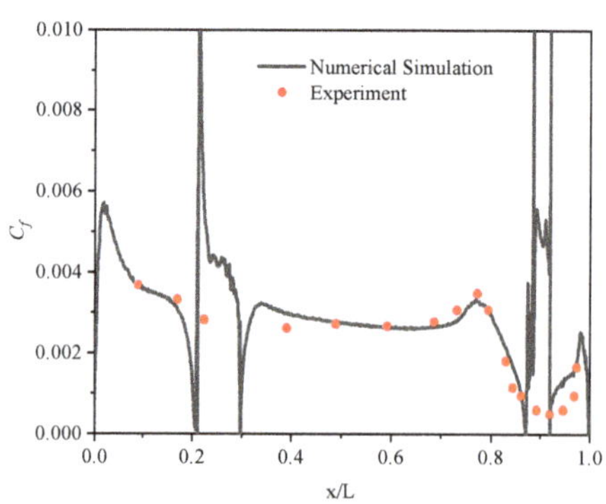

Figure 5. C_f along the upper meridian line.

Through the comparison, it could be found that the numerical calculation in this paper was in good agreement with the experimental results, which verifies the accuracy and feasibility of the numerical method used in this paper.

4. Results and Discussion

4.1. Overview of the Fluid and the Boundary Layer Transition

In this section, the characteristics of the overall and bow flow field of the submarine under three velocities (5.93 kn, 10 kn, 12 kn) were studied. Figure 6 shows the simulation results of the three pressures. It can be seen that the pressure distribution on the surface of the submarine was similar under three velocities. The places where the pressure changed strongly were the bow, the front of the sail, and the front of the tail, which is consistent with the actual flow field. According to Bernoulli's principle, the impact of the fluid on the wall caused the velocity to decrease and the pressure to increase.

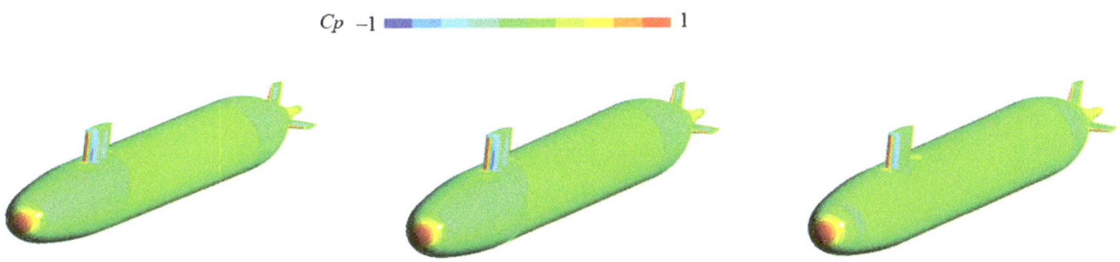

(**a**) 5.93 kn (**b**) 10 kn (**c**) 12 kn

Figure 6. The contours of pressure at three velocities.

The surface distribution of the velocity gradient tensor under three velocities was the second invariant, as seen in Figure 7, where V_A denotes the inflow velocity. The vortex structure around the submarine was clearly visible, which could well capture the horseshoe vortex and top vortex pair at the bottom of the sail and the bottom of the rudder.

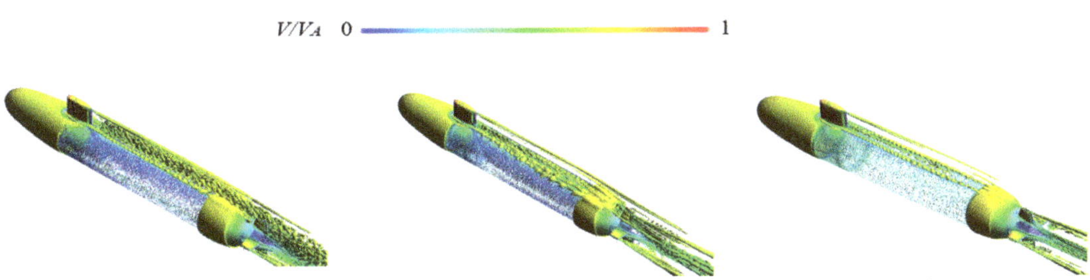

(**a**) 5.93 kn (**b**) 10 kn (**c**) 12 kn

Figure 7. The instantaneous isosurface of the Q-criterion at three velocities.

Table 2 compares the total drag, the drag coefficient, and main parameters of the horseshoe vortex in the sail of the submarine under three velocities, where Re_L is the Reynolds number with the length as the characteristic length; F_x is the drag; C_d is the drag coefficient; A and Γ are the size of the vortex and the flux of the circulation as the flow direction vorticity, respectively. Γ is expressed as

$$\Gamma = \iint\limits_A w_x dA \tag{18}$$

where ω_x is the vorticity component along the x-axis perpendicular to the measurement plane.

Table 2. The drag and structural parameters of the horseshoe vortex at three velocities.

Velocity (kn)	Re_L	F_x (N)	C_d	A (m^2)	Γ (m^2/s^2)
5.93	1.322×10^7	99.6	0.106	8.195×10^{-5}	5.397×10^{-4}
10	2.230×10^7	262.2	0.098	7.902×10^{-5}	1.482×10^{-3}
12	2.676×10^7	370.7	0.096	7.528×10^{-5}	1.815×10^{-3}

It can be seen from Table 2 that as the velocity increased, the drag on the submarine increased, the drag coefficient decreased slightly, and the structure of the horseshoe vortex in front of the sail became smaller, but the strength of the vortex increased.

Extracting the pressure coefficient distribution on the ridge of the longitudinal profile in the submarine, we compared the pressure coefficient of 5.93 kn with [23], which was closer to the literature results. Here, it can be seen that for the SUBOFF standard model, the velocity had essentially no effect on the distribution of the pressure coefficients (Figure 8), and the pressure coefficients for all three velocities reached their lowest values at x/L = 0.025.

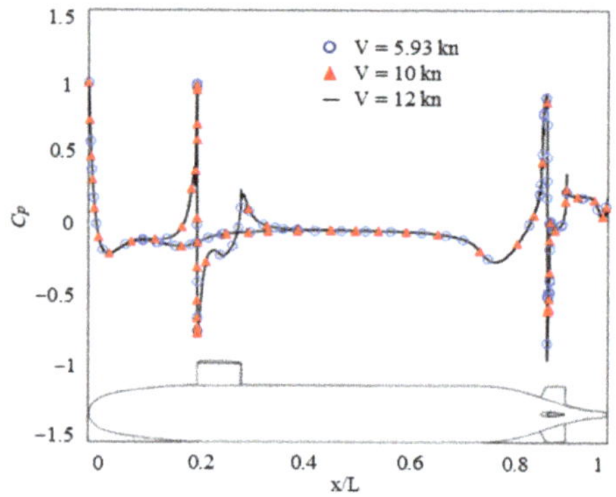

Figure 8. The distribution of the ridge pressure coefficients in the longitudinal profiles.

Table 3 shows the specific location of the turning point calculated by the intermittent factor under the three cases. S is the arc length to the transition point at the bow, and Re_s is the Reynolds number with S as the characteristic length. As can be seen from the table, with the increase in the velocity, the position that reached the critical Reynolds number moved forward, so the turning point position moved forward gradually. The intermittency phenomenon occurred as the boundary layer developed along the flow direction from laminar to fully developed turbulence, that is, during the transition between laminar and turbulent periods. The intermittency factor is typically used to describe the intermittent phenomena. Intermittent phenomena can also occur in the normal upward direction such as in the areas where turbulent patches arise, where the turbulence is fully developed inside the patches, and the laminar is outside the patches. The intermittency factor γ is the time average of the intermittency function $I(x,y,z,t)$, which has a value of 0 in laminar flow and 1 in turbulent flow. The relationship between the two is expressed as

$$\gamma = \frac{1}{T}\int_{0}^{T} I(x,y,z,t)\mathrm{d}t = \gamma(x,y,z,t) \tag{19}$$

Table 3. The parameters of the transition point at different velocities.

Velocity (kn)	S (m)	Re_s
5.93	0.2925	8.880×10^5
10	0.2235	1.144×10^6
12	0.1718	1.056×10^6

Figure 9 depicts the distribution of the bow's intermittent factor, and the position of the transition point of the 1:24 scaling model of SUBOFF at 5.93 kn could be obtained. The arc length of the surface through which the fluid flowed was 0.2925 m at $x = 0.20$ m. Taking the arc length as the characteristic length, the critical Reynolds number $Re_s = 8.880 \times 10^5$ was obtained. The distribution of the intermittency factor allowed us to observe the location where the transient occurred, where the higher the velocity, the more forward the transition location, and Figure 10 shows the distribution of the intermittency factor at different velocities.

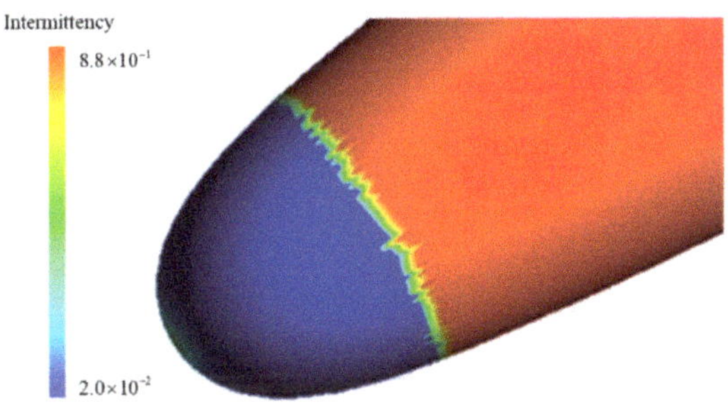

Figure 9. The transition position at the velocity of 5.93 kn.

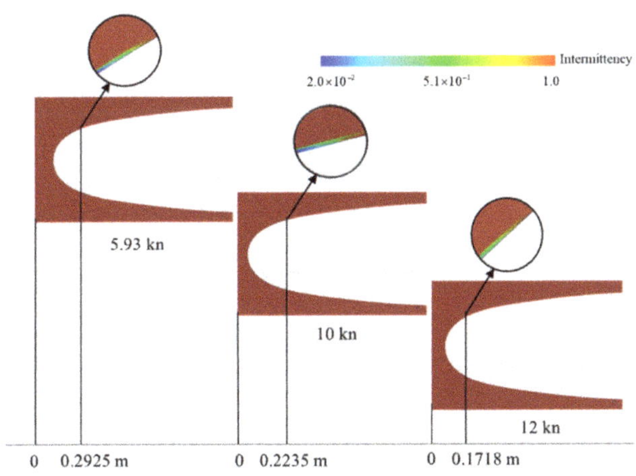

Figure 10. The transition position at three velocities.

Figure 11 shows the time-averaged velocity profiles at the positions of $x = 0.03$ m, 0.087 m, 0.2 m, and 0.5 m, respectively. Based on the previously obtained positions of the transition points, when $x < 0.0976$ m, the boundary layers of the three cases were in laminar flow and the velocity distribution curves had inflection points, as shown in Figure 11a,b. When $x = 0.2$ m (Figure 11c), the case of 5.93 kn was still laminar, while the 10 kn and 12 kn were already after the turning point, and the distribution of the 10 kn and 12 kn velocities along the normal direction at this time also showed the typical turbulent boundary layer velocity profile characteristics. When $x = 0.5$ m (Figure 11d), the boundary layer of all three cases developed into a turbulent state, and the boundary layer was no longer a single laminar structure, but an inner and outer layer, and the velocity distribution profiles of the linear bottom layer, transition layer, logarithmic layer, and outer layer were characterized. This significant change in the velocity pattern is that the velocity profile with flow direction is an important sign of the boundary layer transition phenomenon.

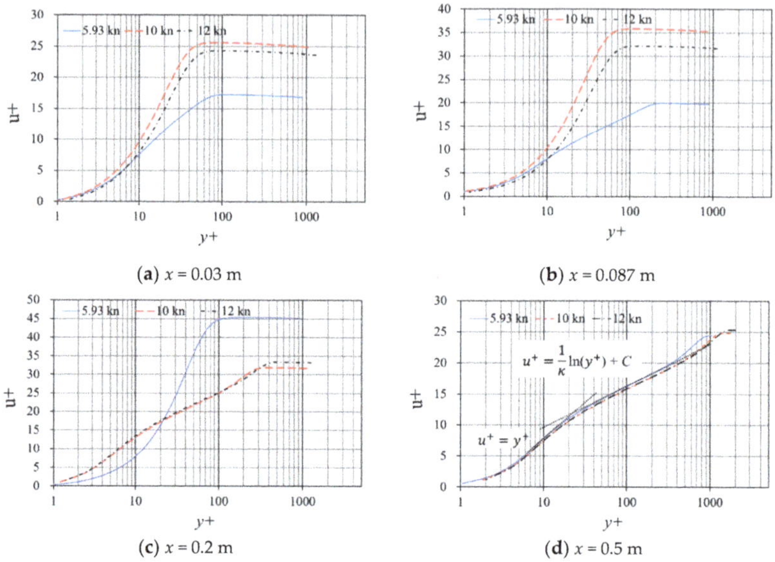

Figure 11. The velocity distribution within the boundary layer at different normal positions of the bow.

Figure 11d shows the distribution of the turbulent boundary layer velocity along the normal direction at $x = 0.5$ m, where the velocity stratification structure of the inner layer was consistent with the wall rate of the turbulent boundary layer at the linear bottom layer $(0 \leq y^+ \leq (5 \sim 8))$, with

$$u^+ = y^+ \tag{20}$$

In the logarithmic layer ($(30 \sim 50)\nu/\mu_\tau \leq y \leq 0.2\delta$), the velocity distribution satisfies the logarithmic law of the form:

$$u^+ = \frac{1}{\kappa}\ln y^+ + C \tag{21}$$

Since there is still no consensus on the choice of coefficients, the calculated results of the three cases were fitted here, and $1/\kappa$ and C are shown in Table 4. Combined with Figure 11d, it is clear that at the same flow position, in the inner layer with $y < 0.2\delta$, the profiles of the boundary layer with various mainstream velocities were very similar and did not change significantly with the change in velocity. While in the outer layer with $y > 0.2\delta$, the data points for different mainstream velocities fell on different profiles, and the appearance of this dispersion implies a change in the turbulence characteristics.

Table 4. The fitting results of $1/\kappa$ and C.

Velocity (kn)	$1/\kappa$	C
5.93	3.1379	1.9797
10	3.1086	1.5708
12	3.1146	1.5411

4.2. Axial Pressure Distribution

As shown in Figure 12, thirty sample points were arranged along the mid-longitudinal profile ridge on the surface of the bow. Figure 13 shows the curves of the pressure fluctuation at several of the selected sample points with time. Due to the small velocity, the pressure fluctuation at different flow locations was not visually distinguishable in the time domain, so Cp' is needed to be used to measure the magnitude of the pressure fluctuation, which represents the strength of the pressure fluctuation, and is expressed as

$$Cp' = \frac{p'_{rms}}{Q_v} \times 100\% \tag{22}$$

where p'_{rms} is the root mean square value of pressure fluctuation; Q_v is the pressure of the fluid velocity.

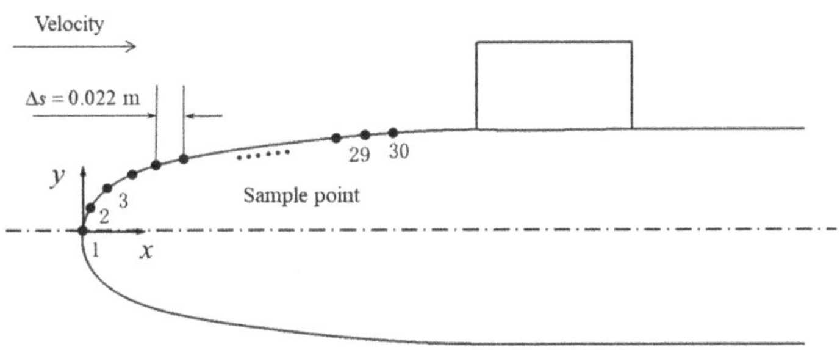

Figure 12. The position of the axial sample points.

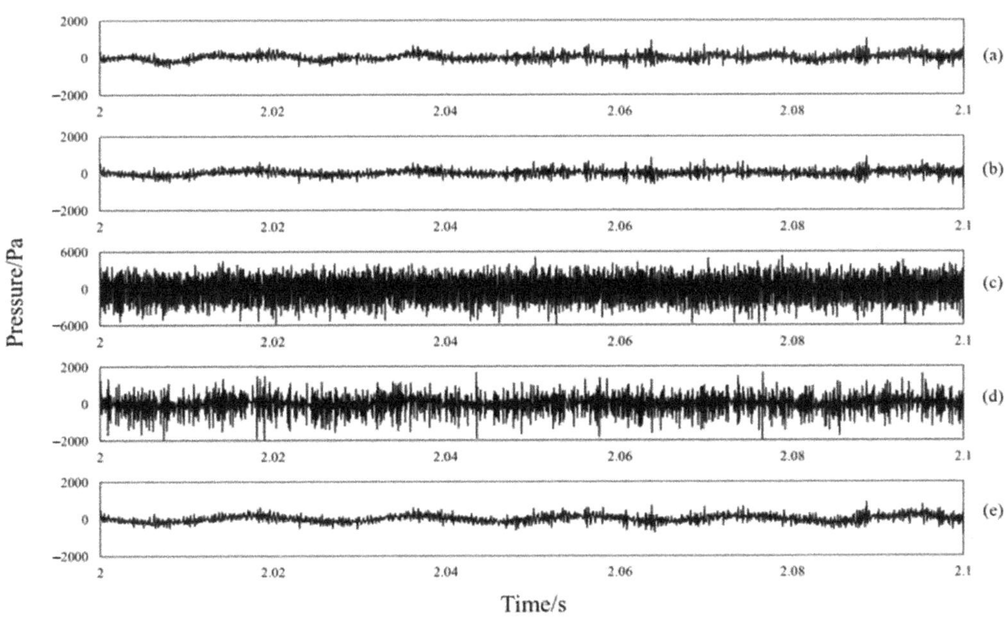

Figure 13. The fluctuation of pressure over time at the velocity of 12 kn. **(a)** $Re_s = 2.689 \times 10^5$. **(b)** $Re_s = 6.738 \times 10^5$. **(c)** $Re_s = 1.011 \times 10^6$. **(d)** $Re_s = 1.482 \times 10^6$. **(e)** $Re_s = 2.561 \times 10^6$.

The pressure fluctuation data at the bow were extracted for the three cases, and the variation curves of the pulsation pressure with times at several sample points at 12 kn are shown in Figure 13. At a higher velocity of 12 kn, it can be observed from the time domain that the amplitude of pulsation pressure was very small in the laminar region, there was a sudden increase in the transient region ($Re_s = 1.011 \times 10^6$), and then there was a certain decrease in the turbulent region, but it was still higher than the laminar.

Figure 14 gives the strength of the pressure fluctuation at different velocities for the first 20 sample points along the axial direction at three velocities. This is in agreement with the time-domain characteristics where the pressure fluctuation is small in the laminar region and has a clear peak at the transient region, after which it decays rapidly and develops to the turbulent phase, where the pressure fluctuation magnitude is maintained at a lower level, but is somewhat larger than in the laminar. This feature of the boundary layer pressure fluctuation is also one of the most effective methods to detect boundary layer transitions. For example, Hu et al. [24] used the pulsation pressure test technique to detect the boundary layer transitions in a 10° cone in a wind tunnel and obtained the evolution of the pulsation pressure coefficient along the bus.

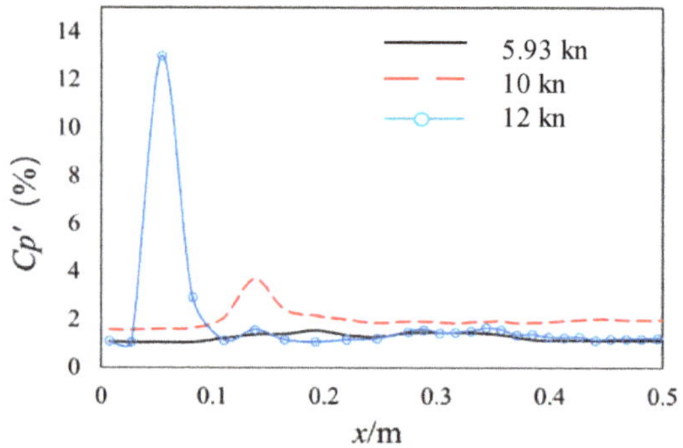

Figure 14. At three velocities.

Comparing the pressure fluctuation evolution law for three velocities, the transition point at the bow moved forward with increasing velocity and the pressure fluctuation coefficient declined, while the strength of the pressure fluctuation in the transient region increased, and the magnitude of the increase grew larger. In conclusion, the pressure fluctuation strength in the transient area may be separated from the laminar and turbulent flow more clearly when the velocity is higher.

4.3. Self-Power Spectral of Wall Pressure Fluctuation

The self-power spectral of the pressure fluctuation at the sample point at different velocities is given in Figures 15–17. According to Parseval's theorem, the total energy of the signal in the time domain was equal to the total energy of the signal in the frequency domain. From the random signal through Fourier transform, and then through the following calculation, we could find its self-power spectrum. In this paper, the frequency characteristics of the pressure fluctuation are represented by the power spectrum G_p and is defined as

$$G_p = 10\lg\frac{G(f)}{p_{ref}^2} \tag{23}$$

where p_{ref} is baseline sound pressure; $p_{ref} = 1\ \mu$ pa; $G(f)$ is the pressure fluctuation power spectral density function calculated with fast Fourier transform.

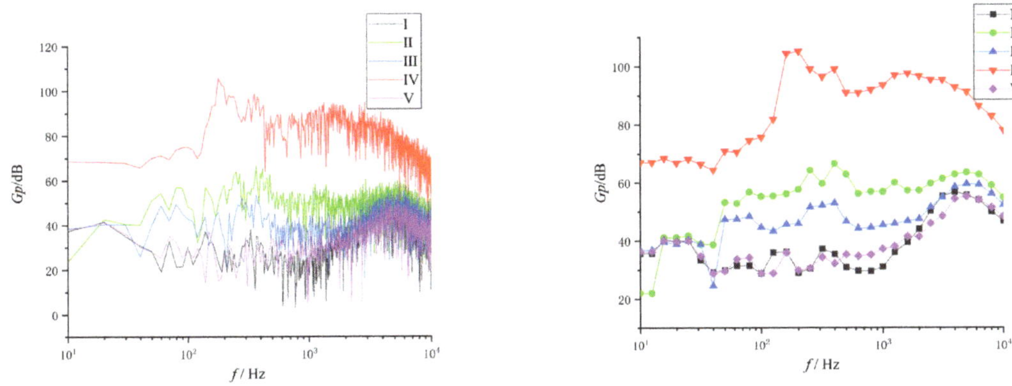

(**a**) Frequency domain characteristic (**b**) Third-octaves

Figure 15. The self-power spectral of the pressure fluctuation at the velocity of 5.93 kn. (I) $Re_s = 1.349 \times 10^5$, (II) $Re_s = 6.060 \times 10^5$, (III) $Re_s = 7.409 \times 10^5$, (IV) $Re_s = 8.760 \times 10^5$, (V) $Re_s = 1.482 \times 10^6$.

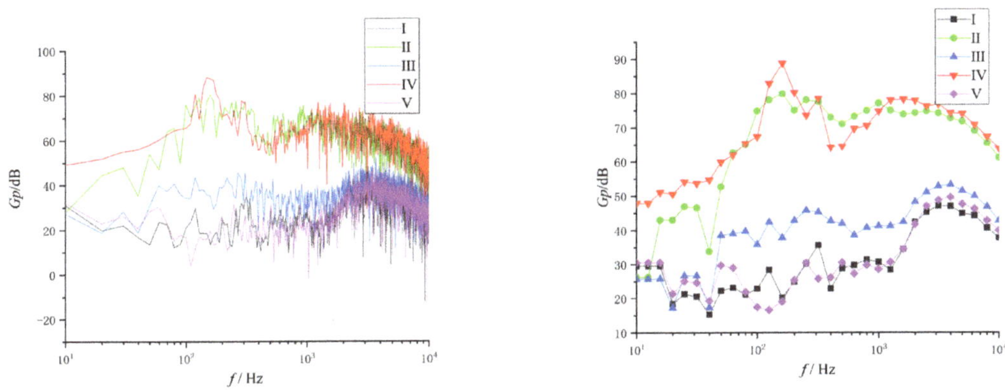

(**a**) Frequency domain characteristic (**b**) Third-octaves

Figure 16. The self-power spectral of the pressure fluctuation at the velocity of 10 kn. (I) $Re_s = 2.275 \times 10^5$, (II) $Re_s = 5.668 \times 10^5$, (III) $Re_s = 7.939 \times 10^5$, (IV) $Re_s = 1.136 \times 10^6$, (V) $Re_s = 1.590 \times 10^6$.

In the low- and medium-frequency range, each velocity case followed the same pattern of the highest spectral level occurring in the transient region, the lower spectral level in the region toward turbulence, and the lowest spectral level in the laminar region. In the frequency range around 100 Hz, a peak gradually appeared during the development from the laminar to the transient, which was associated with the T–S wave generated in the sensible phase and growth in the linear phase. The frequency and spectral level of this peak at each velocity transient point are given in Table 5. It is clear that as the velocity rises, the T–S wave frequency and spectral level increased.

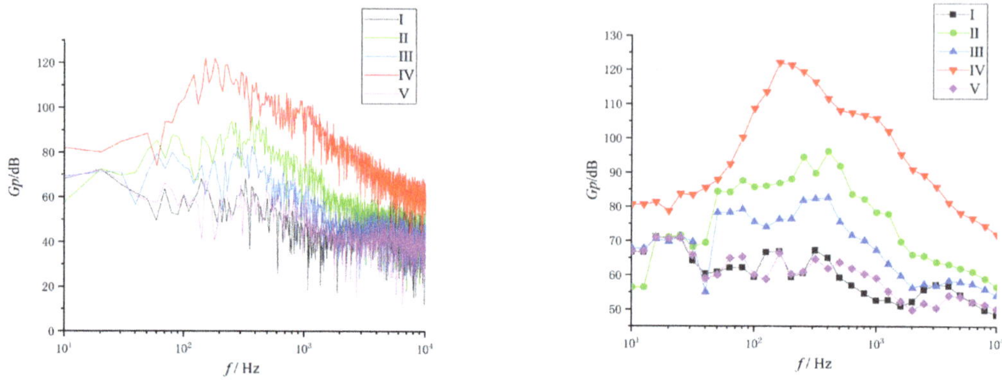

(a) Frequency domain characteristic

(b) Third-octaves

Figure 17. The self-power spectral of the pressure fluctuation at the velocity of 12 kn. (I) $Re_s = 2.730 \times 10^5$, (II) $Re_s = 5.441 \times 10^5$, (III) $Re_s = 8.175 \times 10^5$, (IV) $Re_s = 1.090 \times 10^6$, (V) $Re_s = 1.499 \times 10^6$.

Table 5. The parameters of the T–S wave at different velocities.

	5.93 kn	10 kn	12 kn
f_{TS} (Hz)	93.33	133.33	166.67
Spectrum level (dB)	135.70	141.84	149.92

4.4. Wave-Number Frequency Spectrum of Wall Pressure Fluctuation

The pressure fluctuation was subjected to a two-dimensional Fourier transform from time–space to the wave-number frequency domain in order to examine the movements of the bow turbulent boundary layer vortices and the distribution of energy. With a sampling time step of 0.001 s, only the low-frequency portion of the spectrum, which corresponds to the large-scale eddies, was studied in this paper.

Fourier transform of the correlation function of the turbulent pressure fluctuation time-space signal was used to define the wave-number frequency spectrum. The equation is as follows:

$$\Phi\left(\vec{k},w\right) = \int_{-\infty}^{+\infty} R\left(\vec{\xi},\tau\right) e^{-i(\vec{k},\vec{\xi}+\omega\tau)} d\vec{\xi}\, d\tau \quad (24)$$

By applying a fast Fourier variation to the discrete spatial and temporal signals of turbulent pressure fluctuation and methodically averaging the square of its amplitude, it is typically possible to derive the wave-number frequency spectrum in practice. The formula is:

$$\Phi_m(k,w) = \frac{\left\langle \left| \sum_{n=1}^{N} \sum_{m=1}^{M} W(x_m,t_n) p_m(x_m,t_n) e^{-i(k_x x_m + w t_n)} \Delta x \Delta t \right|^2 \right\rangle}{(2\pi)^2 NMC_w} \quad (25)$$

where p_m is the pressure fluctuation from the mth sensor. N is the number of time nodes and M is the number of sensors. Δx is sensor spacing and Δt is the time step.

$$x_m = (m-1)\Delta x \quad (26)$$

$$t_n = (n-1)\Delta t \quad (27)$$

$$W(x_m,t_n) = W(x_m)W(t_n) \quad (28)$$

where $W(x_m)$ and $W(t_n)$ are window functions. The Hanning window was used in this paper.

$$C_w = \frac{\sum_{n=1}^{N}\sum_{m=1}^{M} W^2(x_m, t_n)}{NM} \tag{29}$$

In the wave-number domain, the range was $-\pi/d$ to π/d with a resolution of $2\pi/\mathrm{Nd}$, where D is the distance between adjacent sampling points and N is the number of sampling points. With an interval of 4.2 mm between adjacent monitoring points, 100 points were placed along the ridge line of the mid-longitudinal profile in the fully developed turbulent area at the bow to increase the resolution of the wave-number domain. According to Figure 18, the first sample point was situated at x = 0.3 m.

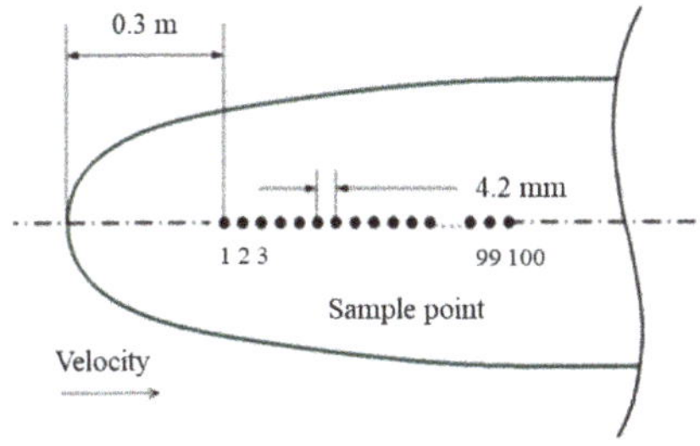

Figure 18. The position of the sample points.

Figure 19 shows the wave-frequency spectrum of the wall pressure fluctuation at three velocities. The acoustic component is what led to the highest value at k = 0. It is clear that the migratory ridge was readily visible. With an increase in frequency, the amplitude lowered, the wave number decreased to the peak value of the migration ridge, and the migration ridge widened. The highest portion of the wavenumber-frequency spectrum, or the distribution of energetic vorticity with the highest energy, had a frequency domain range that expanded with increasing velocity from the standpoint of energy distribution.

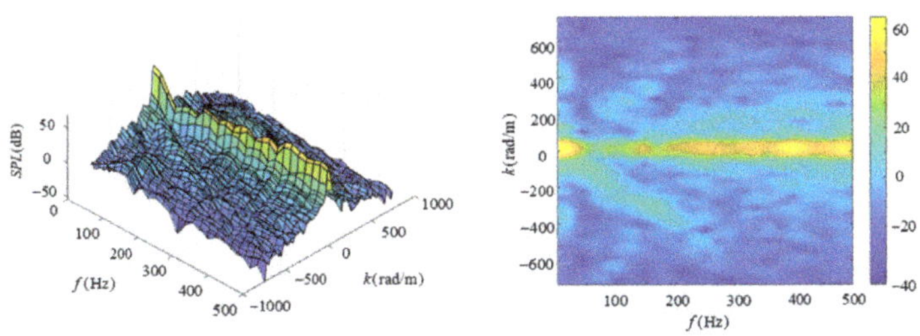

(**a**) 5.93 kn

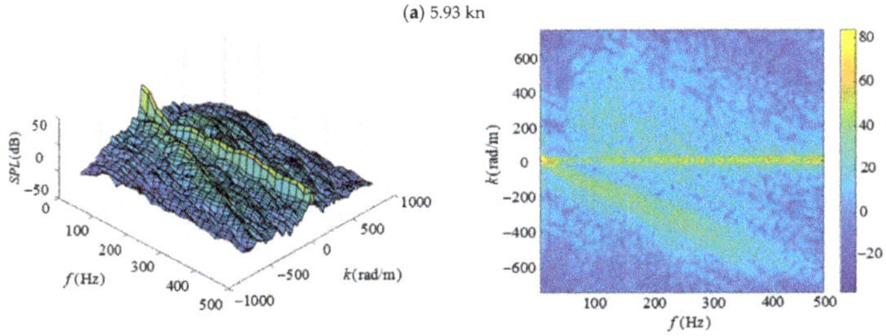

(**b**) 10 kn

Figure 19. *Cont.*

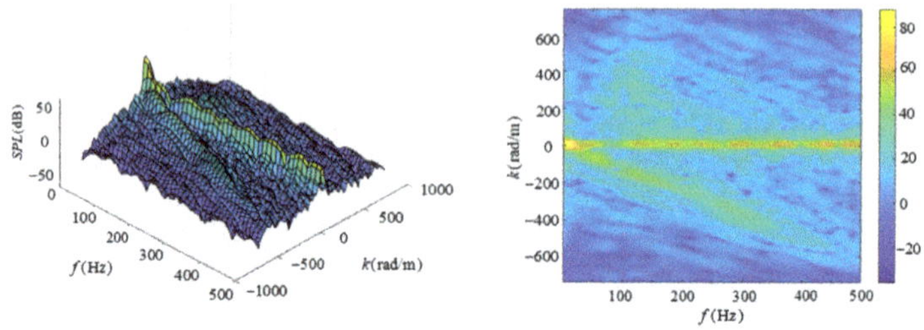

(c) 12 kn

Figure 19. Two views of the wave-number frequency spectrum at different velocities.

Table 6 lists the computed results for the wave-number frequency spectrum parameters. The three velocities had peak spectral levels of 21.79 dB, 64.61 dB, and 83.57 dB, respectively. These values are consistent with the change in the self-power spectral level, which rises with increasing velocity. The migration velocity is one of the crucial physical quantities to examine the turbulent wall pressure, which reflects the motion velocity of the turbulent vortex cluster structure.

Table 6. The parameters of the wave-number frequency spectrum.

Item	5.93 kn	10 kn	12 kn
Spectral peak level (dB)	21.79	64.61	83.57
Frequency range (Hz)	(0,250)	(0,450)	(0,500)
Wavenumber range (rad/m)	(−529,0)	(−680,0)	(−695,0)
Migration rate (m/s)	2.17	3.62	4.49
Dimensionless migration rate	0.71	0.70	0.73

The calculated migration velocities for 5.93 kn, 10 kn, and 12 kn were 2.17 m/s, 3.62 m/s, and 4.49 m/s, respectively, while the corresponding dimensionless velocities were 0.71 m/s, 0.70 m/s, and 0.72 m/s, meaning that the migration velocities rose as the main flow velocities rose, while the corresponding dimensionless velocities hovered around 0.7.

5. Conclusions

For the SUBOFF standard model 5.93 kn, 10 kn, and 12 kn cases, calculations were carried out to determine the flow field structure and pressure fluctuation for various velocities. The following conclusions were drawn after data analysis.

(1) The distribution of pressure coefficients along the ridge of the mid-longitudinal profile and the distribution of pressure and velocity in the flow field of the submarine under various velocities essentially followed the same pattern. The influence of velocities on the flow field was more apparent in the numerical magnitude of pressure and velocity as well as the size and intensity of the vortex structure. With increasing velocity, the submarine experienced an increase in both the drag coefficient and the total drag force. However, the size of the horseshoe vortex at the sail deck's bottom leading edge decreased as the velocity rose, and its intensity increased.

(2) Using the horizontal distance to the stationary point of the bow as the characteristic length, the critical Reynolds number of the submarine's bow was 6.339×10^5. With the higher velocity, the crucial Reynolds number was positioned further ahead and the transient point was located farther forward.

(3) When the transient occurred, the pressure fluctuation amplitude greatly increased and appeared to reach a high peak before swiftly decaying and dropping to a lower level in the turbulent region. The pressure fluctuation at the wall of the submarine's bow was relatively modest in the laminar region. The peak obtained in the transient region rose under conditions of increasing velocity and advanced with it.

(4) As the pulsation energy of each frequency component significantly increased at the transient compared to laminar flow, the magnitude of the pressure fluctuation was reflected in the self-power spectrum. During the development from laminar to the transient, a peak gradually appeared in the frequency range of about 100 Hz, which was associated with the T–S wave generated in the perceptive phase and growing in the linear phase. The T–S wave frequency rose with increasing velocity, and the spectral level followed suit.

(5) In the fully developed turbulent region's wave-number-frequency spectrum, a clear migration ridge with energy focused in the lower frequency range could be seen. The migration ridges increased in frequency, width, and amplitude as they migrated in the direction of lower wave numbers. The migration velocity was the vortex cluster's average movement speed; the larger the velocity, the higher the migration velocity, which was maintained at 0.7.

Author Contributions: Conceptualization, X.H. and X.W.; Data curation, X.H.; Investigation, X.H.; Visualization, X.H.; Writing—original draft, X.H.; Supervision, Q.H. and G.S.; Writing—review and editing, Q.H. and X.W. All authors have read and agreed to the published version of the manuscript.

Funding: This work was supported by the National Natural Science Foundation of China under Project No. 51979226 and by the Fundamental Research Funds for the Center Universities under Project Nos. 3102019HHZY030019 and 3102020HHZY030018.

Institutional Review Board Statement: Not applicable.

Informed Consent Statement: Not applicable.

Data Availability Statement: Not applicable.

Conflicts of Interest: The authors declare no conflict of interest.

References

1. Huang, T.; Liu, H. Measurement of flows over an axisymmetric body with various appendages in a wind tunnel: The DARPA SUBOFF experimental program. In Proceedings of the 19th Symposium on Naval Hydrodynamics, Seoul, Korea, 23–28 August 1992.
2. Saeed, A.; Ali, A.D.; Ali, S. Effects of bulbous bow on cross-flow vortex structures around a streamlined submersible body at intermediate pitch maneuver: A numerical investigation. *J. Mar. Sci. Appl.* **2016**, *15*, 8–15.
3. Manoha, E.; Troff, B.; Sagaut, P. Trailing-edge noise prediction using large-eddy simulation and acoustic analogy. *AIAA J.* **2000**, *38*, 575–583. [CrossRef]
4. Posa, A.; Balaras, E. A numerical investigation of the wake of an axisymmetric body with appendages. *J. Fluid Mech.* **2016**, *792*, 470–498. [CrossRef]
5. Broglia, R.; Posa, A.; Bettle, M. Analysis of vortices shed by a notional submarine model in steady drift and pitch advancement. *Ocean Eng.* **2020**, *218*, 108236. [CrossRef]
6. Ashok, A.; Buren, T.; Smits, A. The structure of the wake generated by a submarine model in yaw. *Exp. Fluids* **2015**, *56*, 123. [CrossRef]
7. Spalart, P. Detached Eddy Simulation. *Annu. Rev. Fluid Mech.* **2009**, *41*, 181–202. [CrossRef]
8. Alin, N.; Fureby, C. LES of the flow past simplified submarine hulls. In Proceedings of the 8th International Conference on Numerical Ship Hydrodynamics, Busan, Korea, 22–25 September 2003.
9. Alin, N.; Bensow, R.; Fureby, C. Current capabilities of DES and LES for submarines at straight course. *J. Ship Res.* **2010**, *54*, 184–196. [CrossRef]
10. Liu, Z.; Ying, X.; Tu, C. Numerical simulation and control of horseshoe vortex around an appendage–body junction. *J. Fluids Struct.* **2011**, *27*, 23–42.
11. Wang, X.; Huang, Q.; Pan, G. Numerical research on the influence of sail leading edge shapes on the hydrodynamic noise of a submarine. *Appl. Ocean Res.* **2021**, *117*, 102935. [CrossRef]
12. Magionesi, F.; Ciappi, E. Characterization of the response of a curved elastic shell to turbulent boundary layer. In Proceedings of the ASME 2010 3rd Joint US-European Fluids Engineering Summer Meeting and 8th International Conference on Nanochannels, Microchannels, and Minichannels, Montreal, QC, Canada, 1–5 August 2010.
13. Bhushan, S.; Alam, M.F.; Walters, D.K. Evaluation of hybrid RANS/LES models for prediction of flow around surface combatant and SUBOFF geometries. *Comput. Fluids* **2013**, *88*, 834–849. [CrossRef]
14. Magionesi, F.; Mascio, A.D. Investigation and modeling of the turbulent wall pressure fluctuation on the bulbous bow of a ship. *J. Fluids Struct.* **2016**, *67*, 219–240. [CrossRef]

15. Li, H.; Huang, Q.; Pan, G. An investigation on the flow and vortical structure of a pre-swirl stator pump-jet propulsor in drift. *Ocean Eng.* **2022**, *250*, 111061. [CrossRef]
16. Dietiker, J.F.; Hoffmann, K.A. Predicting wall pressure fluctuation over a backward-facing step using detached eddy simulation. *J. Aircr.* **2009**, *46*, 2115–2120. [CrossRef]
17. Meng, W.; Moin, P. Computation of trailing-edge flow and noise using large-eddy simulation. *AIAA J.* **2000**, *38*, 2201–2209.
18. Meng, W.; Moreau, S.; Iaccarinoand, G. LES prediction of wall-pressure fluctuations and noise of a low-speed airfoil. *Int. J. Aeroacoustics* **2009**, *8*, 177–198.
19. Menter, F. Stress-Blended Eddy Simulation (SBES)—A new paradigm in hybrid RANS-LES modeling. In *Symposium on Hybrid RANS-LES Methods*; Springer: Cham, Switzerland, 2016.
20. ANSYS. *ANSYS 2020 R1 Fluent User Guide*; ANSYS: Canonsburg, PA, USA, 2020.
21. Groves, N.C.; Huang, T.T.; Chang, M.S. *Geometric Characteristics of DARPA (Defense Advanced Research Projects Agency) SUBOFF Models (DTRC Model Numbers 5470 and 5471)*; David Taylor Research Center Bethesda MD Ship Hydromechanics Dept.: Bethesda, MD, USA, 1989.
22. Liu, H.; Huang, T. *Summary of DARPA SUBOFF Experiment Program Data*; Naval Surface Warface Center, Cardrock Division (NSWCCD): Bethesda, MD, USA, 1999.
23. BULL, P. The validation of CFD predictions of nominal wake for the SUBOFF fully appended geometry. In Proceedings of the Twenty-First Symposium on Naval Hydrodynamics, Washington DC, USA, 24–28 June 1997.
24. Hu, C.; Huang, X. The location of boundary-layer transition detected by pressure fluctuation measurements. *Exp. Meas. Fluid Mech.* **2002**, *2*, 67–71.

Journal of
Marine Science and Engineering

MDPI

Article

Numerical Investigation on Air Film Fusion of Pressure-Equalizing Exhaust around Shoulder Ventilation of Submarine-Launched Vehicle

Yao Shi [1,2,*], Jinyi Ren [1,2], Shan Gao [1,2] and Guang Pan [1,2]

1 School of Marine Science and Technology, Northwestern Polytechnical University, Xi'an 710072, China; rjy@mail.nwpu.edu.cn (J.R.); gsjsy9341@163.com (S.G.); panguang@nwpu.edu.cn (G.P.)
2 Key Laboratory of Unmanned Underwater Projectile, Northwestern Polytechnical University, Xi'an 710072, China
* Correspondence: shiyao@nwpu.edu.cn

Abstract: In order to study the influence of pressure-equalizing exhaust at the shoulder of a submarine-launched vehicle on the surface hydrodynamic characteristics, this paper establishes a numerical calculation method based on the VOF multiphase flow model, the standard RNG turbulence model and the overset mesh technology; the method compares the fusion characteristics of the air film at the shoulder of the underwater vehicle, as well as the distribution of surface pressure along the vehicle's axial direction. The results show that the approximate isobaric zone derived from air film fusion can greatly improve the hydrodynamic characteristics of the vehicle, and the number of venting holes determines the circumferential fusion time of the air film. The greater the number of venting holes, the sooner circumferential fusion starts.

Keywords: submarine-launched vehicle; pressure-equalizing exhaust; multiphase flow; air film fusion

Citation: Shi, Y.; Ren, J.; Gao, S.; Pan, G. Numerical Investigation on Air Film Fusion of Pressure-Equalizing Exhaust around Shoulder Ventilation of Submarine-Launched Vehicle. *J. Mar. Sci. Eng.* **2022**, *10*, 39. https://doi.org/10.3390/jmse10010039

Academic Editor: Alessandro Ridolfi

Received: 19 November 2021
Accepted: 17 December 2021
Published: 31 December 2021

1. Introduction

In recent years, launch modes relying on underwater launch platforms have received increasing attention from various countries for their high stealth, long endurance, and high hitting accuracy. However, the progress of related research and testing has not been smooth, mainly due to the lack of systematic and reliable knowledge of the hydrodynamic load problem when the vehicle is not in the water.

During the movement of an underwater vertically launched vehicle, its structure is completely closed. When the vehicle is constantly close to the free liquid surface, the hydrostatic pressure of the external environment decreases sharply; because of the closeness, once the load of internal pressure exceeds the safety margin of the structure design, the vehicle will be damaged.

The vertical launch process of a submarine-launched vehicle in the water can be roughly divided into three stages: the barrel-out stage, the free navigation stage, and the water-out stage [1]. During the barrel-out stage, the vehicle's energy is gas at high temperature and high pressure, and its speed is increasing; at the same time, cavitation occurs and develops quickly in the low-pressure area at the shoulder of the vehicle. In the free navigation phase, the vehicle is approaching the water surface at decreasing speed. Subsequently, the back jet causes cavitation bubbles to gradually fall off from the end. During the water-out stage, the fluid medium undergoes a sudden change, and the cavitation is continuously collapsed by the interference of the liquid surface. The unsteady, high-frequency, transient-impact load directly acts on the surface of the vehicle [2], which will affect the movement posture and surface of the vehicle. The structure has a serious impact.

In view of the impact load of the vehicle in the process of entering the water, scholars at home and abroad have carried out a series of numerical simulations and experimental testing research and have achieved diverse results. In 1945, Reichardt [3] proposed active ventilated cavitation technology. Later, Knapp et al. [4] directly demonstrated the feasibility of forming cavitation through ventilation technology. Zhang Naimin et al. [5] established a theoretical calculation model of porous exhaust bubble morphology based on the principle of independent cavitation expansion. Ma Guihui et al. [6] studied the effects of single-row exhaust and double-row exhaust combined on the body-fitted air film around the flow field during the vertical water discharge process of the vehicle. Huang Biao [7] carried out research on the dynamic characteristics of unsteady cavitation fluid of a rotating body in a closed water tunnel. Sun Tiezhi [8] carried out research on the influence of the ventilation position on the hydrodynamic characteristics of the submarine-launched vehicle by solving the RANS equation of the mixed medium based on the homogeneous balance flow theory. Gaoshan et al. [9] carried out a refined study on the water exit process of a three-dimensional double-shot salvo based on the VOF multiphase flow model and dynamic mesh technology. Chen Fu et al. [10] simulated the underwater vertical motion process of a vehicle with isobaric exhaust, and analyzed the effect of the exhaust angle on the unsteady development process of the isobaric film and the flow field characteristics of the flow around the vehicle. Bao Wenchun et al. [11] used the VOF multiphase flow model and dynamic grid technology to simulate the bubble development process and pressure distribution on the surface of an underwater launch vehicle. Zhang Ke et al. [12] conducted a numerical study on the exhaust process of an underwater launch vehicle based on the VOF multiphase flow model, and analyzed the development and evolution of cavitation shapes during the navigation and exit stages of the vehicle. Kang Yazhuo et al. [13] simulated the launch process of the vehicle based on the dynamic grid technology and the VOF multiphase flow model, and finally summarized the evolution of bubbles during the whole process of the vertical launch of the vehicle. Yan Guojun et al. [14] obtained the distribution characteristics of the flow field and the pressure on the outer surface of the vehicle through the coupling calculation of the multiphase flow field and the ballistic of the vehicle during the water exit process of a submarine-launched vehicle. Liu Junjun [15] studied and analyzed the influence of various parameters such as the gas pressure, launch speed, and inclination angle of the vehicle on the water outflow of the vehicle. Wang Kelin [16] combined experiments and simulations to study the two-phase flow problem of the double-hole jet flow field under the conditions of multiple flow field parameters and structural parameters. Li Renfeng et al. [17] conducted numerical simulations on the exit and flow field of missiles at different launch speeds. They found that the lateral force produced by the pressure difference between the front and back surfaces of the missile is the reason for the deflection. Zhang Zhao [18] constructed a hybrid mathematical model of the dynamics of underwater-launching high-speed vehicles, and used this model to simulate the dynamics of underwater-launching high-speed vehicles. Based on the RANS equation, Wu Yuyan [19] analyzed the action mechanism of the development of the air mass at the mouth of the launching tube, the internal pressure of the launching tube, and the evolution of the gas–liquid two-phase flow field when the underwater vertically launched vehicle exits the tube. Sun Longquan et al. [20] studied the mechanism of opening a small annular groove at the bottom of the exhaust ring to promote the circumferential fusion of bubbles. They carried out research on the fusion of bubbles by the annular groove under various working conditions and found that the groove can effectively improve the circumferential fusion of the air film. Qiu Yang [21] conducted a numerical simulation study of the multiphase cavitation flow generated by the exhaust gas from the interior cavity of the vehicle, and studied the effect of the exhaust cavitation on the underwater launch process. Duan Lei et al. [22] combined experimental and numerical calculation methods to study the ventilating cavitation flow field around the rotating body. Ren Zeyu et al. [23] used a new type of experimental platform to conduct underwater vertical launch experiments and study the influence of the cavitation number and Froude number on the development and evolution of bubbles. Zhang Subin et al. [24] used numerical simulation methods to study the cavitation flow during the underwater vertical launch of a vehicle.

Recently, great progress has been made in the research of cavitation flow of underwater vertically launched vehicles; however, most of them study supercavitating flow at a method of interface capture. Scholars rarely study the evolution process of local cavitation in ventilation and the influence of local cavitation on the hydrodynamic characteristics of the surface of the vehicle.

The shoulder venting technique can effectively improve the hydrodynamic properties of the vehicle surface. In the initial stage of exhaust, bubbles are attached to the outer surface of the vehicle in a bundle. With continuous movement, the bubbles develop, mix, fuse, and finally fall off, especially when the exhaust bubbles fuse into a large bubble, which can greatly improve the force characteristics of the navigation body and improve the stability of the navigation body's underwater ballistics, as the interior of the bubbles is an approximately isobaric zone.

The hydrodynamic load of high-speed vehicles is the most important basic issue in the underwater vertical launch process; in addition, the stability of cavitation and the characteristics of the collapse load are important objects via which to study this problem. Carrying out a simulation study on the cavitation fusion characteristics of the underwater vehicle shoulder in the circumferential exhaust process can more clearly elucidate the mechanism of the cavitation fusion, which is useful for optimizing the surface structure of the vehicle. It has very important significance to improve the water characteristics and motion stability of the vehicle.

In this paper, based on the VOF multiphase flow model, the overset mesh technique is used to study the air film fusion characteristics of a submerged vehicle. The structure of this paper is as follows: the second part describes the numerical method; the numerical calculation model is established in the third part; the fourth part analyzes the results of the simulation; and the fifth presents the conclusions drawn from the study.

2. Numerical Calculation Method

2.1. Governing Equation

The basic governing equations describing the multiphase flow of underwater vertical launch mainly include the continuity equation, momentum equation, and energy conservation equation. The basic form of the equations is as follows:

Continuity equation:

$$\frac{\partial \rho_m}{\partial t} + \frac{\partial}{\partial x_i}(\rho_m u_i) = 0 \tag{1}$$

Momentum equation:

$$\frac{\partial}{\partial t}(\rho_m u_j) + \frac{\partial}{\partial x_i}(\rho_m u_i u_j) = -\frac{\partial \rho_m}{\partial x_i} + \rho g_i + \frac{\partial}{\partial x_i}\left[(\mu_m + \mu_t)\left(\frac{\partial u_i}{\partial x_j} + \frac{\partial u_i}{\partial x_i}\right)\right] \tag{2}$$

Energy equation:

$$\frac{\partial}{\partial t}(\rho_m T) + \frac{\partial}{\partial x_i}(\rho_m u_i T) = \frac{\partial}{\partial x_i}\left(\frac{k_i}{c_p}\frac{\partial T}{\partial x_j}\right) + S_T \tag{3}$$

2.2. Turbulence Equation

In this paper, the RNG model is used for simulation calculation. The turbulence equations are as follows:

$$\rho\frac{Dk}{Dt} = \frac{\partial}{\partial x_i}\left[\left(\alpha_k \mu_{eff}\right)\frac{\partial k}{\partial x_i}\right] + G_k + G_b - \rho\varepsilon - Y_M \tag{4}$$

$$\rho\frac{D\varepsilon}{Dt} = \frac{\partial}{\partial x_i}\left[\left(\alpha_\varepsilon \mu_{eff}\right)\frac{\partial \varepsilon}{\partial x_i}\right] + C_{1\varepsilon}\frac{\varepsilon}{k}(G_k + C_{3\varepsilon}G_b) - C_{2\varepsilon}\rho\frac{\varepsilon^2}{k} - R \tag{5}$$

2.3. Multiphase Flow Model

This paper uses the VOF multiphase flow model; an equation regarding the phase volume fraction can be established, based on the continuity equation and momentum equation of the mixed phase:

$$\frac{\partial \alpha}{\partial t} + \nabla \cdot (\alpha \mathrm{U}) = 0 \tag{6}$$

The solution of the equation can be divided into the following situations:

When $\alpha = 0$, it means that the phase fluid does not exist in the unit;

When $0 < \alpha < 1$, it means that the unit contains the interface of this phase fluid and other phase fluids;

When $\alpha = 1$, it means that the unit contains only the fluid of this phase.

2.4. Overset Meshing

As shown in Figures 1 and 2, the basic idea of the overset meshing technique is to use the sub-domain mesh to interpolate in the overlapping area, and then achieve the effect of real-time transmission of flow field information. There are three types of points in the overset mesh: in-cavity points, computational points, and interpolation points. The in-cavity points are not involved in the flow field calculation, while the computational points are, and the interpolation points are used for the transmission of flow field information. In this study, the computational domain is divided into a background domain and an overlap domain, and the overset mesh is a combination of the background-region mesh and the overlap-region mesh, which overlap in space but are not connected and exist independently of each other.

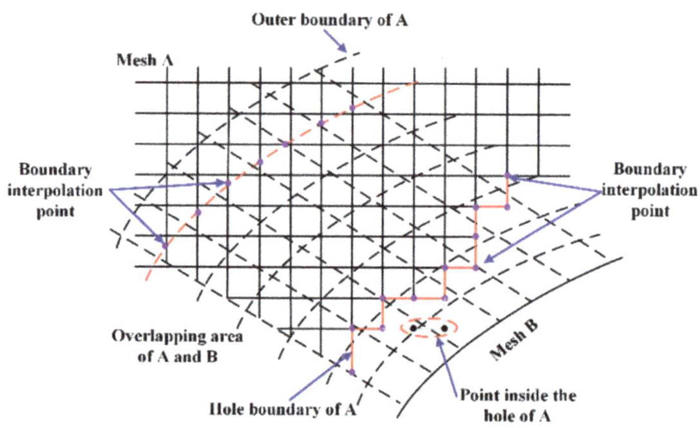

Figure 1. Schematic of overset mesh.

2.5. Numerical Method Verification

In order to verify the effectiveness of the numerical algorithm in this paper, experimental data from the literature [25] are used. The experimental model uses a semi-cylindrical shell with a size of 150 mm × 100 mm × 120 mm (diameter × height × shell thickness), and the speed is 1.2 m/s. The boundary conditions and meshing are shown in Figures 3 and 4.

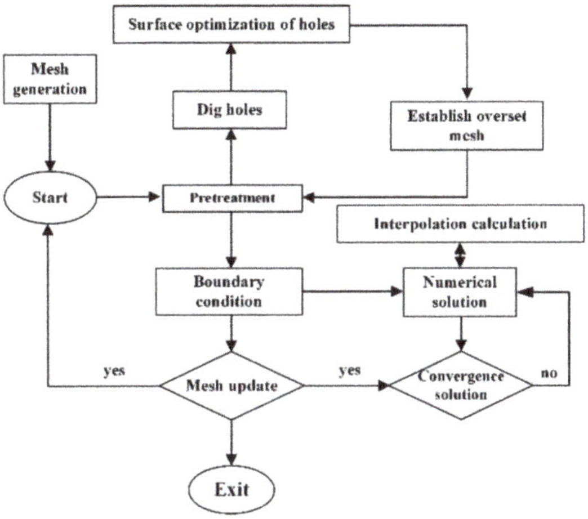

Figure 2. Process flow chart of overset mesh.

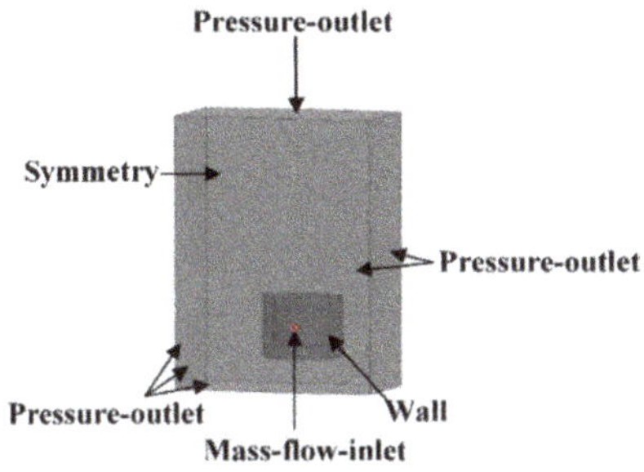

Figure 3. Boundary conditions.

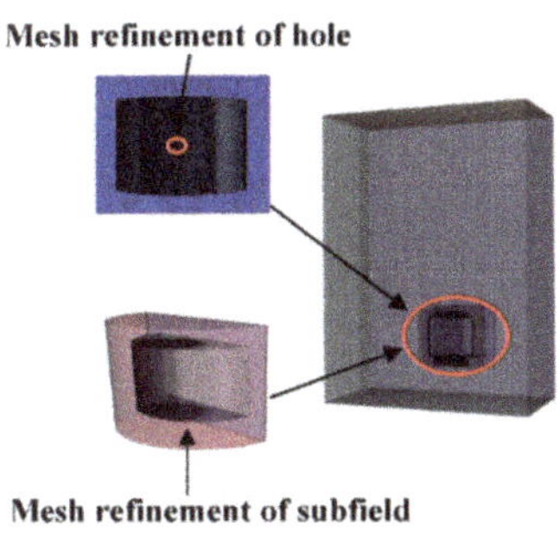

Figure 4. Meshing.

As shown in Figure 5, we used high-speed cameras to capture the growth process of a bubble in the experiment. From the comparison of the maximum thickness of the bubble, it can be found that the bubble has approximately the same shape, with a trapezoid on the top and a spherical shape on the bottom. Since the surface of the cylinder is set as a non-slip wall and there is no contact angle during the numerical calculation, the results of the simulation are somewhat different from the experimental observations, but they are generally close.

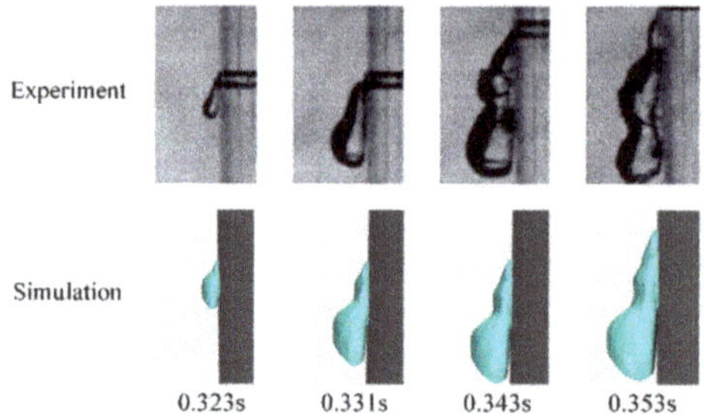

Figure 5. Comparison between numerical and experimental results of cavity.

Tables 1 and 2 show the comparison of the length and width of the air film at different times. The maximum error between the experiment and the numerical calculation is 7.14% and 9.37%, and the error is within the acceptable range, so this research method can be considered effective in studying the size of the bubble.

Table 1. Comparison of bubble length in experiment and CFD.

Time	0.067	0.323	0.331	0.353	0.357	0.360	0.363	0.366
CFD	13	18	23	34	38	42	45	46
Exp	14	19	24	36	40	44	46	48
Error	7.14%	5.26%	4.2%	5.55%	5%	4.55%	2.17%	4.2%

Table 2. Comparison of bubble width in experiment and CFD.

Time	0.067	0.323	0.331	0.353	0.357	0.360	0.363	0.366
CFD	3	5.5	7.5	10	10	11	11	11
Exp	3	6	8	11	11	12	12	12
Error	0	8.33%	9.37%	9.09%	9.09%	9.09%	9.17%	9.17%

3. Numerical Calculation Model

3.1. The Establishment of the Geometric Model

We establish a scaled model of a vehicle with exhaust. As shown in Figure 6, the diameter of the vehicle (D) is 40 mm, the length (L) is 230 mm, and the venting hole's diameter (d) is 0.6 mm.

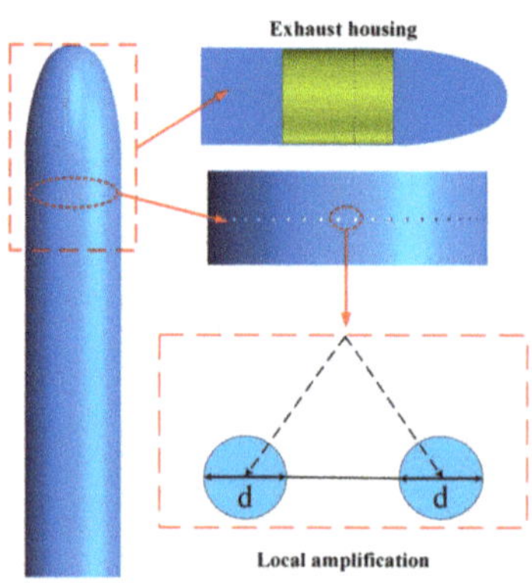

Figure 6. Geometric model.

3.2. Computational Domain and Boundary Condition Division

As shown in Figures 7 and 8, the computational domain is divided into a background domain and an overlapping domain. The top of the background domain is the pressure outlet, and the vehicle part of the overlap domain is the overset mesh. All other surfaces are non-slip wall, except for the symmetry plane.

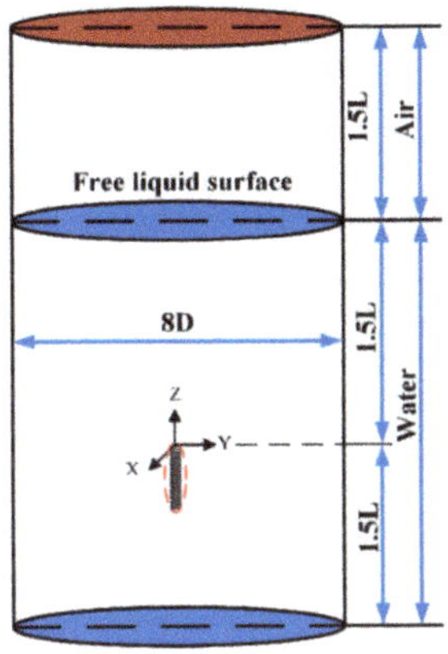

Figure 7. Computational domain.

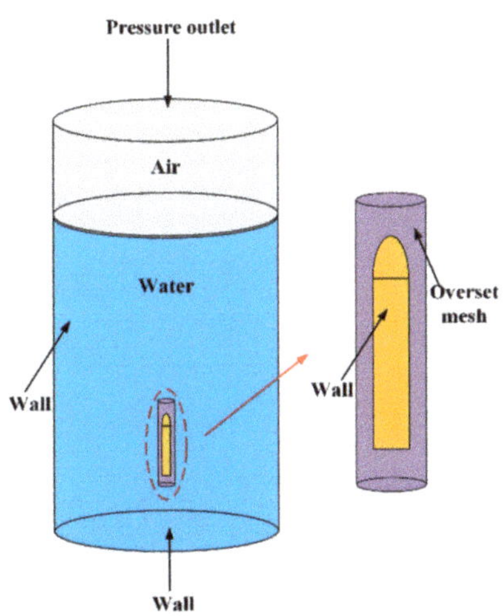

Figure 8. Boundary conditions.

The notations used in this paper are as follows: density of water ρ_w, body diameter D, length L, velocity v, time T^* et al. In order to facilitate the research and analysis, the physical quantities of pressure, time, cavity length, and cavity diameter are treated as dimensionless in Table 3:

Table 3. Non-dimensionalization of physical quantities.

Pressure	Time	Cavity Length	Cavity Diameter
$\overline{p} = \frac{p^*}{1/2\rho_w v^2}$	$\overline{T} = \frac{T^*}{D/v}$	$\overline{L} = \frac{L^*}{L}$	$\overline{D} = \frac{D^*}{D}$

3.3. Meshing

As shown in Figure 9, the STAR-CCM+ software is used for meshing. In order to ensure the symmetry of the initial bubble as much as possible, the background mesh and the sub-domain mesh are trimmed volume mesh.

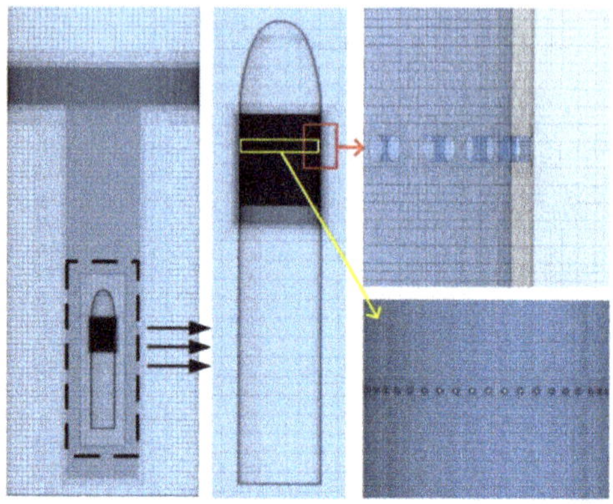

Figure 9. Schematic of numerical mesh.

It is necessary to refine the mesh near the motion area of the vehicle and the surface of the water. In addition, in order to reduce the workload and improve the calculation efficiency, we carried out research on the half-mode. The number of meshes in the background domain is around 4.2 million, and that in the sub-domain is around 5.3 million, which meet the calculation requirements of the overset mesh while ensuring calculation accuracy.

3.4. Mesh Independence Verification

In order to verify the influence of the number of grids on the simulation results, we analyzed two different numbers of numerical grids under the first layer height of 0.05 mm. Based on the grid independence approach, coarse grids N_{coarse}= 9.5M and fine grids N_{fine}= 14.3M are generated. The grid convergence index (GCI) is used for reporting.

The error of coarse grid is defined as:

$$E_{coarse} = \left| \frac{r^p \varepsilon}{1 - r^p} \right| \tag{7}$$

$$\varepsilon = f_2 - f_1 \tag{8}$$

$$r = \left(\frac{N_{fine}}{N_{coarse}}\right)^{1/d} \tag{9}$$

In this study, f_1 and f_2 represent the minimum dimensionless pressure on the surface of the vehicle on the coarse and fine grids, p is the formal order of accuracy of the algorithm ($p = 2$), r is the refinement factor between the coarse and fine mesh, and is the dimension of the simulation. The GCI of the coarse grid is defined as:

$$GCI_{coarse} = F_s |E_{coarse}| \tag{10}$$

where F_s is the safety factor and, in this study, $F_s = 3$.

In this study, the minimum dimensionless pressure on the surface of vehicle is used for the irrelevant verification of the different grid numbers. The calculation results are shown in Table 4. From Table 4, the calculation results show that the results of the two kinds of grids are similar, and the GCI of the coarse mesh result is within 2%, which meets the calculation requirements.

Table 4. Analysis of physic coefficients under different grid numbers.

Physic Coefficients	Coarse	Fine	E_{coarse}	GCI_{coarse}
Min dimensionless pressure	0.496	0.497	0.0042	1.26%

4. Analysis of Numerical Results

4.1. Analysis of Air Film Fusion Characteristics

Figure 10 shows the fusion process of the air film when the speed of the vehicle is 3.71 m/s, the number of venting holes in a single row is 50, and the diameter is 0.6 mm.

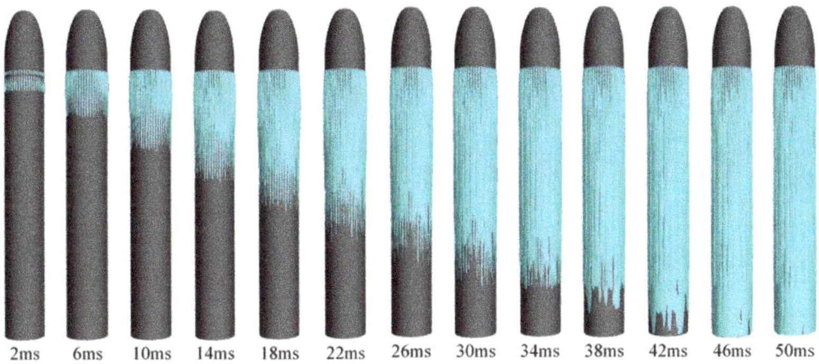

Figure 10. Evolution of exhaust air film at different times.

As shown in the Figures 11 and 12, the evolution process of the air film can be roughly divided into three stages: the initial growth stage, the mixing and fusion stage, and the late stage of air film fusion.

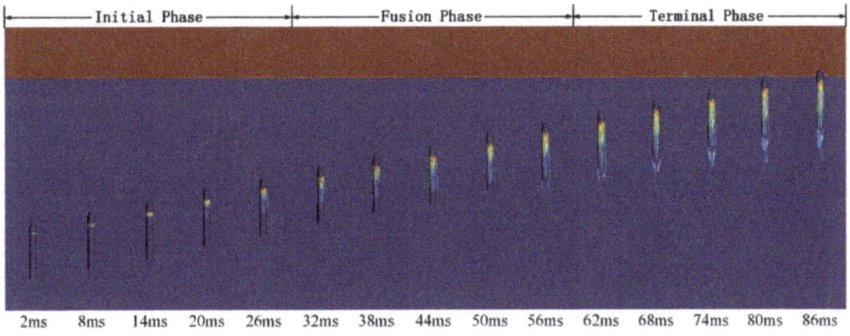

Figure 11. Three stages of air film development.

In the initial stage, the vehicle leaves the launch tube and moves to the water surface; the external hydrostatic pressure in the area where the venting holes are located decreases continuously, resulting in the ideal gas in the shoulder cavity being discharged from the venting holes. As for the air film, the upper part is elongated and the lower part becomes spherical. It is obvious that multiple rows of air film with a bundle shape are attached to the surface of the vehicle side-by-side.

In the mixing and fusion stage, the hydrostatic pressure decreases faster, and the pressure difference between the inside and outside of the cavity increases continuously. At the same time, the gas flow rate of the venting holes increases, and the initial side-by-side bundle-shaped air film expands significantly in both axial and circumferential directions under the supplement of sufficient high-pressure gas; the shape of the air film gradually becomes strip-like, completely covering the area below the venting holes.

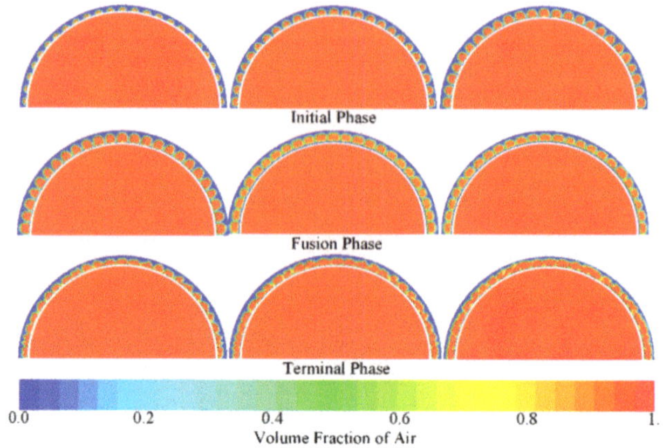

Figure 12. Distribution of phase volume fraction.

In the late stage of fusion, the fused air film continues to develop down to the bottom of the vehicle to cover the surface below the shoulder and eventually develops from a strip-like to sheet-like structure. When the head of the vehicle reaches the free liquid surface, the air film covering the surface of the vehicle begins to fall off, and, finally, the vehicle is completely out of the water.

4.2. Research on the Characteristics of Surface Pressure of Vehicle

In order to study the influence of the air film on the surface pressure characteristics of the vehicle, as shown in Figure 13, the pressure distribution on the axial centerline of the vehicle's surface with or without venting holes was examined.

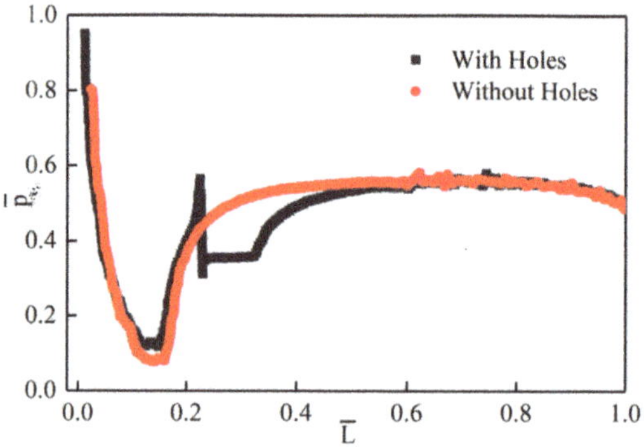

Figure 13. Distribution of surface pressure along vehicle.

From the distribution of the graph in Figure 14, it can be seen that the existence of venting holes will only affect the area covered by the air film, while the pressure distribution in other areas on the surface of the vehicle is approximately the same. The fusion of the air film causes a low-pressure area at the shoulder of the vehicle and a peak in pressure at the end of the film. The pressure fluctuation at the rear of the vehicle is due to the fusion of the air film wrapped in the upper region of the center of mass, while the lower region of the center of mass is not completely covered and the bubble is partially dislodged.

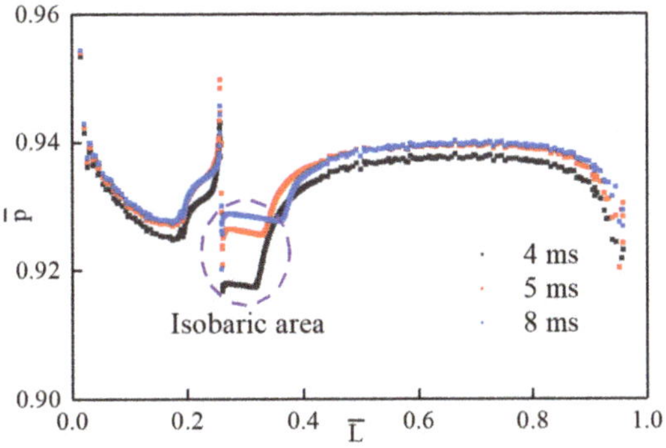

Figure 14. Schematic of the exterior of vehicle.

It can be seen from Figure 15, during the movement of the vehicle, the peak pressure appears at the very front of the head, while a low-pressure region appears in the shoulder region as well as near the tail. The appearance of low pressure in the region of the shoulder comes from the generation of the air film near the venting holes, and the low pressure in the region of the tail is due to the flow separation of the fluid in the tail of the vehicle. The pressure cloud map near the venting hole shows that the top of the hole is a high-pressure area, and the pressure decreases along the incoming flow direction to be similar to that around the hole. In the Figure 16, from the cross-sectional pressure distribution, the pressure increases gradually along the radial direction, with the air film as the center. The generation of the air film makes the pressure near it lower than the local hydrostatic pressure, so it can well improve the flow field environment around the vehicle.

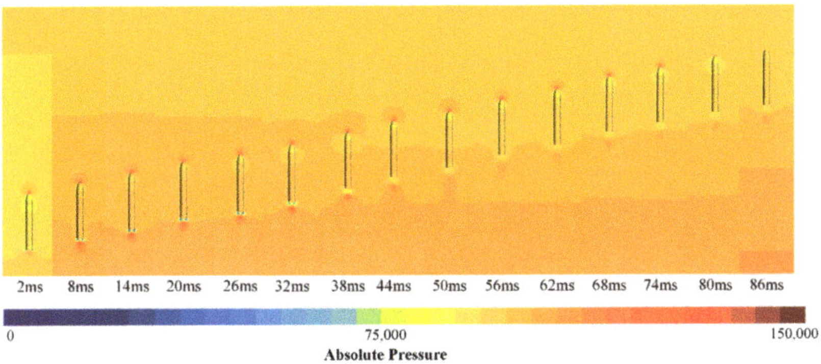

Figure 15. Pressure distribution in the computational domain at different moments.

As time goes on, the vehicle keeps moving upward, the depth of water decreases gradually, and the pressure of the local flow field also decreases. The low-pressure area at the shoulder expands continuously, while the high-pressure area at the head decreases gradually. In the movement to T = 80 ms, the high pressure in the head completely disappears; the reason for this is that the head of the vehicle has been out of the water so is no longer subject to water pressure but standard pressure. After the head of vehicle is out of the water, the low-pressure area in the shoulder completely disappears; the reason for this is that the air film passes through the water surface, and surrounding media from water to air. The internal equilibrium of the attached bubble is broken due to the pressure difference between the internal and external areas and the complex force acting on the surface of the vehicle, and it eventually collapses.

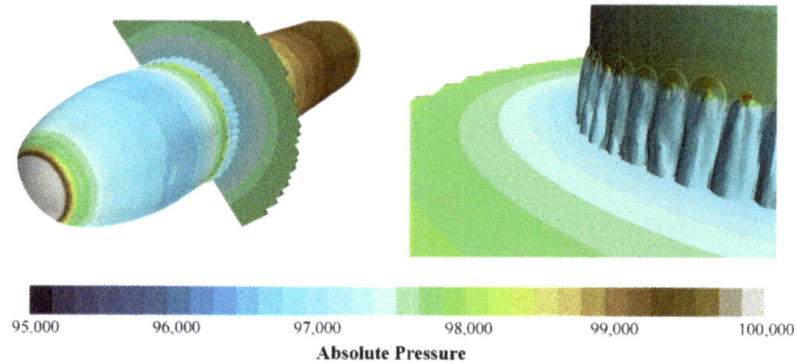

Figure 16. Distribution of pressure near the air film.

4.3. Effect of the Number of Holes on Air Film Fusion and Pressure Distribution

In this section, the evolution process of the bubble and the pressure distribution characteristics of the double row of holes with the spacing of 6 mm, 7 mm, 8 mm, 9 mm, and 10 mm at the velocity of 3.71 m/s is considered.

In order to study the effect of double-row pores on the evolution of the vacuole morphology, the length and thickness of the vacuole were measured at different moments, and the evolution is shown in Figure 17.

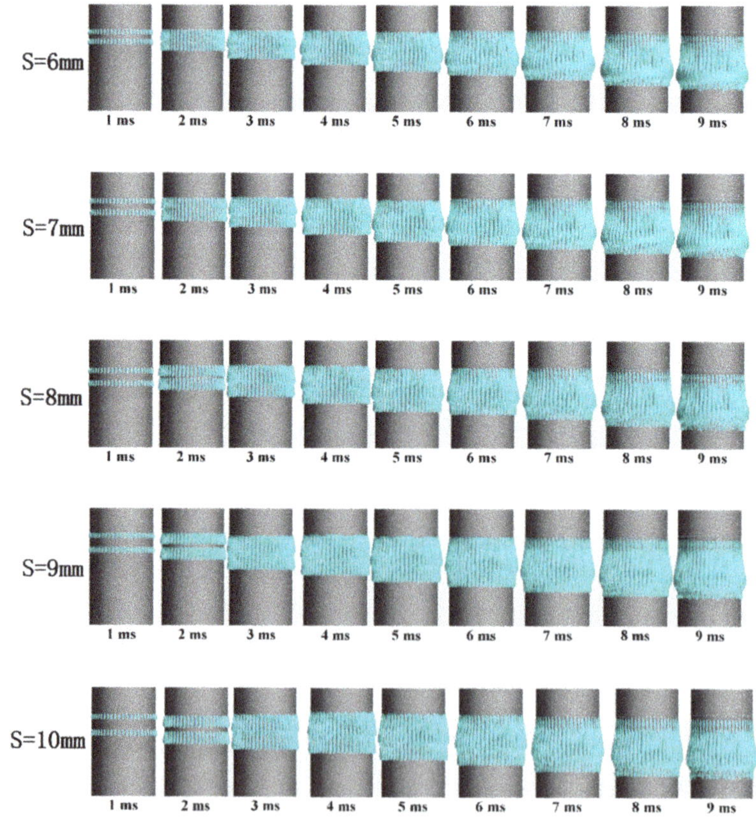

Figure 17. Three-dimensional shape of the air film at different times.

As can be seen from Figures 18 and 19, with the development of time, the maximum length of the air film grows linearly as a whole, and the maximum diameter grows faster in the first period, slows down in the later period, and finally stays the same. At the same time, the larger the spacing between the double-row holes, the larger the maximum length of the air film and the smaller the maximum diameter. As for the structure, the exhaust area of the double-row holes is twice as large as that of the single-row holes, and the larger exhaust area determines the higher expansion rate of the air film. Due to the same volume of gas in the cavity, the excessive consumption of gas in the initial stage also determines the slow rate in the later stage, and the maximum diameter growth rate slows down gradually, as seen in the graph.

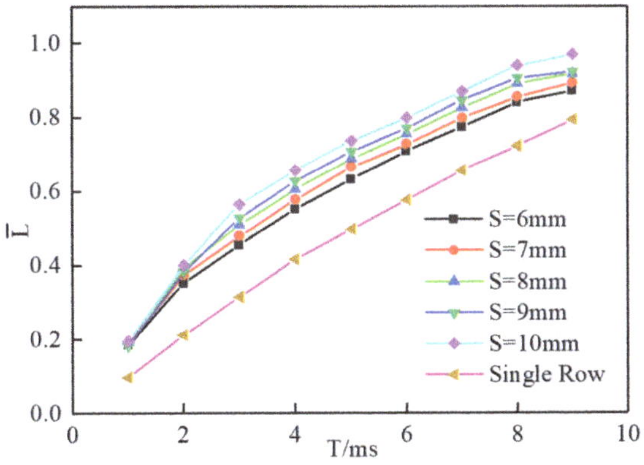

Figure 18. Evolution of length.

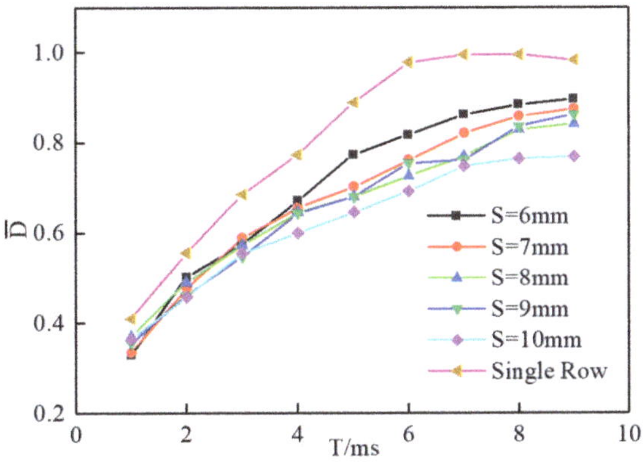

Figure 19. Evolution of diameter.

In order to demonstrate the difference in the effect of the air film on the surface pressure distribution of the vehicle in different cases, the distribution of the model surface pressure along the axial direction is given in Figure 20.

In the early stage of exhaust, the double-row hole solution is significantly better than the single-row hole solution for the expansion rate of the approximate isobaric zone in the axial direction; meanwhile, in the late stage of exhaust, the growth rate of the approximate isobaric zone with larger axial dimensions slows down, and the magnitude of the strong inverse pressure gradient caused by the hysteresis effect at the rear of the air film decreases. Generally speaking, axial coverage is better than that of the single-row hole. In the early stage of exhaust, an approximate isobaric zone with a small axial pressure gradient is formed between the two rows of holes, which can effectively improve the hydrodynamic characteristics of the vehicle. In addition, the bubbles produced by the first row of holes cover the surface of the second row; compared with the single row, the surface of the second row is no longer a high-speed transverse water flow, but a low-pressure, low-density gas, and the shearing effect of the axial incoming flow on the second row is weaker, which is more conducive to the fusion of the air film.

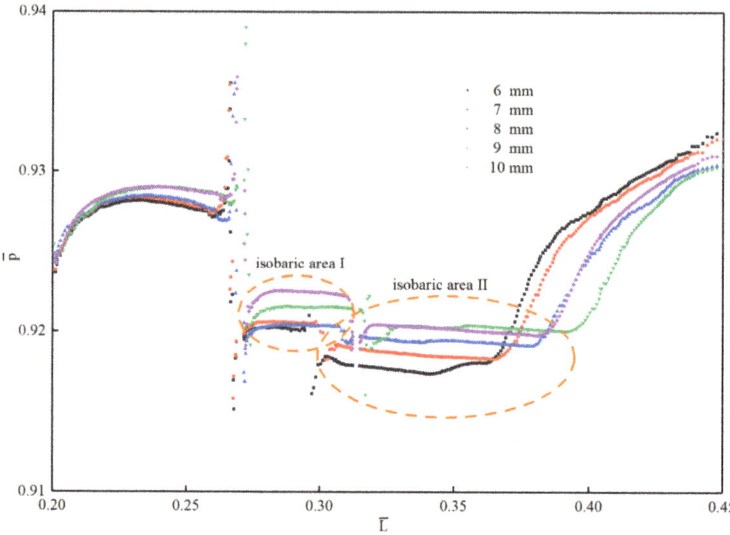

Figure 20. Distribution of surface pressure along vehicle's axial direction.

5. Conclusions

Based on the VOF multiphase flow theory, standard RNG turbulence model, and overset mesh technology, this paper establishes a numerical calculation model for the pressure equalization exhaust of a submarine-launched vehicle and simulates the evolution and development of the air film on the shoulder of the vehicle under different exhaust conditions. The conclusions are as follows:

(1) Pressure-equalizing exhaust is the passive exhaust process of gas under the joint action of the incoming shear force and the hydrostatic pressure drop of the holes. The gas morphology of the holes develops from a bundle-like to a strip-like structure, and finally develops into a sheet-like structure and achieves complete fusion.

(2) The fusion film forms an isobaric zone in the coverage area, which reduces the amplitude of the surface pressure of the vehicle and effectively improves the hydrodynamic characteristics of the vehicle.

(3) Compared with the single row of holes, the double row has a greater speed of axial advance, radial expansion, and circumferential fusion, and the approximate isobaric zone inside the film also develops faster, so the double row is more conducive to improving the hydrodynamic characteristics of the vehicle.

Author Contributions: Conceptualization and methodology, all authors; design of simulation ideas, J.R. and S.G.; setting of simulation cases, sorting and analysis of data, visualization of results and writing-original draft, J.R.; writing-review and editing, Y.S. and J.R.; management of project, supporting of funding and equipment, Y.S. and G.P. All authors have read and agreed to the published version of the manuscript.

Funding: The research was supported by the National Natural Science Foundation of China (Grant No. 52171324) and the Fundamental Research Funds for the Central Universities (Grant No. 3102019JC006).

Data Availability Statement: All data and models generated or used during the research period appear in the submitted manuscript.

Acknowledgments: We are grateful to the National Natural Science Foundation of China (Grant No. 52171324) and the Fundamental Research Funds for the Central Universities (Grant No. 3102019JC006) for their financial support of our research work.

Conflicts of Interest: The authors have no conflict of interest and unanimously agree to submit the manuscript to the journal.

References

1. Wang, Y.W.; Huang, C.G. Research progress on hydrodynamics of high-speed vehicles in the underwater launching process. *J. Adv. Mech.* **2018**, *48*, 259–298.
2. Wang, Y.W.; Huang, C.G.; Du, T.Z.; Fang, X.; Liang, N.G. Mechanism analysis about cavitation collapse load of underwater vehicles in a vertical launching process. *J. Chin. J. Theoretical Appl. Mech.* **2012**, *44*, 39–48.
3. Reichardt, H. The Laws of Cavitation Bubbles at Axially Symmetrical Bodies in A Flow. *Kais. Wilhelm Inst. Stromungsforsch.* **1945**, *766*, 322–326.
4. Knapp, R.T.; Daily, J.W.; Hammitt, F.G. Cavitation. *J. Fluid Mech.* **1972**, *54*, 189–191.
5. Zhang, N.M.; Zhao, Y.; Wei, H.P.; Cheng, S.H. Research on fusion characteristics of porous exhaust bubbles in underwater vertical launch vehicle. *J. Ship Mech.* **2018**, *22*, 135–143.
6. Ma, G.H.; Chen, F.; Yu, J.Y. Research on the effect of the number of stomata on the flow around the exhausted vehicle. *J. Eng. Thermophys.* **2018**, *39*, 1945–1951.
7. Huang, B.; Wang, G.Y.; Quan, X.B.; Zhang, M.D. Study on the unsteady cavitating flow dynamic characteristics around a 0-caliber ogive revolution body. *J. Exp. Fluid Mech.* **2011**, *25*, 22–28.
8. Sun, T.Z.; Wei, Y.J.; Wang, C.; Cao, W. Analysis of the effect of ventilation positions on hydrodynamic characteristics of submarine-launched vehicle. *J. Beijing Univ. Aeronaut. Astronaut.* **2013**, *39*, 1303–1308.
9. Gao, S.; Pan, G. Characteristics Research on Unsteady Flow Field during Salvo of Submarine-launched Vehicles. *J. Digital Ocean. Underw. Warf.* **2020**, *3*, 271–275.
10. Chen, F.; Ma, G.H.; Cheng, S.H.; Jiang, S. Effect of Straight or Inclined Hole Exhaust on Flow around Underwater Vehicle: Part 1-Flow Field Structure. *J. Eng. Thermophys.* **2016**, *37*, 507–513.
11. Bao, W.C.; Quan, X.B.; Wei, H.P. Numerical Simulation on the Flow Dynamics of Underwater Vehicle Launching with Exhaust. *J. Missiles Space Vehicles.* **2014**, *21*, 14–18.
12. Zhang, K.; Liu, G.T. Three-dimensional Numerical Simulation on Flow Field of Underwater Vehicle Launching with Air Exhaust. In Proceedings of the the 14th National Hydrodynamics Conference and the 28th National Hydrodynamics Conference, Changchun, China, 8 August 2017; Volume 7, pp. 100–106. Available online: https://kns.cnki.net/kcms/detail/detail.aspx?dbcode=CPFD&dbname=CPFDLAST2017&filename=SLDX201708002012&uniplatform=NZKPT&v=tDXf9Ep8FtMAzLJpFwtYF_hgiCscmTTdN_mIZgdWM4h_5ieTIFLA8YBa_eFtqbK4YrvN9iPVPZA%3d (accessed on 25 November 2021).
13. Kang, Y.Z.; Wang, C.C.; Wu, Q.; Huang, B.; Kong, D.C. Numerical Simulation on Vertical Launch of Vehicle with Pressure-equalizing. In Proceedings of the the 29th National Hydrodynamics Symposium, Zhenjiang, China, 25 August 2018; Volume 6, pp. 187–192. Available online: https://kns.cnki.net/kcms/detail/detail.aspx?dbcode=CPFD&dbname=CPFDLAST2019&filename=SLDX201808001021&uniplatform=NZKPT&v=fat6wacm5-FMkZ1XaeObTnEuVKxgflFJv0_6-zTbTLak7UbHZvDUAo3vKQSQP36189W24fZQy6M%3d (accessed on 25 November 2021).
14. Yan, G.J.; Liang, X.X.; Zhang, J.; Quan, X.B.; Wei, H.P. Investigation on the Vertical Launching Process of the Underwater Vehicle by Couple the Flow Field and the Hydro-ballistics. *J. Energy Conserv. Technol.* **2019**, *37*, 307–312.
15. Liu, J.J. The Research of the Head-Ventilated Projectile on the Process of Water-Exit. Ph.D. Thesis, Harbin Engineering University, Harbin, China, 2020.
16. Wang, K.L. Research on Gas-liquid Multiphase Flow Characteristics of Pinhole Jet Flow Field. Ph.D. Thesis, Harbin Institute of Technology, Harbin, China, 2019.
17. Li, R.F.; Hu, X.L.; Wang, Z.H.; Le, G.G. Effect of Flow Speed on Tube-exit of Submarine-launched Missile. *J. Fire Control Command Control.* **2019**, *44*, 98–101, 106.
18. Zhang, Z. Study on Dynamic Simulation of Submarine High Speed Voyage Based on Mixed Mathematical Model. *J. Ship Sci. Technol.* **2020**, *42*, 31–33.
19. Wu, Y.Y. Analysis of Flow Field Characteristics of Submarine Launched Vehicle during Launching Out of Cylinder. Ph.D. Thesis, Harbin Institute of Technology, Harbin, China, 2020.
20. Sun, L.Q.; Yan, H.; Ma, G.H.; Zhao, J.P. Analysis of the Promoting Effect of Annular Groove on the Coalescence of Ventilated Cavity. *J. Chin. J. Theor. Appl. Mech.* **2021**, *53*, 386–394.
21. Qiu, Y. Numerical Simulation Research on the Influence of Cavity Exhaust to the Process of Vehicle Underwater Launch. Ph.D. Thesis, Harbin Institute of Technology, Harbin, China, 2013.
22. Duan, L.; Wang, G.Y.; Fu, X.N. Research on Unsteady Characteristics of Ventilated Cavitating Flow in the Form of Gas-leakage by Toroidal Vortex. *Acta Armam.* **2014**, *35*, 712–718.

23. Ren, Z.Y.; Sun, L.Q.; Li, Z.P.; Xiao, W. Experimental Study on the Cavitation Development and Collapse Characteristics of Underwater Vehicle. *J. Astronaut. Syst. Eng. Technol.* **2021**, *5*, 42–49.
24. Zhang, S.B.; Lu, C.J.; Chen, X. Investigation of Ventilated Cavity Shapes of a High-speed Underwater Vehicle. *J. Shanghai Jiaotong Univ.* **2012**, *46*, 329–334.
25. Wei, Q.D. Research on Experimental Characteristics of Underwater Porous Exhaust. Ph.D. Thesis, Harbin Engineering University, Harbin, China, 2016.

Journal of
Marine Science and Engineering

MDPI

Article

Numerical Simulation of Cavitation and Damping Force Characteristics for a High-Speed Supercavitation Vehicle

Rui Lu [1,2], Guang Pan [1,*], Kun Tan [2] and Shaoping Yin [2]

1 School of Marine Science and Technology, Northwestern Polytechnical University, Xi'an 710071, China; gaopengcheng@mail.nwpu.edu.cn
2 Xi'an Institute of Optics and Precision Mechanics, Xi'an Hi-Tech Industrial Development Zone, Xi'an 710119, China; huangqiaogao@nwpu.edu.cn (K.T.); qianzhe540@mail.nwpu.edu.cn (S.Y.)
* Correspondence: panguang@nwpu.edu.cn

Abstract: In this study, an attempt has been made to investigate the supercavitation and hydrodynamic characteristics of high-speed vehicles. A homogeneous equilibrium flow model and a Schnerr–Sauer model based on the Reynolds-averaged Navier–Stokes method are used. Grid-independent inspection and comparison with experimental data in the literature have been carried out to verify the accuracy of numerical methods. The effect of the navigation speed and angle of attack on the cavitation morphology and dynamic characteristics has been investigated. It has been demonstrated that the angle of attack has a remarkable influence on the wet surface and hydrodynamic force, whereas navigation speed has little effect on the position force of the vehicle under the circumstance of no wet surface. The hydrodynamic force changes periodically with the swing of the vehicle, but its maximum is greater than that for the direct navigation state at the same attack angle. Moreover, the damping effect obviously affects the hydrodynamic force amplitude and movement trend.

Keywords: high-speed supercavitation vehicle; wet surface; hydrodynamic force; damping force; periodic swing

Citation: Lu, R.; Pan, G.; Tan, K.; Yin, S. Numerical Simulation of Cavitation and Damping Force Characteristics for a High-Speed Supercavitation Vehicle. *J. Mar. Sci. Eng.* **2021**, *9*, 1171. https://doi.org/10.3390/jmse9111171

Academic Editors: Alon Gany, Peng Du, Haibao Hu, Xiaopeng Chen and Abdellatif Ouahsine

Received: 19 September 2021
Accepted: 18 October 2021
Published: 25 October 2021

1. Introduction

Supercavitating vehicles have received much attention in recent years due to their advantages of high speed, energy savings, and environmental friendliness [1,2]. The supercavitating vehicle can avoid direct contact with liquid, reduce its surface wettability and viscous resistance during sailing, and increase sailing speed. However, due to the fast-moving speed of the supercavitating vehicle, a small area of wetting can produce a huge force, accompanied by a nonnegligible damping force. Therefore, there is important theoretical and practical significance in studying changes in the supercavitation shape and the damping force characteristics of underwater vehicles.

Numerous investigations on supercavitation underwater vehicles have been conducted. Semenenko [3] proposed the basic physical properties and computational methods of artificial cavitation (ventilation) flows. Karn [4] revealed the effect of internal flow physics on the characteristics observed during supercavity formation and closed mode transitions. Ahn [5] analyzed the general characteristics of supercavitation and experimentally observed the size of supercavitation. Based on the slender body theory, Vasin [6–9] explored the supercavitation flow field under subsonic, transonic, and supersonic conditions; analyzed the cavitation state and stress characteristics under corresponding conditions; and reached a conclusion different from that under common low-speed conditions, which has greatly developed the research on the insurmountable sound speed in water. Logvinovich [10] proposed the principle of independent expansion of cavitation sections in 1969. The expansion law of cavitation on any fixed section formed by a high-speed moving object is the same and has nothing to do with the movement of the cavitating object before or after the moment. Vasin [11] and Serebryakov [12] derived the section development equation based on this principle.

Fu et al. [13] studied the supercavitation resistance characteristics of a rotating body based on the Kubota cavitation model. Yi et al. [14] studied the change in the drag coefficient of a supercavitating vehicle at speeds of 300 m/s to 900 m/s and explored the relationship between the drag coefficient and cavitator diameter. The results showed that in the supercavitation shape, the drag coefficient of the vehicle at the head of the disk cavitator was inversely proportional to the area of the cavitator, and increasing the slenderness ratio of the vehicle was conducive to the drag reduction effect of the supercavitation. Based on the relationship between cavitation shape, drag characteristics, and cavitation number, Xiong et al. [15,16] studied the influence of attack angle on the cavitation shape and drag characteristics under small angles. Li [17] investigated the cavitation morphology and tail beat effect of underwater vehicles at high speed (1000 m/s). The numerical simulation was in good agreement with the experimental results. Li [18] conducted a detailed study on the variation trend of the supercavitation shape of underwater vehicles with cavitation number, attack angle, and rudder angle at small speed (100 m/s) and analyzed the force variation characteristics of underwater vehicles.

The present research focuses on supercavitating vehicles under 100 m/s, while studies on supercavitation vehicles with high subsonic, transonic, or even supersonic speeds are rarely involved. Therefore, in this paper, the CFD (Computational Fluid Dynamics) commercial software FLUENT 18.2 (Ansys Inc., Canonsburg, PA, USA) is used to carry out the numerical simulation of direct navigation and periodic swing of a high-speed supercavitation vehicle. The effects of the attack angle, navigation speed, and swing motion on the cavitation shape and hydrodynamic force are analyzed. Then, the damping force characteristics are discussed in detail.

2. Numerical Calculation Model

The high-speed supercavitation vehicle is a typical slender body, as shown in Figure 1. The front body is composed of two directly transitional cones, the middle is a cylinder, and the tail is equipped with six wings. The computing domain and boundary conditions are shown in Figure 2. Different cavitation number conditions are obtained by adjusting the incoming flow velocity. Based on the finite volume method, the fully implicit multigrid technique [19,20] is used to solve the momentum equation, continuity equation, and energy equation at the same time, and then scalar equations such as turbulence are solved. The transient term of the governing equation is discretized by a second-order implicit Euler scheme.

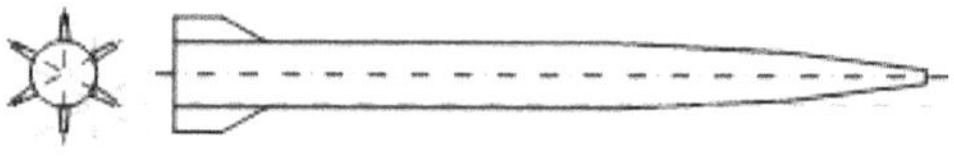

Figure 1. The appearance of the ultrahigh-speed vehicle.

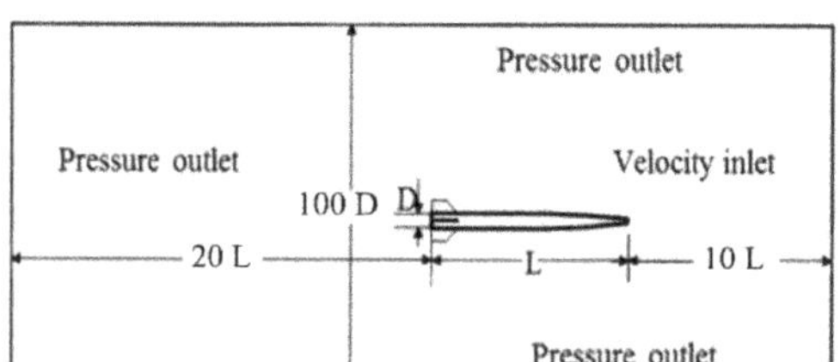

Figure 2. Calculation domain and boundary condition settings.

The computational mesh around the underwater vehicle is shown in Figure 3. The multiblock structured grids were used. The calculation domain is divided into two parts.

The grid of the inner region is dense, and the outflow region is sparse, thus reducing the total number of grids and improving the calculation efficiency. These two regions are connected through the interface. The grid refinement algorithms are conducted in the boundary layer near the wall and the water–vapor interface. Moreover, adaptive mesh refinement according to the pressure and velocity gradient is utilized to accurately capture the evolution of the cavitation shape during the calculation.

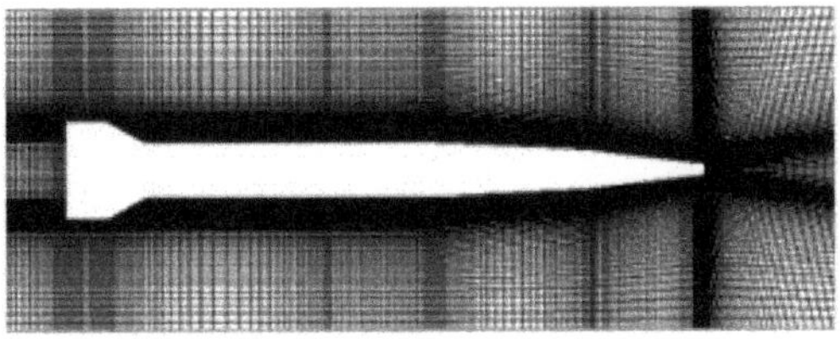

Figure 3. Grids around the vehicle.

3. Cavitation Flow Control Equation

3.1. Homogeneous Equilibrium Flow Model

Cavitation multiphase flow is involved in high-speed supercavitation. According to the published literature, the homogeneous equilibrium flow model is widely used in numerical simulations and performs well in predicting natural cavitation. Generally, there are two models, the mixture model and the VOF (Volume of Fluid) model, which are suitable for the natural cavitation and ventilated supercavity, respectively. For the mixture model, there is a unified velocity field and pressure field in the unit volume, and no relative velocity between the two phases is considered. The flow field density is determined by the volume fraction and density of the two phases. In this paper, natural cavitation is studied, so the mixture homogeneous equilibrium flow model is adopted. The main equations involved include the continuity equation, momentum equation, and energy equation.

(1) The continuity equation:

$$\frac{\partial \rho}{\partial t} + \nabla \bullet (\rho u) = 0 \tag{1}$$

$$\rho = \sum_{k=1}^{n} \alpha_k \rho_k \tag{2}$$

$$u = \frac{1}{\rho} \sum_{k=1}^{n} \alpha_k \rho_k u_k \tag{3}$$

where ρ is the fluid mixing density; u is the mixing speed; and n is the number of phases. Since noncondensable gas is considered, $n = 3$, α_k, ρ_k, and u_k are the volume fraction, density, and velocity of the k-th phase.

(2) The momentum equations:

$$\frac{\partial (\rho u)}{\partial t} + \nabla \bullet (\rho u u) = -\nabla p + \nabla \bullet \tau_{ij} + S \tag{4}$$

$$\tau_{ij} = \mu \left[\left(\frac{\partial u_i}{\partial x_j} + \frac{\partial u_j}{\partial x_i} \right) - \frac{2}{3} \delta_{ij} \frac{\partial u_k}{\partial x_k} \right], \quad \delta_{ij} = \begin{cases} 1 & (i \neq j) \\ 0 & (i = j) \end{cases} \tag{5}$$

$$\mu = \sum_{k=1}^{n} \alpha_k \mu_k \tag{6}$$

where p is the pressure, and S is the source term. The term τ_{ij} represents the shear stress, which is written as (5). μ is the hybrid viscosity, and its expression is (6). μ_k is the dynamic viscosity of the k-th phase.

(3) The volume fraction equation of water vapor phase:

$$\frac{\partial \rho_v \alpha_v}{\partial t} + \nabla \bullet (\rho_v \alpha_v u) = \dot{m}^+ + \dot{m}^- \tag{7}$$

where ρ_v is the water vapor density, and α_v is the water vapor volume fraction. The term $\dot{m}^+, \dot{m}^-$ represents the condensation rate and vaporization rate of water vapor, which are described in the cavitation model.

(4) The energy equation:

$$\frac{\partial}{\partial t} \sum_{k=1}^{n} (\alpha_k \rho_k E_k) + \nabla \sum_{k=1}^{n} (\alpha_k v_k (\rho_k E_k + p)) = \nabla \left[k_{eff} \nabla T \right] + S_E \tag{8}$$

$$E_k = h_k - \frac{p}{\rho} + \frac{v_k^2}{2} \tag{9}$$

$$k_{eff} = \sum_{k=1}^{n} \alpha_k (k_k + k_l) \tag{10}$$

where the term E_k represents the internal energy of the k-th phase; for a compressible flow, $E_k = h_k$; and h_k is the apparent enthalpy of the k-th phase. k_{eff} is the effective thermal conductivity, k_l is the turbulent thermal conductivity, and T is the local temperature.

3.2. Viscous Model

The flow around the high-speed supercavitation vehicle has strong turbulence characteristics. Therefore, the viscous model plays an important role in the simulation of turbulence. In this paper, the realizable k-ε model based on the RANS (Reynolds-Averaged Navier-Stokes) method is adopted. Moin [21] pointed out that the standard k-ε model may have negative positive stress when the time-average strain rate is large. Shin [22] combined C_μ and the strain rate in the turbulence formula to mathematically constrain the normal stress to ensure that the flow conforms to the physical laws and proposed a realizable k-ε model that is more in line with the actual situation. The modeled transport equations for k and ε in the realizable k-ε model are

$$\rho \frac{Dk}{Dt} + \frac{\partial(\rho k u_j)}{\partial x_j} = \frac{\partial}{\partial x_j} \left[\left(\mu + \frac{\mu_t}{\sigma_k} \right) \frac{\partial k}{\partial x_j} \right] + G_k + G_b - \rho\varepsilon - Y_M + S_k \tag{11}$$

$$\rho \frac{D\varepsilon}{Dt} + \frac{\partial(\rho \varepsilon u_j)}{\partial x_j} = \frac{\partial}{\partial x_j} \left[\left(\mu + \frac{\mu_t}{\sigma_t} \right) \frac{\partial \varepsilon}{\partial x_j} \right] + \rho C_1 S\varepsilon - \frac{\rho C_2 \varepsilon^2}{k + \sqrt{v\varepsilon}} + C_{1\varepsilon} \frac{\varepsilon}{k} C_{3\varepsilon} G_b + S_\varepsilon \tag{12}$$

$$C_1 = max\left[0.43, \frac{\eta}{\eta+1}\right], \quad \eta = S\frac{k}{\varepsilon}, \quad \mu_t = \rho C_\mu \frac{k^2}{\varepsilon}, \tag{13}$$

In these equations, G_k represents the generation of turbulence kinetic energy due to the mean velocity gradients. G_b is the generation of turbulence kinetic energy due to buoyancy. Y_m represents the contribution of the fluctuating dilatation in compressible turbulence to the overall dissipation rate, and the term incompressible flow is 0. C_2 and $C_{1\varepsilon}$ are constants. σ_k and σ_ε are the turbulent Prandtl numbers for k and ε, respectively. S_k and S_ε are user-defined source terms. μ_t is the turbulent viscosity coefficient. In the realizable k-ε model, C_μ is a function of the strain rate and curl, and its value is 0.09 on the viscous bottom layer.

3.3. Cavitation Model

Cavitation models are widely used in CFD calculations. This type of model uses mass source terms to simulate the mass conversion between water and vapor, as well as the unsteady characteristics of flow. The Schnerr and Sauer model proposed in 2001 [23] is used in this article. This model relates the volume fraction of the vapor phase to the number of cavitations per unit volume of fluid. The expressions are

$$\begin{cases} \dot{m}^{+} = \frac{\rho_l \rho_v}{\rho_w} \alpha_{nuc}(1-\alpha_{nuc}) \frac{3}{R_B} \sqrt{\frac{2}{3}\frac{p_v - p}{\rho_l}} & p < p_v \\ \dot{m}^{+} = \frac{\rho_l \rho_v}{\rho} \alpha_{nuc}(1-\alpha_{nuc}) \frac{3}{R_B} \sqrt{\frac{2}{3}\frac{p - p_v}{\rho_l}} & p > p_v \end{cases} \tag{14}$$

$$\begin{cases} R_B = \left(\frac{\alpha_{nuc}}{1-\alpha_{nuc}} \frac{3}{4\pi} \frac{1}{n}\right)^{\frac{1}{3}} \\ \alpha_{nuc} = \frac{\frac{4}{3}\pi R_B^3 N_B}{1+\frac{4}{3}\pi R_B^3 N_B} \end{cases} \tag{15}$$

where $\dot{m}^{+}$ is the evaporation rate. $\dot{m}^{-}$ is the condensation rate. ρ_l and ρ_v are the water and vapor densities, respectively. ρ_w is the water density. α_{nuc} is the gas core volume fraction. R_B is the gas core cavity diameter. n is the number of cavities per unit volume.

4. Verification of the Calculation Model

The numerical simulation method established above is evaluated for the supercavitation flow around a flat-headed conical vehicle (as shown in Figure 4). The experiment was conducted by Hrubes [24]. The test uses a 30 mm caliber underwater gun to launch at a depth of 4 m. The animation of the water entry process is taken by a high-speed camera. The laser is used to measure the velocity of a vehicle. In this paper, the simulation is carried out for the condition with a water depth of 4 m at room temperature and an incoming flow velocity of 970 m/s corresponding to a cavitation number of $\sigma = 2 \times 10^{-4}$, which is defined as follows:

$$\sigma = \frac{P_\infty - P_v}{\frac{1}{2}\rho V^2} \tag{16}$$

where P_∞ and P_v represent the environment pressure and the saturate pressure, respectively. ρ is the density of the water, and V is the velocity of the main flow.

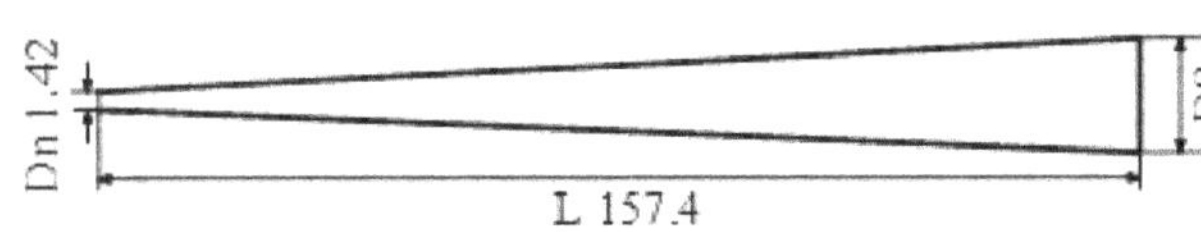

Figure 4. Experimental model of high-speed vehicle.

Figure 5 shows the comparison of the density distribution between the experimental and numerical results. The vehicle has been enveloped by the cavitation, showing an obvious state of supercavitation. The simulation result is in good agreement with the test result. Figure 6 shows the cavitation radius at different positions from the head of the cavitator calculated by the test, simulation, and semiempirical formula. The simulation of the supercavitation shape in this paper has good accuracy and high reliability. Therefore, the above numerical methods are further used in the later simulation of supercavitation flow.

In addition, the drag coefficient obtained by using five different grids is shown in Figure 7. The numbers of grids are 3.5×10^5, 4.3×10^5, 5.4×10^5, 6.6×10^5 and 7.3×10^5, respectively. The drag coefficient C_d is defined as follows:

$$C_d = \frac{F_d}{\frac{1}{2}\rho V^2 S} \tag{17}$$

where F_d is the drag force, ρ is the density of the water, and V is the velocity of main flow. S is the area of the cavitator.

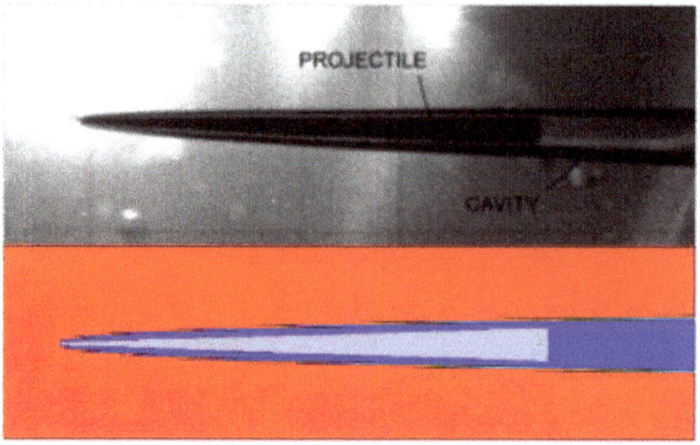

Figure 5. Comparison diagram of the velocity v = 970 m/s test and simulation cavitation flow field.

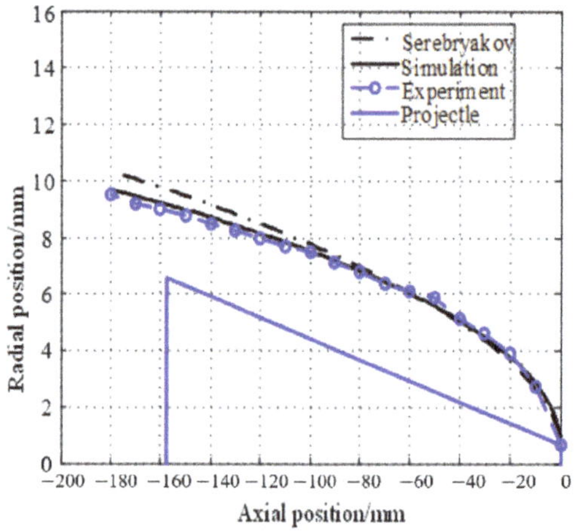

Figure 6. Comparison of simulation, experiment, and theoretical calculation of cavity shape.

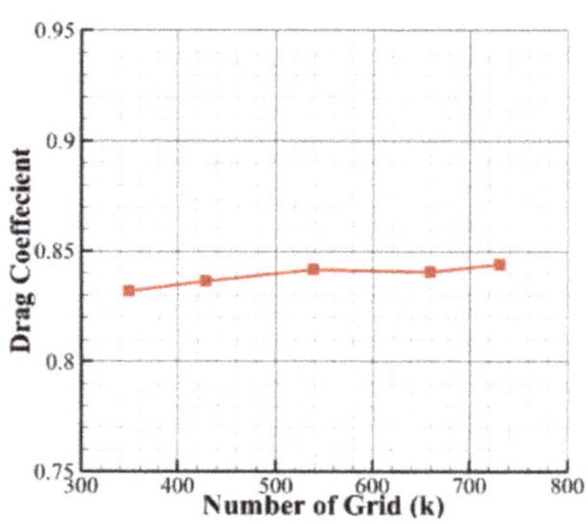

Figure 7. The drag coefficient obtained by using different density grids.

It can be observed that with the number of grids increasing, the drag coefficient rises slightly. When the number is larger than 540,000, the variety of the drag coefficient is within

0.5%. Therefore, the number of grids in this paper is set to 540,000, and the numerical simulation is validated.

5. Cavitation Evolution Characteristics of High-Speed Vehicles

Due to the high speed of the vehicle, such as with the ejection tube used for testing in Ukraine and the United States, the velocity is more than 1000 m/s, and the trajectory is relatively stable for horizontal straight navigation. However, for the condition with an attack angle, the projectile trajectory is extremely easy to be made unstable. Even a small attack angle below 1° would significantly affect the projectile motion and cavitation shape. Therefore, the flow field and cavitation shape of a projectile at high speed at different attack angles are studied and analyzed in detail in this paper. The incoming flow velocity considered ranges from 300 m/s to 1500 m/s, and the corresponding Reynolds number $Re = \frac{VD}{v}$ ranges from about 450,000 to 2,250,000. The attack angle ranges from 0° to 3°. Due to the max diameter being small, the gravity effect is not considered in the numerical simulation. Therefore, the Froude number is not involved in this paper.

Figures 8–11 show the density distribution on the projectile surface at different speeds and attack angles, as well as the comparison of the isosurface with the contour density surface of 500 kg/m^3 (which is defined as the cavitation interface in this paper). The redder the color is, the greater the density is. In the local cavitation state, the cavitation number has a great influence on the cavitation shape. When supercavitation occurs (as shown by the equal density surface of 500 kg/m^3), the shape of the cavitation around the projectile under different cavitation numbers is basically the same in the cases with the same attack angle.

When the speed is the same (300m/s), the attack angle has a significant effect on the shape of the cavity and its positional relationship with the projectile. With an increasing attack angle, the offset of the back surface of the cavitation increases significantly. When the attack angle is 0°, the projectile is cavitation-wrapped. As the attack angle increases to 1°, the wing is slightly wetted, indicating that the cavitation part is punctured, and a secondary cavitation appears near the wetted part of the wing. When the attack angle increases to 2° (as shown in Figure 10), the wing on the water-facing surface has been fully wetted, and the wetted area has transitioned to the body. The bottom cavities are fully pierced by the wing, but at the tip of the wing, they are still in a state of cavitation. In addition, the conical transition zone of the second part of the projectile also appears partially wet. When the attack angle is 3°, the conical transition zone of the first part of the projectile is wetted on a large scale, secondary cavitation appears, and the shape of the cavitation is more complicated. Therefore, as the attack angle increases, the cavitation on the water-facing surface is more severely compressed by the incoming flow, and the offset of the cavitation to the back surface becomes increasingly serious.

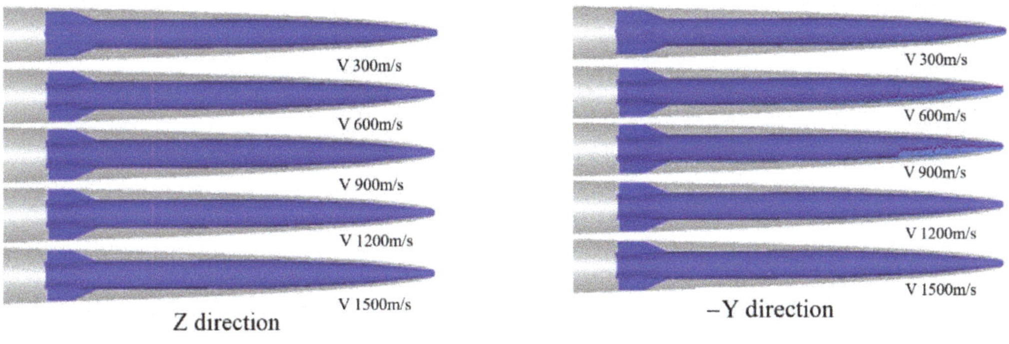

(a) (b)

Figure 8. Cavitation morphology with different flow speeds in the directions of (**a**) Z and (**b**)–Y, respectively. (attack angle 0°).

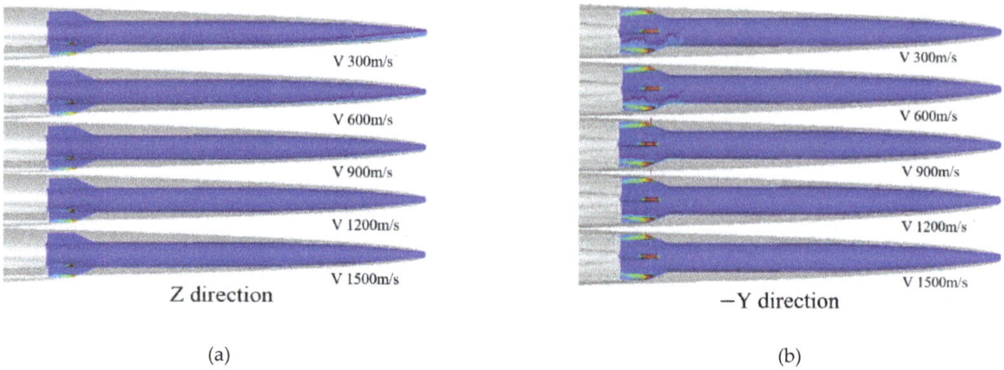

Figure 9. Cavitation morphology with different flow speeds in the directions of (**a**) Z and (**b**)–Y, respectively. (attack angle 1°).

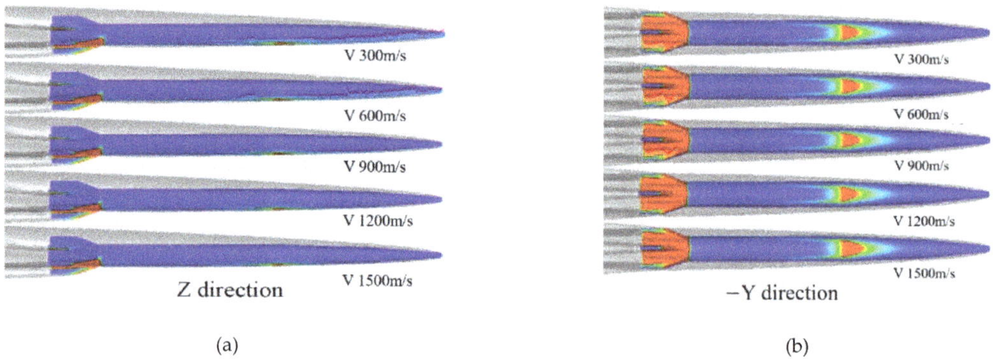

Figure 10. Cavitation morphology with different flow speeds in the directions of (**a**) Z and (**b**)–Y, respectively. (attack angle 2°).

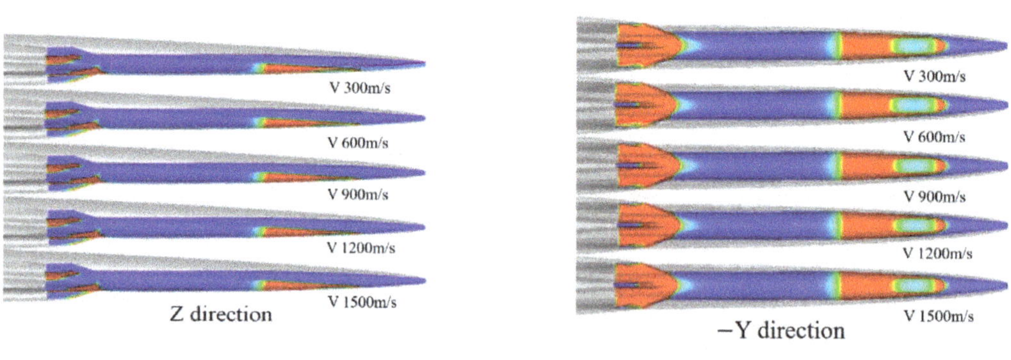

Figure 11. Cavitation morphology with different flow speeds in the directions of (**a**) Z and (**b**)–Y, respectively. (attack angle 3°).

To quantitatively measure the magnitude of the offset, Figure 12a shows the comparison of the cavitation interface on the back water surface at different attack angles. From 0° to 3.0°, the upper boundary of the cavitation in the longitudinal plane continuously removes wards of nearly 9 mm. From the given radial offset distribution of the cavitation axis relative to the axis when the attack angle is 0° at each attack angle, it can be seen that at the same position, the larger the attack angle is, the greater the offset is. However, from the amount of change, the tendency of the increase slows down as the attack angle increases. At the same attack angle, the distance from the cavitator is also positively correlated with the offset, and the farther it is from the cavitator, the more violently the offset increases.

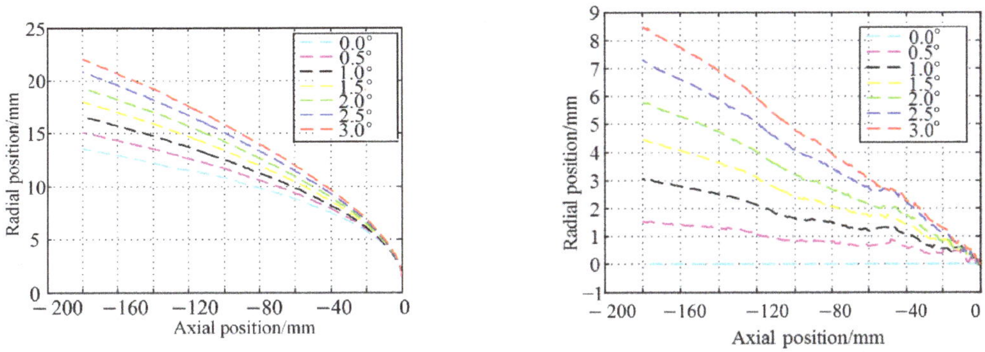

(a) (b)

Figure 12. (**a**) Upper half of the cavitation shape at different attack angles (v = 300 m/s). (**b**) The offset of the cavitation axis in the longitudinal plane at different angles of attack (v = 300 m/s).

Figure 13 gives the comparison of the cavitation morphology of different axial cross-sections. The cross-section of the cavitation is a regular circular shape 100 mm behind the cavitation when the attack angle is 0°. With an increasing attack angle, the cross-sectional shape gradually changes to an ellipse. When the angle of attack increases to 2.0° and 2.5°, the shape is not smooth, and a pit appears. When the attack angle reaches 3°, the strong effect of the attack angle and secondary cavitation severely flattens and elongates the cross-section of the cavity, and an irregular contour appears, making the edge of the cavity more complicated.

Downstream of the projectile (160 mm behind the cavitator), a small attack angle (such as 0.5°) also causes the wing tip of the missile to wet and an accompanied secondary cavitation. Furthermore, at an attack angle of 3.0° the secondary cavitation is more pronounced. Therefore, the edge shape of the downstream cavitation is also obviously affected by the wing.

In summary, as the attack angle increases, the wetness of other parts starts from the bottom wing tip to the wing root, the tail of the projectile, the second part of the tapered transition zone, and the first part except for the foremost cavitator. In this process, every time a part of the wetted area is increased and the conditions for generating cavitation are reached, secondary cavitation will appear here, causing the flow field to become more complex, and the cross-sectional profile of the cavitation is more irregular. Flat and elongated cavities are added to the ellipsoid, and this situation becomes more serious as the attack angle increases.

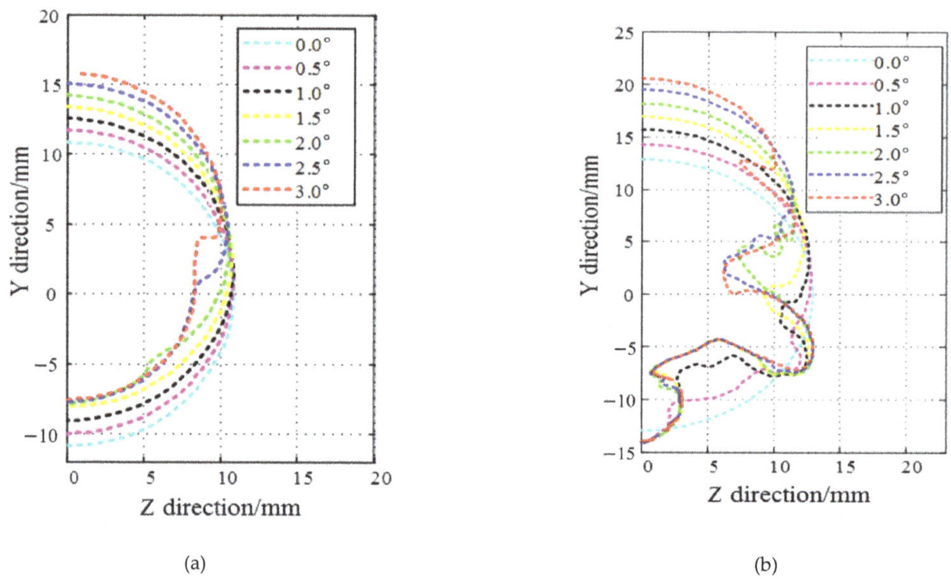

Figure 13. Cavitation cross-sectional shape at different positions of the cavitator with different attack angles. (**a**) Cavitation at 100 mm behind the cavitator; (**b**) cavitation at 160 mm behind the cavitator.

6. Damping Characteristics Analysis of Ultrahigh-Speed Vehicles

Ignoring the inertial force, the hydrodynamic forces of the ultrahigh-speed vehicle when rotating or swinging in a fixed period include the position force and damping force. First, we conduct a numerical simulation study on the direct navigation conditions of ultrahigh-speed vehicles and obtain the position force characteristics in the longitudinal plane at different speeds and angles of attack. Comparing the direct sailing conditions with the oscillating conditions of the vehicle, the damping force characteristics of the ultrahigh-speed vehicle at a given angular velocity are obtained.

6.1. Position Force Characteristics of Ultrahigh-Speed Vehicles

The variation in the hydrodynamic coefficient with the attack angle at different speeds under the direct navigation condition is shown in Figure 14. For a high speed with supercavitation, the vehicle is wrapped by cavitation, and the drag coefficient and lift coefficient are slightly affected by the speed when the attack angle is constant. With the change in attack angle, the hydrodynamic force of the projectile changes sharply. At the same speed, when the attack angle increases, the wetted state of the missile body and wing changes significantly (as shown in Figures 7–10 above), so the fluid force (including resistance, lift, and pitch moment) changes accordingly. For example, as the attack angle changes from 0° to 3°, the drag coefficient increases sharply from 0.83 to 2.9, and the lift coefficient increases from 0 to 5.7.

In Figure 14c, since the wetted area at the tail of the wing is large, the lift generated on the after part of the projectile is larger than that generated by the cavitator and the conical transition zone at the front part, and an obvious negative value of the pitching moment coefficient is observed, resulting in a pitch down moment. Therefore, in the range of the attack angle, the given vehicle presents static stability characteristics, which are the excellent characteristics required by the supercavitation vehicle.

The changes in lift and pitching moment coefficients with attack angle and speed in Figure 14b,c are basically consistent with the changes in drag coefficient, which are all caused by the wetted area. The difference is that due to the larger wetting area of the tail of the wing projectile, the lift generated by the wing missile is larger than that generated by the cavitator and the tapered transition zone of the front end, which results in a significant negative value of the pitching moment coefficient, that is, a bowing moment. Within the range of the attack angle under study, the given ultrahigh-speed vehicle exhibits static and stable characteristics, which are the excellent characteristics required by the ultrahigh-speed vehicle.

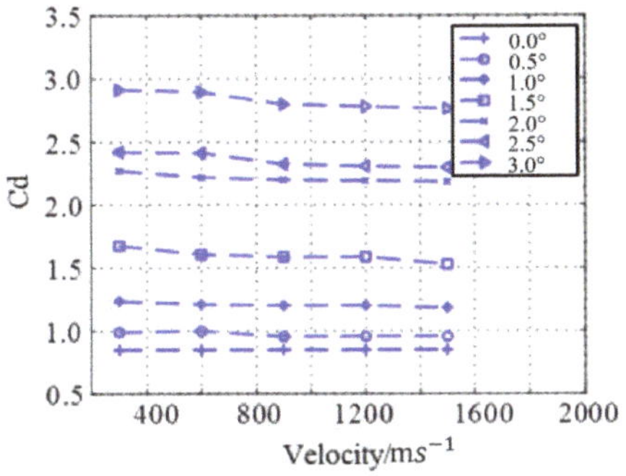

(a)

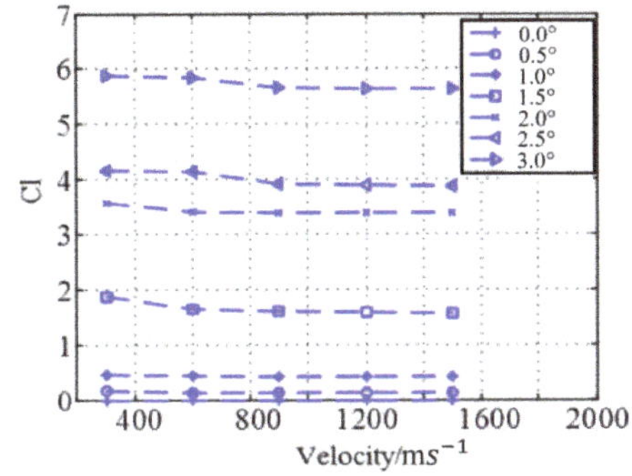

(b)

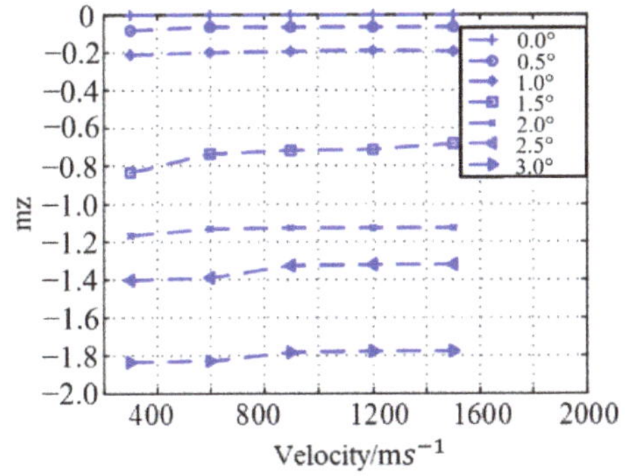

(c)

Figure 14. The influence of speed and attack angle on the fluid force coefficient, (**a**) drag coefficient, (**b**) lift coefficient, and (**c**) pitching moment coefficient.

6.2. Damping Characteristics of Ultrahigh-Speed Vehicle

Numerical simulations of the supercavitation vehicle oscillating periodically around the z-axis are carried out to study the damping characteristics. The swing of the supercavitation vehicle is shown in the Figure 15, The working conditions are as follows: sailing speed of 300 m/s, swing amplitude of 3°, and swing period of $T = 4$ ms. The swing angle is calculated by Equation (18):

$$\alpha(t) = -\frac{3\pi}{180}\sin(500\pi t) \tag{18}$$

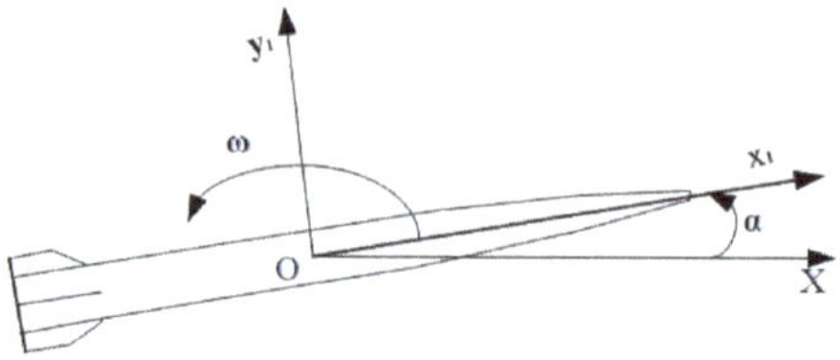

Figure 15. Schematic diagram of the swing of the ultrahigh-speed vehicle.

The user-defined function (UDF) provided by FLUENT is used for secondary development, and the projectile angular velocity is given to simulate the periodic swing. The time step is 0.02 ms, corresponding to 1/200 of the swing period. The damping force characteristics at the corresponding angular velocity and velocity are obtained by subtracting the hydrodynamic force at the corresponding attack angle and velocity from the direct navigation condition.

The coefficients of the high-speed supercavitation vehicle under periodic swing conditions are shown in Figures 16–18. For comparison, the force and torque coefficients under constant direct navigation conditions are also given. Generally, when the high-speed vehicle swings periodically, the hydrodynamic force changes correspondingly. From Figure 16, as the attack angle of the vehicle first increases to the maximum value and then decreases, the resistance coefficient increases from small to large and then decreases accordingly. However, before the attack angle reaches the maximum, the maximum resistance coefficient is observed in advance. In Figure 17, the change in the lift characteristic is more complex than that of the drag coefficient. At the same angle, the lift coefficient is different in the process of raising the head and lowering the head. In addition, similar to the variation in the drag coefficient, the lift coefficient reaches the maximum value in advance before the attack angle reaches the maximum value.

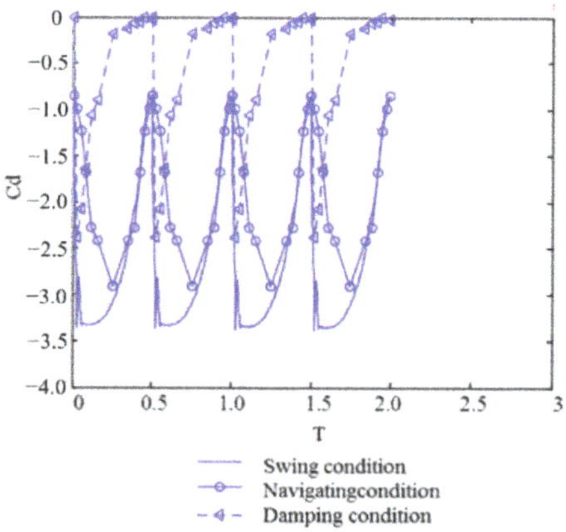

Figure 16. Damping characteristics curves.

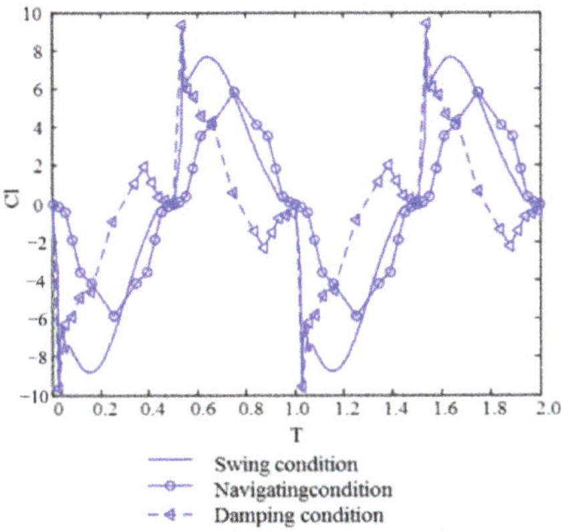

Figure 17. Curves of lift characteristic.

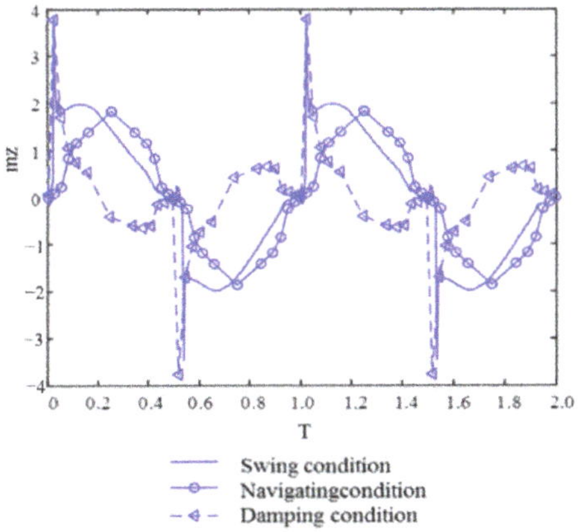

Figure 18. Curves of pitching moment characteristic.

From the change in pitching torque shown in Figure 18, at the beginning of the swing of the high-speed vehicle (or at a small attack angle), the shoulder is not wet, and the tail is slightly wet. During the continuous head-up motion of the vehicle, the damping force presents a positive effect on the motion, and then the torque coefficient quickly reaches the maximum value. When the head lowers, due to the hysteresis of the damping force, part of the position torque is offset. As a result, the change process of the torque is also relatively gentle.

Overall, the force and moment coefficients of the high-speed vehicle increase with increasing attack angle. In the swing process, due to the damping force, the variation of the force and moment coefficients with the attack angle is obviously different. When the vehicle gradually rises from zero, the damping force increases sharply to quickly reach the extreme value greater than that in direct navigation. When moving back from the maximum attack angle, the position force is partly offset by the damping force, thus delaying the variation of the force. Therefore, for the high-speed vehicle involved in the wet state, the damping term will have an important impact on the magnitude and the variation trend of the hydrodynamic force.

7. Conclusions

In this paper, the hydrodynamic characteristics of a high-speed vehicle under direct navigation at different speeds and angles of attack are simulated. Then, the damping force of the ultrahigh-speed vehicle during the periodic swing is analyzed. It can be concluded that:

(1) When there is no wet surface on the surface of an ultrahigh-speed vehicle, its position force characteristics are less affected by the navigation speed. At the same velocity, with the increase of the attack angle, the wet area of the tail increases gradually, and then the fluid force is also enlarged. At the same attack angle as the wet surface, the drag coefficient decreases slightly with increasing velocity.

(2) The hydrodynamic force changes roughly periodically according to the swing of the high-speed vehicle, whereas the maximum value is greater than that during the direct navigation state with the same attack angle.

(3) The damping term has an obvious impact on the amplitude and change trend of the hydrodynamic force. When the high-speed vehicle gradually rises from a zero attack angle, the damping force plays a driving role. When it reaches the maximum attack angle and swings low, the damping force acts as an obstacle, and a hysteresis effect is observed for the force and torque coefficients.

Author Contributions: Conceptualization, R.L.; methodology, R.L.; software, R.L.; validation, R.L., G.P.; formal analysis, R.L.; investigation, R.L.; data curation, R.L.; writing—original draft preparation, R.L.; writing—review and editing, K.T.; visualization, R.L.; supervision, G.P. and S.Y.; project administration, G.P. and S.Y.; funding acquisition, G.P. All authors have read and agreed to the published version of the manuscript.

Funding: This work was supported by the National Natural Science Foundation of China (Grant No. 51879220), the National Key Research and Development Program of China (Grant No. 2020YFB1313201), and Fundamental Research Funds for the Central Universities (Grant No. 3102019HHZY030019 and 3102020HHZY030018).

Institutional Review Board Statement: Not applicable.

Informed Consent Statement: Not applicable.

Data Availability Statement: The data that support the findings of this study are available within the article.

Conflicts of Interest: The authors declare no conflict of interest.

References

1. Vanek, B.; Bokor, J.; Balas, G. High-speed supercavitation vehicle control. In Proceedings of the AIAA Guidance, Navigation, and Control Conference and Exhibit, Keystone, CO, USA, 21–24 August 2006; p. 6446.
2. Ceccio, S.L. Friction Drag Reduction of External Flows with Bubble and Gas Injection. *Annu. Rev. Fluid Mech.* **2010**, *42*, 183–203. [CrossRef]
3. Semenenko, V.N. *Artificial Supercavitation. Physics and Calculation*; RTO AVT Lecture Series on "Supercavitating Flows"; A St. Joseph Corporation Company: Brussels, Belgium, 2001.
4. Karn, A.; Arndt, R.; Hong, J. An experimental investigation into supercavity closure mechanisms. *J. Fluid Mech.* **2016**, *789*, 259–284. [CrossRef]
5. Ahn, B.K.; Lee, T.K.; Kim, H.T.; Lee, C.S. Experimental investigation of supercavitating flows. *Int. J. Nav. Archit. Ocean. Eng.* **2012**, *4*, 123–131. [CrossRef]
6. Vasin, A.D. Some Problems of Supersonic Cavitation Flows. Available online: http://caltechconf.library.caltech.edu/82/ (accessed on 17 October 2021).
7. Vasin, A.D. Application of the Slender Body Theory to Investigation of the Developed Axially Symmetric Cavitation Flows in a Subsonic Stream of Compressible Fluid. *Izv. Vyssh. Uchebn. Zaved. Mat.* **2001**, *3*, 122–129. [CrossRef]
8. Vasin, A.D. *Supercavities in Compressible Fluid*; RTO EN-016; RTO AVT Lecture Series on "Supercavitating Flows"; von Kármán Institute VKI: Brussels, Belgium, 2001.
9. Vasin, A.D. Calculation of axisymmetric cavities downstream of a disk in a supersonic flow. *Fluid Dyn.* **1997**, *32*, 513–519.
10. Logvinovich, G.V. *Hydrodynamics of Free-Boundary Flows*; Naukova Dumka: Kyiv, Ukraine, 1969; p. 208.

11. Vasin, A.D. *The Principle of Independence of the Cavity Sections Expansion (Logvinovich's Principle) as the Basis for Investigation on Cavitation Flows*; Project-Nerd: Arvada, CO, USA, 2001; Volume 2, pp. 161–162.
12. Serebryakov, V.V. Ring model for calculation of axisymmetric flows with developed cavitation. *J. Hydromech.* **1974**, *27*, 25–29.
13. Fu, H.; Lu, C.J.; Li, J. Numerical Research on Drag Reduction Characteristics of Supercavitating Body of Revolution. *J. Ship Mech.* **2004**, *8*, 1–7.
14. Yi, W.J. Research on drag characteristics of natural supercavitation profile for high speed bodies. *Ship Sci. Technol.* **2009**, *31*, 38–42.
15. Xiong, T.H.; Yi, W.J.; Wu, J.J. Analysis on Supercaviting Flow Characteristics of Underwater High-Speed Projectile. *J. Gun Launch Control.* **2008**, *3*, 6–9.
16. Xiong, T.H.; Yi, W.J.; Wu, J.J. Investigation on Characteristic of Supercaviting Flow around Underwater Highspeed Vehicle at Small Angle of Attack. *Ship Sci. Technol.* **2009**, *31*, 38–42.
17. Li, D.X. Numerical analysis and experiment on high-speed projectile of underwater supercavity. *J. Sichuan Ordnance* **2011**, *32*, 25–29.
18. Li, Y.T. Calculation of Hydrodynamic Characteristics of Supercavity Vehicle with CFD Method. *Torpedo Technol.* **2015**, *23*, 262–268.
19. Triki, A. Multiple-grid finite element solution of the shallow water equations: Water hammer phenomenon. *Comput. Fluids* **2014**, *90*, 65–71. [CrossRef]
20. Kandula, J.; Sri, P.U.; Reddy, P.R.; Gugulothu, S.K. Numerical and experimental cavitation assessment of near-wake characteristics of hydrodynamic performance characteristics of cavitating flow with and without ultrasonic transducers-ScienceDirect. *Measurement* **2021**, *173*, 108591. [CrossRef]
21. Moin, P. Progress in large eddy simulation of turbulent flows. *AIAA Pap.* **1997**, *970749*, 1–11.
22. Shih, T.H.; Liou, W.W.; Shabbir, A.; Yang, Z.; Zhu, J. A new k-ϵ eddy viscosity model for high reynolds number turbulent flows. *Comput. Fluids* **1995**, *24*, 227–238. [CrossRef]
23. Schnerr, G.H.; Sauer, J. Physical and numerical modeling of unsteady cavitation dynamics. In Proceedings of the 4th International Conference on Multiphase Flow, New Orleans, LA, USA, 27 May–1 June 2001.
24. Hrubes, J.D. High-speed imaging of supercavitating underwater projectiles. *Exp. Fluids.* **2001**, *30*, 57–64. [CrossRef]

Journal of
Marine Science and Engineering

MDPI

Article

Propulsion Performance and Wake Dynamics of Heaving Foils under Different Waveform Input Perturbations

Pengcheng Gao [1,2], Qiaogao Huang [1,2,*] and Guang Pan [1,2]

1 School of Marine Science and Technology, Northwestern Polytechnical University, Xi'an 710072, China; gaopengcheng@mail.nwpu.edu.cn (P.G.); panguang@nwpu.edu.cn (G.P.)
2 Key Laboratory of Unmanned Underwater Vehicle, Northwestern Polytechnical University, Xi'an 710072, China
* Correspondence: huangqiaogao@nwpu.edu.cn

Abstract: A numerical simulation is used to investigate the effects of adding high frequency and low amplitude perturbations of different waveforms to the sinusoidal-based signal of the heaving foil on the propulsion performance and wake structure. We use the adjustable parameter k to achieve a heaving motion of various waveform cycle trajectories, such as sawtooth, sine, and sq3e. Adding a perturbation of whatever waveform is beneficial in increasing the thrust of the heaving foil, especially by adding a square wave perturbation with a frequency of 10 Hz, pushes the thrust up to 10.49 times that without the perturbation. However, the addition of the perturbation signal brings a reduction in propulsion efficiency, and the larger the perturbation frequency, the lower the efficiency. The wake structure of the heaving foil behaves similarly under different waveform perturbations, all going through some intermediate stages, which eventually evolve into a chaotic wake with the increase in the perturbation frequency. However, a lower frequency square wave perturbation can destabilize the heaving foil wake structure. This work further explains the effect of different waveform perturbation signals on the base sinusoidal signal and provides a new control idea for underwater vehicles.

Keywords: heaving foil; different waveform perturbation; hydrodynamic characteristics; vortex structure

Citation: Gao, P.; Huang, Q.; Pan, G. Propulsion Performance and Wake Dynamics of Heaving Foils under Different Waveform Input Perturbations. *J. Mar. Sci. Eng.* **2021**, 9, 1271. https://doi.org/10.3390/jmse9111271

Academic Editor: Kostas Belibassakis

Received: 27 October 2021
Accepted: 11 November 2021
Published: 15 November 2021

1. Introduction

Motion is the fundamental property and the existing mode of matter. From running cheetahs to flying swallows and swimming carp, and even to bacteria, moving organisms can be commonly observed. Birds and fish, which usually overcome drag by flapping their wings or fins, can drive themselves with fewer power costs and greater maneuverability after more than 500 million years of natural selection.

Over the past century, researchers have made numerous efforts to learn about the physical mechanisms at the back of flapping propulsion and apply them to bionic vehicles, such as the pioneering work of Garrick [1] and Gray [2], and the theoretical investigations presented by Lighthill [3] and Wu [4]. As the saying goes: three parts land, seven parts ocean. With the over-exploitation of land resources, humans gradually shifted their focus to the ocean, which has more resources, and thus bionic underwater vehicles were born. With the development of technology, scholars have analyzed the skeletal–muscular structure of marine organisms through technical means such as X-ray scanning and dissection, and also used high-speed cameras to photograph the movement of different fishes, summarizing a large number of kinematic equations, which have greatly contributed to the development of bionic vehicles. [5–8]

Most commonly studied today is the replacement of fins/wings of fish and birds by a simplified model (flapping foil), while simulating biological motion by forcing an excitation signal on the leading edge of the foil. Previously, smooth and periodic excitation signals (sinusoidal, non-sinusoidal, asymmetric functions, etc.) were often chosen to control the foil to accomplish a pure pitch motion, pure heave motion, or a combination of both. Mao et al. [9] used a computational fluid dynamics (CFD) approach to study the hydrodynamic characteristics of 3D flapping foil with different bias angles in pitch motion, and his results can be used to improve control systems for the heave and pitch motions of vehicles. Wu [10] conducted a numerical study on the self-propulsion problem of rigid–flexible composite plates, in which he applied forces to the connection points to keep the rigid part in a given pitch motion, while the deformation of the flexible part was consequent. The effect of the stiffness of the flexible plate was investigated by varying the length ratio of the flexible and rigid parts to simulate the composite plate. In the experiments of Cros [11], a transverse harmonic displacement is applied at the fixed leading edge of the flexible plate. The excitation frequency and air velocity are the two control parameters in their experiment. Their experimental results demonstrate that the resonance phenomenon exhibited by the cantilever plate in the airflow is consistent with the linear theoretical predictions of Eloy [12] and Michelin [13].

However, the simple excitation signals mentioned above cannot accurately represent the mechanisms that generate motor output in biological systems. The reasons for this are mainly the following: Firstly, when fish move, their fins and bodies are deformed, thus generating multiple waves [14,15]. Secondly, the biological locomotor signal is not formed by a single signal, it is generated by a combination of central modulation signals and organ feedback signals [16–18]. Finally, the marine environment in which fish live is complex and variable. Therefore, fish must adapt their movements to the changing ocean currents for efficient propulsion. Lehn et al. [19] conducted perturbation experiments in a closed recirculating water tank, where he added a high frequency, low amplitude sinusoidal perturbation signal to the base signal of a sinusoidal heave motion, and the experimental results showed that the addition of the perturbation resulted in a substantial increase in thrust. Inspired by Lehn et al. [19], we add the nonsinusoidal perturbation signal to the sinusoidal base signal and investigate the effects of different waveforms of the perturbation signal on the hydrodynamic characteristics of the heaving foil using numerical simulations.

In this paper, we give the description of the problem and the methodology in Section 2, where the schematic representation of the flow configuration and the kinematics of heaving foil are described. We also define the relevant parameters and validate the numerical method in this section. Then, we discuss the wake structure and hydrodynamics in Section 3. The last section concludes the paper with a summary of our results.

2. Problem Definition and Methodology

2.1. Problem Definition

Figure 1 depicts this problem through a flow schematic. A two-dimensional NACA0012 foil is placed in a uniform flow with constant velocity U in the x-direction. The control signal $h(t)$ consists of the base signal $h_b(t)$ and the perturbation signal $h_p(t)$. The base signal is a sinusoidal signal, which has a heaving amplitude (A) of 0.25 times the chord length (c) and a fixed frequency (f_b) of 1 Hz. The heaving amplitude (0.1 A) of the perturbation signal is 0.025 times the chord length (c) and the frequency (f_p) ranges from 0 to 10 Hz with increments of 2 Hz, but its waveform is not fixed, and we will give a detailed definition of the perturbation signal later.

Different from Lehn et al. [19], the adjustable parameter k is used to achieve perturbation signals of various waveform cycle trajectories, such as sawtooth, sine, and square. According to the work of Lu [20], we define the nonsinusoidal heaving perturbation signal of the foil as follows:

$$\begin{cases} h_p(t) = \frac{h_{p0}\sin^{-1}[-k\sin(2\pi f_p t)]}{\sin^{-1}(-k)}, -1 \le k < 0 \\ h_p(t) = h_{p0}\sin(2\pi f_p t)\ , \ k = 0 \\ h_p(t) = \frac{h_{p0}\tanh[k\sin(2\pi f_p t)]}{\tanh k}\ , \ k > 0 \end{cases} \tag{1}$$

where $h_p(t)$ denotes the instantaneous heaving perturbation amplitude, h_{p0} is the maximum heaving amplitude, f is the heaving frequency, and t is the instantaneous time. In Equation (1), we regulate the trajectory of the heaving perturbation signal through an adjustable parameter k. As k increases, the heaving perturbation signal changes from a sawtooth wave to a sine wave and finally to a square wave. In Figure 2, we use the k adopted in this study to show the waveform shapes of various perturbation signals.

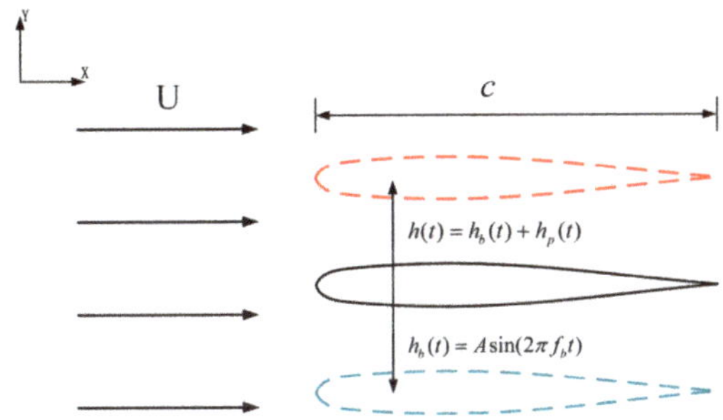

Figure 1. Schematic diagram of the flow field.

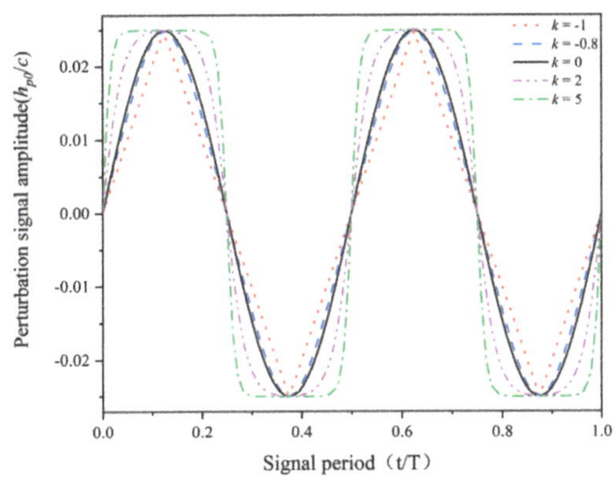

Figure 2. Variation of the perturbation signal waveform in one period at different k.

We present several dimensionless parameters to describe the motion of the foil. The non-dimensional heave amplitude h, Reynolds number Re, and non-dimensional frequency $\overline{f}$ are defined:

$$h = \frac{A}{c} \tag{2}$$

$$\text{Re} = \frac{Uc}{v} \tag{3}$$

$$\overline{f} = \frac{2\pi fc}{U} \tag{4}$$

where v is the kinematic viscosity of the fluid, U is the incoming flow velocity, which has been described earlier, A is the heaving amplitude of the base signal, c is the chord length, and here f is the base frequency (f_b).

To analyze the forces produced by the heaving foil, we use the symbols C_T and C_L to represent the instantaneous thrust coefficient and the instantaneous lift coefficient of the heaving foil, respectively. Where s is the span length. Hence, the time-averaged coefficients over a motion period are calculated as:

$$\begin{cases} C_T = \frac{F_T}{1/2\rho U^2 cs} \\ C_L = \frac{F_L}{1/2\rho U^2 cs} \\ C_{T-mean} = \frac{1}{T}\int_0^T C_T(t)dt \\ C_{L-mean} = \frac{1}{T}\int_0^T C_L(t)dt \end{cases} \quad (5)$$

According to the definition, it can be easily found that $C_{T-mean} > 0$ produces thrust while $C_{T-mean} < 0$ produces drag. From the previous study [21], we computed the instantaneous power coefficient, C_P, as $C_P(t) = C_L(t)\dot{h}(t)$. Therefore, the time-averaged input power coefficient is calculated as:

$$C_{P-mean} = \frac{1}{T}\int_0^T C_L(t)\frac{\dot{h}(t)}{U}dt \quad (6)$$

Thus, the propulsion efficiency of the heaving foil is determined as:

$$\eta = \frac{C_{T-mean}}{C_{P-mean}} \quad (7)$$

2.2. Numerical Approach

Numerical simulation of the unsteady flow field around the heaving foil is performed with the software Fluent 20.0. The finite volume method (FVM) is used to discretize the Navier–Stocks equations. The PISO (Pressure-Implicit with Splitting of Operators) algorithm is used to solve the coupled velocity–pressure solution. Momentum discretization is performed using the second-order windward method and derivatives are calculated on the basis of Green-Gauss nodes. The foil motion is achieved by user-defined functions (UDF).

The computational domain consists of two parts: the inner domain and the outer domain. The total size of the computational domain is a rectangle of $20c \times 12c$, and the inner domain is $10c \times 3c$. The inner domain uses an unstructured grid to reduce computational resources by the dynamic diffusion grid method and avoid obvious changes in the grid, and the outer domain uses a structured grid.

The boundary conditions of the calculation area are set as follows: the inlet boundary is the velocity inlet; the outlet boundary is the pressure outlet, and the no-slip boundary conditions are used for the foil's surface and the upper and lower boundaries.

2.3. Validations

To verify the grid-independence, three grid sizes (coarse, medium, and fine) were selected, and the number of grids in grid 1 to 3 was 9×10^4, 2×10^5, and 3×10^5. The C_T curves of the heaving foil at $h = 0.25$, $f_b = 1$ Hz, and $f_p = 0$ Hz were used to verify the grid-independence. The convergence validation of grids 1 to 3 is shown in Figure 3a. There is no significant difference in the C_T variation under the three grid scales, indicating that the numerical calculation is grid convergent, and the subsequent results are analyzed using the calculation results of grid 2.

To verify the reliability of the calculation method, the results of this paper are compared with those of Heathcote [22] and Ashraf [23]. Heathcote conducted experiments on a NACA0012 airfoil with an amplitude $h = 0.175$ at different frequencies $\overline{f}$. To determine whether turbulence effects play an important role in the flow, Ashraf compared his calculations with the Sparlart–Allmaras turbulence model and the experimental results of Heathcote. It was shown that the NS (Navier-Stokes) laminar simulation results coincided with the turbulence simulation results and were similar to the experimental data when the product of the dimensionless amplitude h and the dimensionless frequency $\overline{f}$ was less than 1.5 ($\overline{f}h < 1.5$). From Figure 3b, we can see that the calculation results of the method in this paper are in good agreement with the experimental results and are better than the calculation results of Ashraf. A time-step refinement study has also been performed by varying the number of time steps per cycle. For this study, 500, 800, 1000, and 1500 steps were used, using a medium grid for each cycle. As shown in Figure 3c, increasing the resolution of the time step from 1000 to 1500 steps per cycle does not change the thrust coefficient. Therefore, in our study, we use 1000 steps for further simulations.

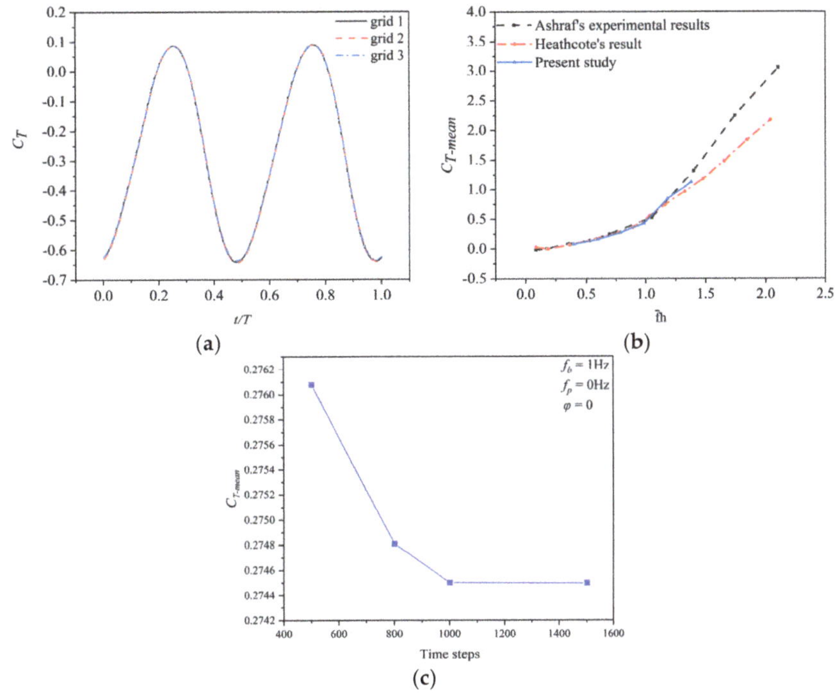

Figure 3. Validation for present numerical simulation: (**a**) grid independence verification, (**b**) calculation method validation and (**c**) validation for present numerical simulation.

3. Result and Discussion

To investigate the physical mechanism behind the effect of different waveform perturbation signals on the heaving foil, this section shows the average thrust coefficient, propulsion efficiency, and wake structure of the heaving foil for different waveforms of perturbation signals.

3.1. Force Measurement

From the preceding parameter settings, we can see that the base frequency f_b has been kept constant at 1 Hz and the perturbation frequency f_p is in variation. Therefore, the calculation period is determined by the base frequency, which means that one calculation period is 1 s and the total calculation time is 10 s. Note that all calculations are averaged over the last two cycles.

The variation of the thrust coefficient of the heaving foil with the perturbation frequency for different waveform perturbations is shown in Figure 4. The comparisons between square and sine, and sawtooth and sine waves are carried out in Figure 4a,b, respectively. The 0 Hz perturbation frequency means that the input signal is only the base signal.

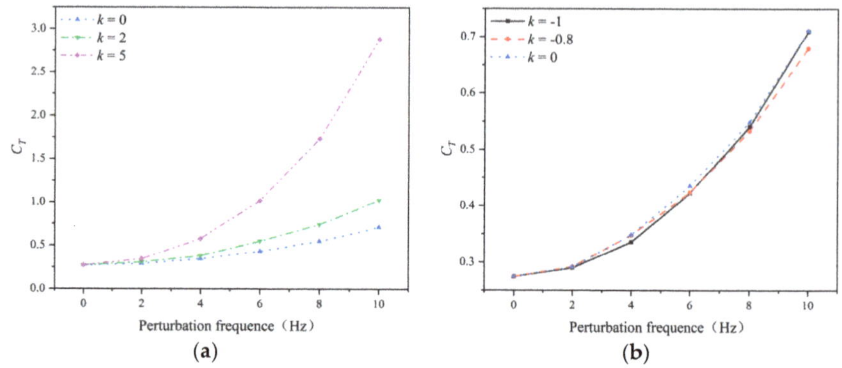

Figure 4. Variation of thrust coefficient: (**a**) a square wave vs. a sine wave and (**b**) a sawtooth wave vs. a sine wave.

The thrust coefficient of the heaving foil gradually improves as the perturbation frequency increases, which is consistent with the conclusion reached in the experiment of Lehn et al. [19]. We also observe that the sine wave perturbation signal brings a slightly better thrust enhancement than the sawtooth wave, while the square wave perturbation signal brings a substantial thrust enhancement, and we can observe that at k = 5 and f_p = 10 Hz, the thrust is directly raised to 1049.89% of that without the addition of perturbation. However, we did not observe the peak thrust as in the experiment of Lehn et al. [19], perhaps because the stiff foil used in his experiment is defined only relative to the flexible foil he used and is not an absolutely rigid body.

For the purpose of quantifying the effect of perturbation frequency on the thrust coefficient, we attempted to fit a response function from the data to derive the corresponding thrust coefficient based on the input perturbation parameters. We fitted a polynomial of the fourth order to the existing thrust coefficient and perturbation frequency curves to obtain the following equation.

$$\begin{cases} y = 1.0807 \times 10^{-5}x^4 - 1.3154 \times 10^{-4}x^3 + 0.0048x^2 - 0.002x + 0.2748, \quad k = -1 \\ y = 2.8125 \times 10^{-5}x^4 - 5.2407 \times 10^{-4}x^3 + 0.0067x^2 - 0.0024x + 0.2745, \quad k = -0.8 \\ y = 3.9583 \times 10^{-5}x^4 - 7.544 \times 10^{-4}x^3 + 0.0086x^2 - 0.0061x + 0.2747, \quad k = 0 \\ y = 6.6406 \times 10^{-6}x^4 - 5.4456 \times 10^{-5}x^3 + 0.0073x^2 + 4.7447 \times 10^{-4}x + 0.2747, k = 2 \\ y = 1.3750 \times 10^{-4}x^4 - 9.2546 \times 10^{-4}x^3 + 0.0222x^2 - 0.0066x + 0.2749, \quad k = 5 \end{cases} \tag{8}$$

where y represents the value of the thrust coefficient and x represents the value of the perturbation frequency. To test the accuracy of the empirical equation we obtained, we numerically simulated that the value of the thrust coefficient under a 12 Hz sinusoidal wave perturbation is 0.9429, which was calculated from our fitted equation as 0.9505, resulting in an error of 0.81%. This indicates that the fitted equation is predictive and yields a universal relationship between the perturbation parameters and the thrust coefficients.

Figure 5 illustrates the variation pattern of propulsion efficiency with perturbation frequency for heaving foil under five perturbation waveforms. As can be seen from the figure, the propulsion efficiency of the heaving foil tends to decrease as the perturbation frequency increases, regardless of which waveform perturbation is added. The propulsion efficiency performance of heaving foil under the perturbation of five waveforms can be roughly classified into three levels: excellent (k = −0.8, 0), good (k = 2), and poor (k = −1, 5). No increase in propulsion efficiency similar to that obtained experimentally by Lehn et al. [19] was found in our numerical simulations, but there is a large number of previous research [24] showing differences in the propulsive performance of flexible and rigid foils. Therefore, we have reason to believe that the reason for this difference is because the research object of this paper is a rigid foil, and the effect of flexibility on the foil motion performance is not considered.

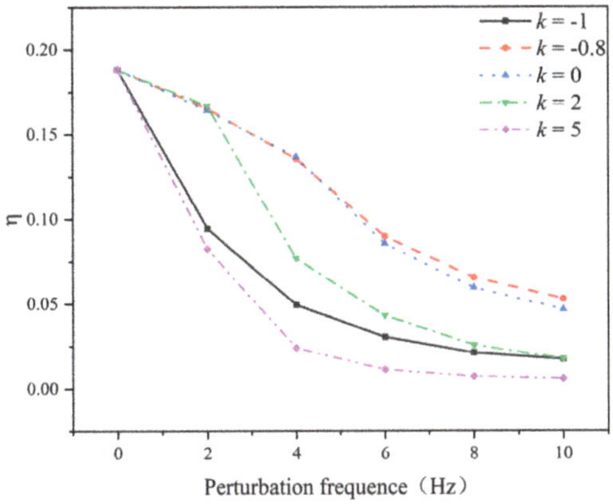

Figure 5. Variation of propulsion efficiency.

To explore the physical mechanism of the efficiency decrease, we show the variation of the input power coefficient C_P with the perturbation frequency f_p at different base frequencies f_b in Figure 6. It can be observed that the input power coefficient almost always increases, while the perturbation frequency gradually increases. Since the increase of the perturbation frequency makes the foil move more distance at the same time, it inevitably leads to the increase of the input work, which causes the increase of the input power coefficient. Since the increase of input power coefficient is much larger than the increase of thrust coefficient, the propulsion efficiency of foil decreases with the increase of perturbation frequency. Meanwhile, we see that the input power coefficient at $k = -1$ is much larger than $k = 0, -0.8$, which explains why the propulsion efficiency is low when the perturbation signal is a sawtooth wave.

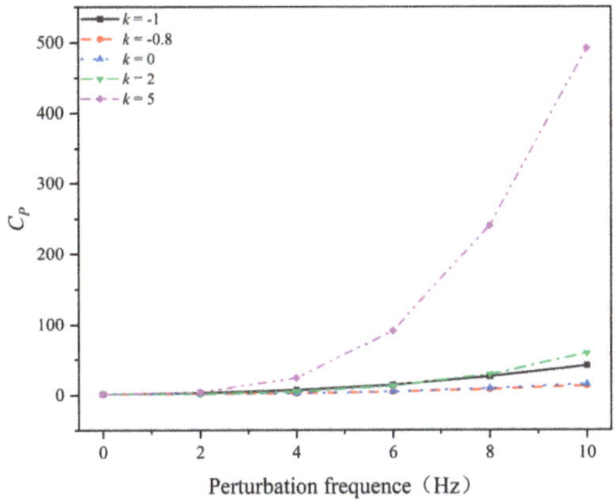

Figure 6. Variation of input power coefficient.

Considering the thrust and propulsion efficiency performance of the heaving foil under different perturbations, it seems that this can give us some inspiration for the drive mechanism of the bionic vehicle. First, the control signal of the vehicle trys to avoid the use of the sawtooth wave signal. The thrust gain from the sawtooth wave is not significantly different from the sine wave, but the sawtooth wave brings a significant reduction in propulsion efficiency. Second, it can consider sacrificing part of the propulsion efficiency and adding a square wave signal to the drive signal to promote the vehicle, gaining great thrust. Next, to ensure high propulsion efficiency for a long voyage, a sinusoidal signal (or a signal with $k = -0.8$) can be added to the drive signal. Last, the waveform signal with $k = 2$ seems to be the most cost-effective choice, which provides a larger thrust without a very drastic decrease in propulsion efficiency.

3.2. Wake Structures

In this section, we study the wake structure as well as the wake evolution to better understand the foil propulsion properties, and we summarize all the wake structures into a (k, f_p) map.

It can be seen from Figure 7 that the wake evolves similarly when the waveform parameter k is less than 0, in other words, when the perturbation signal is in the sawtooth to sine wave range. The transition from rbvk wake in the absence of perturbation to an asymmetric wake then goes through a complex wake phase and finally evolves to a chaotic wake. Even when the perturbation signal is a square wave, the wake still behaves as an asymmetric wake at $f_p = 2$ Hz, and the difference starts to appear when the perturbation frequency reaches 4 Hz. Also, this confirms the thrust curve, where the difference of thrust is small when $k = -1, -0.8, 0$; the difference in thrust also occurs at $f_p = 4$ Hz for $k = 2, 5$.

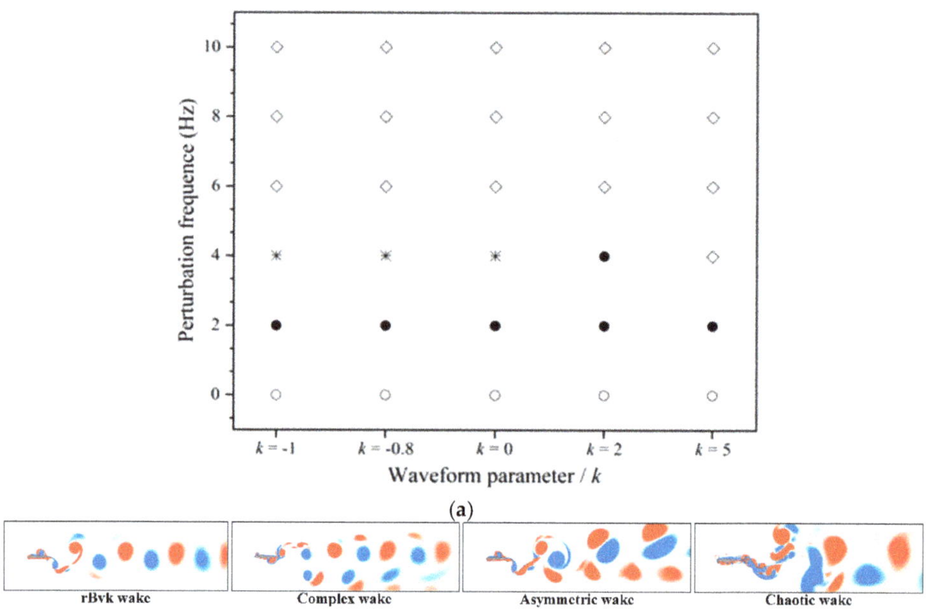

Figure 7. The numerical points in the (k, f_p) plane for the heaving foil. ○: reverse Bénard–von Kármán (rBvk) wake; •: asymmetric wake; ✻: complex wake; ◇: chaotic wake. (**a**) Represents the (k, fp) map and (**b**) presents examples of the wake structures.

Because the flow structure is similar, this paper only shows the wake structure of the heaving foil at different perturbation frequencies at five different moments ($0\,T$, $1/4\,T$, $1/2\,T, 3/4\,T, T$) in one computational period for $k = -1, 2, 5$, with the initial motion direction all along the positive y-axis.

From Figure 8, it can be found that at $f_b = 1$ Hz, the wake structure behaves as an rBvk wake at $f_p = 0$ Hz. When f_p increases to 2 Hz, the wake structure is presented as asymmetric wake and the vortex pair deflects upward. When f_p increases to 4 Hz, the wake structure is not chaotic and sheds four vortices respectively in one cycle; when the perturbation frequency continues to increase to 6 Hz, the trailing edge structure starts to become chaotic. As can be seen from the Figure 4b, the thrust coefficient is increased when the perturbation frequency reaches 4 Hz. The reason for this phenomenon is that the number of wake vortex pairs increases and is accompanied by an increase in vortex strength when the perturbation frequency is 4 Hz. In particular, when the perturbation frequency is greater than 6 Hz, vortices start to fall off at the leading edge of the foil and converge into the wake, further enhancing the thrust coefficient.

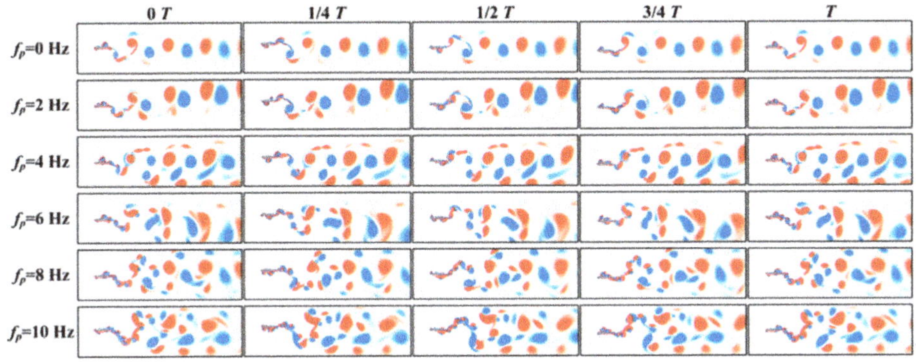

Figure 8. Instantaneous wake structure for foil at $k = -1$.

The wake evolution at $k = 2$ is shown in Figure 9. When a perturbation of 2 Hz is added, the wake structure also deflects and appears as asymmetric wake, and the degree of deflection is greater than $k \leq 0$. Unlike the sine and sawtooth wave perturbations, the wake structure remains as asymmetric wake when the perturbation frequency increases to 4 Hz and becomes chaotic when the perturbation frequency reaches 6 Hz. Because of this pattern of change in wake structure, no increase in thrust coefficient is observed at a perturbation frequency of 4 Hz as at $k = -1$. Instead, a significant increase in thrust coefficient is observed at $f_p = 6$ Hz. In addition, it can be observed that when the perturbation frequency is higher than 6 Hz, the vortex density and vortex intensity are significantly increased relative to those at $k = -1$, which also leads to a significant increase in the heaving foil thrust coefficient.

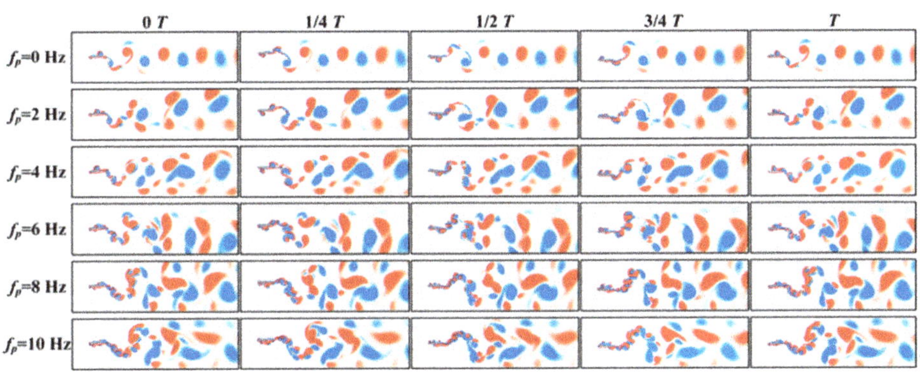

Figure 9. Instantaneous wake structure for foil at $k = 2$.

Figure 10 shows the wake structure when $k = 5$, in other words, when the perturbation signal is a square wave. Unlike the two evolutionary modes mentioned previously, the wake structure becomes directly chaotic when the perturbation frequency increases to 4 Hz. This fits with the phenomenon in Figure 4a that the thrust coefficient produces a large enhancement when the perturbation frequency reaches 4 Hz when $k = 5$. Furthermore, by comparing the wake structure diagrams for $k = -1, 2, 5$, it can be observed that the vortex strength is greatest when $k = 5$. This also explains why the square wave perturbation brings the most significant thrust improvement for the same perturbation amplitude and perturbation frequency.

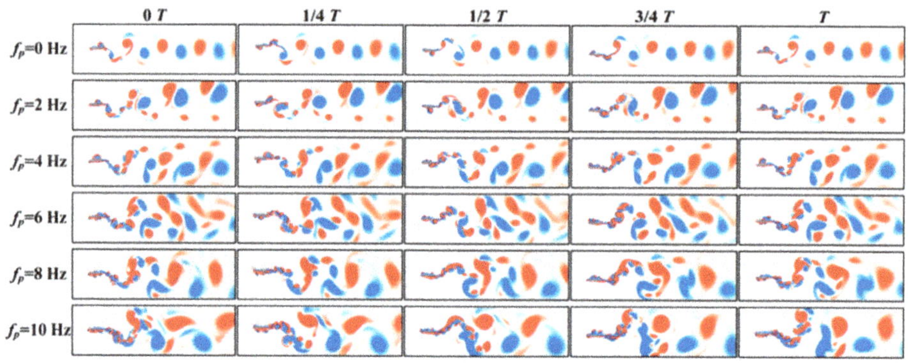

Figure 10. Instantaneous wake structure for foil at $k = 5$.

From the above analysis, it can be observed that although we add different perturbation waveforms, there is no significant effect on the wake structure of the heaving foil at perturbation frequencies below 4 Hz, which means that a higher perturbation frequency is beneficial for the perturbation signal to affect the flow structure. At the same time, this can also trigger us to think about why the effect of different waveforms of perturbation on the heaving foil wake structure is only different when the perturbation frequency is more or equal to 4 Hz. Our preliminary speculation is that this perturbation frequency value is related to the parameters, such as the base frequency we selected and the amplitude ratio between the base signal and the perturbation signal, which will also be the focus of our subsequent research.

4. Conclusions

A systematic numerical investigation of the fluid dynamics around the heaving foil, including the time-averaged thrust coefficient, time-averaged input work coefficient, propulsion efficiency, and the wake structure generated by the foil, was conducted to better understand the effects of different waveform perturbations on the heaving foil force generation and wake structure, and to provide recommendations for underwater vehicle motion parameter settings from a hydrodynamic perspective. The simulation results are concluded as follows:

1. The addition of any waveform perturbation increases the axial thrust, especially when a square wave ($k = 5$) perturbation with a frequency of 10 Hz is applied, which directly raises the thrust to 10.49 times higher than when no perturbation is added. A response function is obtained which gives the relationship between the input perturbation parameters and the thrust coefficients.
2. However, the propulsion efficiency decreases gradually with the increase of the perturbation frequency, especially when a sawtooth wave perturbation ($k = -1$) is applied, which brings limited thrust gain but a rapid decrease in the propulsion efficiency.
3. The wake structure is roughly the same for different waveform perturbations, all of which first go through an asymmetric wake phase (at $f_p = 2$ Hz). With increasing perturbation frequency, adding sine wave perturbation ($k = 0$) and sawtooth wave perturbation ($k = -1, -0.8$) leads to a complex wake phase (at $f_p = 4$ Hz) and then evolves to a chaotic wake; adding square wave perturbation (k =2) keeps the asym-

metric wake at f_p = 4 Hz and then evolves to a chaotic wake; adding square wave perturbation (k = 5) leads to a direct evolution from an asymmetric wake to a chaotic wake.

4. The simulation results have a guiding meaning for the setting of the vehicle motion parameters. The control signal of the vehicle tries to avoid the use of sawtooth wave signal. It can consider sacrificing part of the propulsion efficiency and adding a square wave signal to the drive signal to promote the vehicle gaining greater thrust. To ensure high propulsion efficiency for a long voyage, a sinusoidal signal (or a signal with $k = -0.8$) can be added to the drive signal. The waveform signal with k = 2 seems to be the most cost-effective choice, which provides a larger thrust without a significant decrease in propulsion efficiency.

Author Contributions: Conceptualization, P.G.; methodology, P.G.; software, P.G.; validation, P.G., Q.H., and G.P.; formal analysis, P.G.; investigation, P.G.; data curation, P.G.; writing—original draft preparation, P.G.; writing—review and editing, Q.H.; visualization, P.G.; supervision, G.P. and Q.H.; project administration, G.P. and Q.H.; funding acquisition, G.P. and Q.H. All authors have read and agreed to the published version of the manuscript.

Funding: This work was supported by the National Natural Science Foundation of China (Grant No. 51879220), the National Key Research and Development Program of China (Grant No. 2020YFB1313201), and Fundamental Research Funds for the Central Universities (Grant No. 3102019HHZY030019 and 3102020HHZY030018).

Institutional Review Board Statement: Not applicable.

Informed Consent Statement: Not applicable.

Data Availability Statement: The data that support the findings of this study are available within the article.

Conflicts of Interest: The authors declare no conflict of interest.

References

1. Garrick, I.E. Propulsion of a flapping and oscillating airfoil. *NACA Rep.* **1937**, *567*, 419–427.
2. Gray, J. Studies in animal locomotion: VI. The propulsive powers of the dolphin. *J. Exp. Biol.* **1936**, *13*, 192–199. [CrossRef]
3. Lighthill, M.J. Hydromechanics of aquatic animal propulsion. *Annu. Rev. Fluid Mech.* **1969**, *1*, 413–446. [CrossRef]
4. Wu, T.Y. Fish swimming and bird/insect flight. *Annu. Rev. Fluid Mech.* **2011**, *43*, 25–58. [CrossRef]
5. Shiau, J.; Watson, J.R.; Cramp, R.L.; Gordos, M.A.; Franklin, C.E. Interactions between water depth, velocity and body size on fish swimming performance: Implications for culvert hydrodynamics. *Ecol. Eng.* **2020**, *156*, 105987. [CrossRef]
6. Downie, A.T.; Illing, B.; Faria, A.M.; Rummer, J.L. Swimming performance of marine fish larvae: Review of a universal trait under ecological and environmental pressure. *Rev. Fish Biol. Fish.* **2020**, *30*, 93–108. [CrossRef]
7. Russo, R.S.; Blemker, S.S.; Fish, F.E.; Bart-Smith, H. Biomechanical model of batoid (skates and rays) pectoral fins predicts the influence of skeletal structure on fin kinematics: Implications for bio-inspired design. *Bioinspiration Biomim.* **2015**, *10*, 046002. [CrossRef] [PubMed]
8. Tytell, E.D.; Leftwich, M.C.; Hsu, C.Y.; Griffith, B.E.; Cohen, A.H.; Smits, A.J.; Hamlet, C.; Fauci, L.J. Role of body stiffness in undulatory swimming: Insights from robotic and computational models. *Phys. Rev. Fluids.* **2016**, *1*, 073202. [CrossRef]
9. Mao, L.; Wang, H.; Li, Y.; Yi, H. Force model of flapping foil stabilizers based on CFD parameterization. *Ocean Eng.* **2019**, *187*, 106151. [CrossRef]
10. Wu, W. Study on the self-propulsion of the rigid-flexible composite plate. *Fluid Dyn. Res.* **2021**, *53*, 045501. [CrossRef]
11. Cros, A.; Castro, R.F.A. Experimental study on the resonance frequencies of a cantilevered plate in air flow. *J. Sound Vib.* **2016**, *363*, 240–246. [CrossRef]
12. Eloy, C.; Souilliez, C.; Schouveiler, L. Flutter of a rectangular plate. *J. Fluids Struct.* **2007**, *23*, 904–919. [CrossRef]
13. Michelin, S.; Llewellyn Smith, S.G. Resonance and propulsion performance of a heaving flexible wing. *Phys. Fluids* **2009**, *21*, 071902. [CrossRef]
14. Root, R.G.; Courtland, H.W.; Shepherd, W.; Long, J.H. Flapping flexible fish. In *Animal Locomotion*, 1st ed.; Taylor, G.K., Triantafyllou, M.S., Tropea, C., Eds.; Springer: Berlin/Heidelberg, Germany, 2010; pp. 141–159.
15. Root, R.G.; Liew, C.W. Computational and mathematical modeling of the effects of tailbeat frequency and flexural stiffness in swimming fish. *Zoology* **2014**, *117*, 81–85. [CrossRef]
16. Grillner, S. Neural control of locomotion in lower vertebrates: From behavior to ionic mechanisms. *Neural Control. Rhythm. Mov. Vertebr.* **1988**, *1*, 1–40.

17. Pearson, K.G. Common principles of motor control in vertebrates and invertebrates. *Annu. Rev. Neurosci.* **1993**, *16*, 265–297. [CrossRef]
18. Pearson, K.G. Generating the walking gait: Role of sensory feedback. *Prog. Brain Res.* **2004**, *143*, 123–129. [PubMed]
19. Lehn, A.M.; Thornycroft, P.J.; Lauder, G.V.; Leftwich, M.C. Effect of input perturbation on the performance and wake dynamics of aquatic propulsion in heaving flexible foils. *Phys. Rev. Fluids* **2017**, *2*, 023101. [CrossRef]
20. Lu, K.; Xie, Y.H.; Zhang, D. Numerical study of large amplitude, nonsinusoidal motion and camber effects on pitching airfoil propulsion. *J. Fluids Struct.* **2013**, *36*, 184–194. [CrossRef]
21. Read, D.A.; Hover, F.S.; Triantafyllou, M.S. Forces on oscillating foils for propulsion and maneuvering. *J. Fluids Struct.* **2003**, *17*, 163–183. [CrossRef]
22. Heathcote, S.; Gursul, I. Jet switching phenomenon for a periodically plunging airfoil. *Phys. Fluids* **2007**, *19*, 027104. [CrossRef]
23. Ashraf, M.A.; Young, J.; Lai, J.C.S. Oscillation frequency and amplitude effects on plunging airfoil propulsion and flow periodicity. *AIAA J.* **2012**, *50*, 2308–2324. [CrossRef]
24. Gao, P.; Huang, Q.; Pan, G.; Zhao, J. Effects of flexibility and motion parameters on a flapping foil at zero freestream velocity. *Ocean Eng.* **2021**, *242*, 110061. [CrossRef]

Journal of
Marine Science and Engineering

MDPI

Article

Effects of Blade Number on the Propulsion and Vortical Structures of Pre-Swirl Stator Pump-Jet Propulsors

Han Li [1], Qiaogao Huang [1,2,*], Guang Pan [1,2], Xinguo Dong [3] and Fuzheng Li [1]

1 School of Marine Science and Technology, Northwestern Polytechnical University, Xi'an 710072, China; lihan_nwpu@mail.nwpu.edu.cn (H.L.); panguang@nwpu.edu.cn (G.P.); jinglelfz@mail.nwpu.edu.cn (F.L.)
2 Key Laboratory of Unmanned Underwater Vehicle, Northwestern Polytechnical University, Xi'an 710072, China
3 Science and Technology on Water Jet Propulsion Laboratory, Marine Design and Research Institute of China, Shanghai 200011, China; dongxinguosh@126.com
* Correspondence: huangqiaogao@nwpu.edu.cn

Abstract: Reducing the noise of the underwater propulsor is gaining more and more attention in the marine industry. The pump-jet propulsor (PJP) is an extraordinary innovation in marine propulsion applications. This paper inspects the effects of blade number on a pre-swirl stator pump-jet propulsor (PJP) quantitatively and qualitatively. The numerical calculations are conducted by IDDES and ELES, where the ELES is only adopted to capture the vortical structures after refining the mesh. The numerical results show good agreement with the experiment. Detailed discussions of the propulsion, the features of thrust fluctuation in time and frequency domains, and the flow field are involved. Based on the ELES results, the vortices in the PJP flow field and the interactions between the vortices of the stator, rotor, and duct are presented. Results suggest that, though changing the blade number under a constant solidity does not affect the propulsion, it has considerable effects on the thrust fluctuation of PJP. The wakes of the stator and rotor are also notably changed. Increasing the stator blade numbers has significantly weakened the high-intensity vortices in the stator wake and, hence, the interaction with the rotor wake vortices. The hub vortices highly depend upon the wake vortices of the rotor. The hub vortices are considerably broken by upstream wake vortices when the load per rotor blade is high. In summary, the blade number is also vital for the further PJP design, particularly when the main concerns are exciting force and noise performance.

Keywords: pump-jet propulsor; hydrodynamics; blade number; vortices

Citation: Li, H.; Huang, Q.; Pan, G.; Dong, X.; Li, F. Effects of Blade Number on the Propulsion and Vortical Structures of Pre-Swirl Stator Pump-Jet Propulsors. *J. Mar. Sci. Eng.* **2021**, *9*, 1406. https://doi.org/10.3390/jmse9121406

Academic Editor: Kourosh Koushan

Received: 11 November 2021
Accepted: 6 December 2021
Published: 9 December 2021

1. Introduction

Nowadays, the pump-jet propulsor (PJP) has been widely equipped in torpedos and submarines owing to its high efficiency and outstanding noise resistance. PJP is a multi-component underwater propulsor consisting of a duct and a stator and rotor vane cascades inside the duct. The stator can be installed either upstream or downstream of the rotor, whose installed position categorizes the PJPs into two main forms: the pre-swirl stator form and the rear stator form. The PJP hydrodynamics sets the characteristic performance of the underwater vehicles, and hence the PJP is receiving increased attention.

The propulsion features and flow of PJPs have been widely studied through both numerical approaches and experiments. Suryanarayana et al. [1] evaluated the PJP performance in a wind tunnel and then Suryanarayana et al. [2] evaluated the PJP performance again in a cavitation tunnel and considered the trim angle effects. Later, Suryanarayana et al. [3] obtained the cavitation performance of a PJP. Results show that the tip clearance between the rotor blade tip and the duct is the easiest cavitation place.

Since computers have become powerful, investigating the PJP via the numerical method is the main strategy. Many aspects of the PJP investigation have been studied via computational fluid dynamics (CFD), including the performance [4–9], flow characteristics [10–12], cavitation [13,14], and noise [15]. Besides, the panel method based on potential flow theory is an alternative to obtain the performance [16,17], which plays an important role in quickly obtaining PJP performance and optimization. In CFD, the Reynolds-averaged Navier–Stokes (RANS) approach is preferred to obtain the PJP performance and flow [4–10]. However, the RANS approach is not expert in modeling the turbulence structures as the Reynolds-averaged method smoothes many details. The large eddy simulation (LES) and even the direct numerical simulation (DNS) are outstanding in modeling the complex turbulence and have been successfully applied in modeling the flow around propellers [18–20] and a propeller blade [21]. However, it is still a challenge when employing the LES to model the whole flow field of a complex propulsor due to the mesh requirement of LES. Up to now, the hybrid RANS/LES is a compromise to model the interesting flow region, which has been employed in underwater propulsors for performance and flow details [11,12,22]. Li et al. [11] comparatively investigated the PJP wake vortical structure with different hybrid RANS/LES approaches, which suggests that the improved delayed detached eddy simulation (IDDES) and stress-blended eddy simulation (SBES) have an outstanding ability to model the wake. Then, they revealed the underlying mechanism of the wake instability [12]. Pan et al. [14] obtained the PJP cavitation performance in the hull condition via the RANS approach with the shear stress transport (SST) $k - \omega$ turbulence model. The results illustrated that the cavitation begins at the tip clearance region. The cavitation in tip clearance leakage flow enhances the leakage flow vorticty [23] and accelerates the expansion and contraction of the tip vortex [24]. These studies further suggest that controlling the tip clearance leakage flow is important. Sun et al. [15] investigated the feasibility of using a serrated trailing edge duct in a PJP for noise reduction. Su et al. [25] concluded that the force fluctuation of the rotor dominates the vibro-acoustics response caused by PJP.

As PJP is a multi-component propulsor, most investigations on PJP performance and flow are conducted focusing on the effects of these component parameters. The rotor is the working component and there is a tip clearance between the rotor blade tip and duct. The tip clearance forms a characteristic flow in the PJP flow field and significantly affects the PJP performance [6,7,9]. Ji et al. [26] assessed the effects of thickening and raking the rotor blade tip on attenuating the tip clearance flow. Decreasing or thicking and raking tip clearance considerably weakens the intensity of the tip vortices. Yu et al. [8] investigated the effects of stator incidence angle, stator-rotor spacing, and stator chord length on PJP performance, which shows that the first two parameters considerably affect the rotor force fluctuations. Huang et al. [9] studied the effects of duct main parameters on PJP performance. The expansion ratio of the duct outlet and the incidence angle of the duct also significantly affects the tip clearance flow.

The tip clearance and duct also exist in a ducted propeller. According to the investigations on the ducted propeller, the duct existence delays the wake constraction [27] and changes the tip vortex trajectory [28]. The bio-inspired blade also weakens the effects of low pressure in tip clearance leakage flow on the blade suction side [29]. For a propeller, the tip flow characterizes its wake topology. Without holding the propeller expansion area ratio, directly changing the blade number notably changes the performance [30–32] while the tip vorticity is not notably changed [31] as the load on per blade is not changed. Besides, the blade trailing wakes in propeller wake are restrained by increasing the skew angle [32,33].

The stator and rotor as vane cascades in a PJP, their blade numbers characterize the wake topologies of the stator and rotor, also the tip clearance leakage flow for the rotor, hence affecting the interactions between the stator, rotor, and duct, as well as the PJP wake features. As introduced before, the effects of blade number under constant solidity on the PJP performance and flow vortical structure are not considered. So, in the present work, the effects of blade number under constant solidity are investigated. In the following content, the research models and methodology are discussed in Section 2, followed by the computational setting overview in Section 3. Finally, detailed analysis and discussion are presented in Section 4, and the conclusion in Section 5.

2. Research Object

In the present work, the considered PJP model is a pre-swirl stator form, as shown in Figure 1. The basic PJP model has an eight-bladed stator (Z_s = 8) and a six-bladed rotor (Z_r = 6). Figure 2 presents the sketch of the flow velocity into the stator and rotor vane cascades at radius r, where V_∞, n, and α are the oncoming flow velocity of the PJP, the revolution speed of the rotor, the angle of attack of the rotor blade section, respectively. The symbol V denotes the velocity, where the subscripts 'u' and 'm' indicate the circumferential and meridional components of velocity, and the subscripts '1', '2', and '3' indicate the axial locations. The rotor diameter D_r ($D_r = 2R_r$) is 0.1664 m and the tip clearance between the tip and the duct inside is 0.001 m. The rotor blade is zero skew and rake. The pitch angle of the rotor blade β_r in the radial direction is given in Table 1. The stator axial length L_s is constant in the radial direction while the pre-swirl angle of the stator blade β_1 is variable, as presented in Table 1. The stator and rotor sections are the NACA66 and NACA16 series, respectively. The blade-to-blade distance of the stator t_s, the blade-to-blade distance of the rotor t_r, and the chord length of the rotor blade section L_r are sketched in Figure 2. Particulars of the duct section are given in Table 2. Table 3 gives the blade numbers of the stator and rotor of the derived PJP models. The solidities of the stator and rotor are held constant as same as the basic PJP model (S8R6). Hence, the product of the chord length and blade numbers are held constant according to the solidity definition. The PJP is mainly working behind the submarines, and hence the rudder wakes are even number. So, the blade with an even number is not considered. The stator and rotor of the derived PJP models are shown in Figure 3.

In Figure 1, two inertial reference frames are shown: a cartesian coordinate system $Oxyz$ and a cylindrical coordinate system $O\theta rz$. The PJP advance direction is the z negative direction. The rotor rotates around the z axis, and the rotation direction is right-handed. The axis direction of the cylindrical coordinate system $O\theta rz$ is oriented towards the z axis.

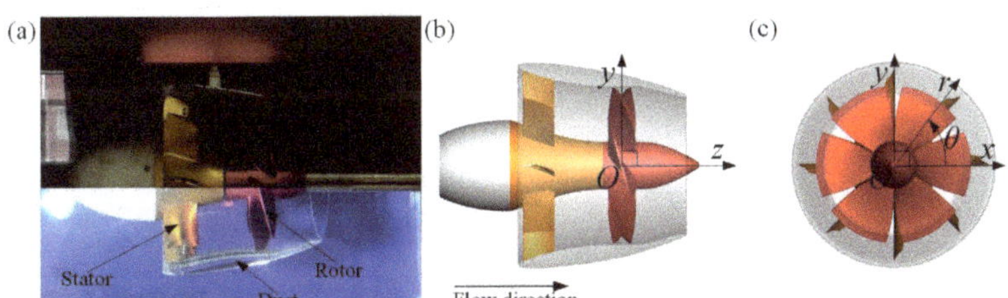

Figure 1. The PJP model in experimental configuration (**a**) and numerical configurations (**b**,**c**).

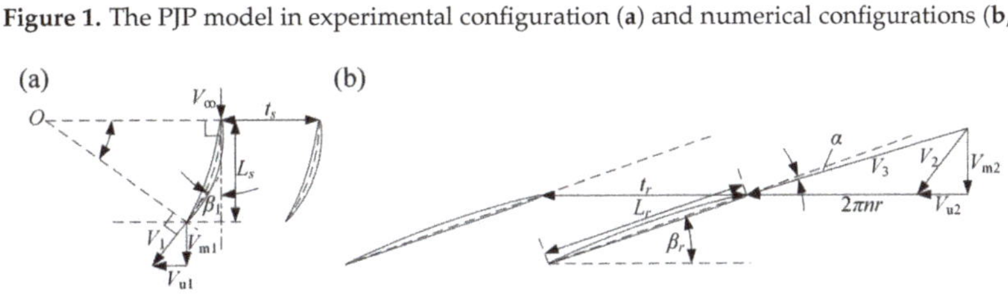

Figure 2. The sketch of the velocity across the blades: (**a**) stator blades; (**b**) rotor blades.

Table 1. The radial distribution of the stator pre-swirl angle β_1 and rotor pitch angle β_r, as shown in Figure 2.

r/R_r	0.5	0.6	0.7	0.8	0.9	1.0	1.1	1.2
β_1 (rad)	0.2009	0.2716	0.3146	0.3311	0.3246	0.2967	0.2399	0.1705
r/R_r	0.3	0.4	0.5	0.6	0.7	0.8	0.9	1.0
β_r (rad)	0.8161	0.7061	0.6075	0.5287	0.4649	0.4012	0.3351	0.2658

Table 2. The duct section, where y_i and y_o denote the duct inside and outside the y coordinate.

z/D_r	0.000	0.0250	0.050	0.075	0.100	0.200	0.300	0.400
y_i/D_r	0.6393	0.6156	0.6053	0.5952	0.5866	0.5582	0.5361	0.5182
y_o/D_r	0.6393	0.6548	0.6567	0.6588	0.6592	0.6561	0.6445	0.6307
z/D_r	0.500	0.600	0.700	0.800	0.900	1.000	1.025	1.050
y_i/D_r	0.5076	0.5060	0.5060	0.4990	0.4762	0.4468	0.4421	0.4387
y_o/D_r	0.6151	0.5978	0.5758	0.5488	0.5168	0.4795	0.4681	0.4582

Table 3. The PJP model with different rotor and stator blade numbers (Z_r and Z_s).

PJP Model ID	**S8R6 1**	**S9R7 2**	**S11R7 3**	**S11R9 4**	**S13R7 5**	**S13R9 6**	**S13R11 7**
Z_r	6	7	7	9	7	9	11
Z_s	8	9	11	11	13	13	13

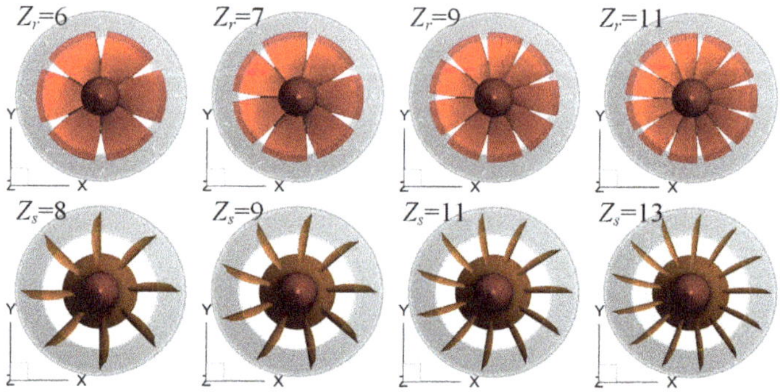

Figure 3. The rotor and stator under different blade numbers.

The following dimensionless coefficients are used to describe the PJP propulsion performance:

$$J = \frac{V_\infty}{nD_r};\ K_{Tr} = \frac{T_r}{\rho n^2 D_r^4};\ K_{Qr} = \frac{Q_r}{\rho n^2 D_r^5};\ K_{Ts} = \frac{T_s}{\rho n^2 D_r^4};\ K_{Qs} = \frac{Q_s}{\rho n^2 D_r^5};\ K_{Td} = \frac{T_d}{\rho n^2 D_r^4}, \tag{1}$$

where J is the advance coefficient. T_r, T_s, and T_d are the thrust of the rotor, stator, and duct, respectively, and their corresponding dimensionless coefficients are K_{Tr}, K_{Ts}, and K_{Td}. The total thrust coefficient of PJP is defined as $K_T = K_{Tr} + K_{Ts} + K_{Td}$. Q_r and Q_s are the torques of rotor and stator, respectively, and similarly, their corresponding dimensionless coefficients are K_{Qr} and K_{Qs}. Unless otherwise specified, all quantities are in dimensionless form using the rotor diameter D_r, the velocity $n\pi D_r$, and the fluid density ρ = 998.2 kg/m^3 as reference values. The dynamic viscosity of the fluid is μ = 0.001003 kg/(m·s).

3. Computational Overview

3.1. Numerical Method

As discussed in Section 1, the hybrid RANS/LES [34,35] is preferred to model the flow structures in PJP investigations. So, the IDDES [11,22,36] is employed to obtain the PJP propulsion and flow as it is a further developed formulation in the hybrid RANS/LES family. In the RANS portion, the SST $k-\omega$ [37] turbulence model is employed. Considering that in the IDDES strategy, the wall boundary regions are modeled by RANS, the transition model [38] with cross-flow effect is included in the RANS portion during the IDDES calculation. Moreover, to better capture the vortical structures and reduce the RANS portion effects, the embed LES (ELES) is adopted to model the flow structures. The ELES is also a hybrid RANS/LES form, and its main difference with IDDES is that the LES region can be manually assigned. In ELES, the Wall-Adapting Local Eddy-viscosity (WALE) [39] subgrid-scale model is employed to model the turbulence where the scale is smaller than the filtering scale. In this work, the body force and gravitational acceleration are ignored.

3.2. Mesh and Computational Setting

All meshes are hexahedral structural, the details of which are presented in Figure 4. For IDDES and ELES calculations, the y+ on the model surfaces is controlled close to 1. The computational fluid domain is shown in Figure 5. The whole computational domain is a cylinder with a diameter of $10D_r$ and a length of $30D_r$. The PJP is located at $10D_r$ downstream of the inlet. The domain was divided into four subdomains for the convenience of generating mesh and modeling the rotor revolution, including the rotor domain, stator domain, outer0 domain, and outer1 domain. These divided subdomains are connected by interior interfaces. Details of mesh count are given in Table 4. The discussion on the effects of mesh density on the prediction of PJP performance is not presented in the current investigation as the meshing strategy and mesh density are based on Refs. [9,11], where more information about the discussion on mesh density can be found. Moreover, the total mesh count is more than that in previous work [9,11]. The mesh of outer0 and outer domains are further refined for better capturing the vortices in ELES calculation, where the calculations in the rotor, stator, and outer0 domains are in LES.

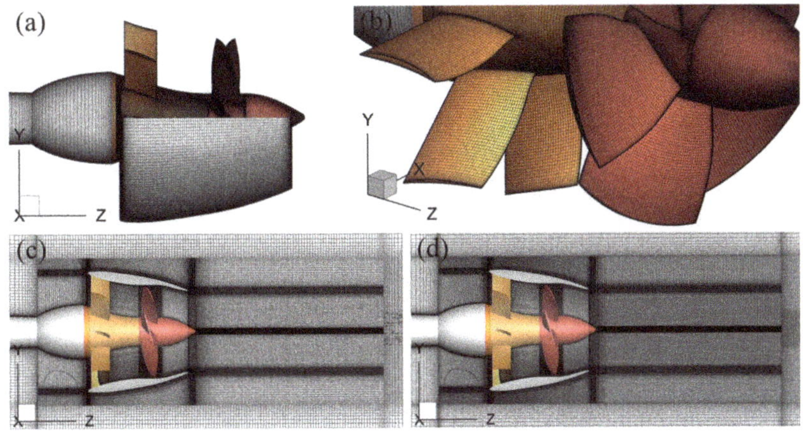

Figure 4. The PJP mesh (**a–c**), and the further refined mesh (**d**).

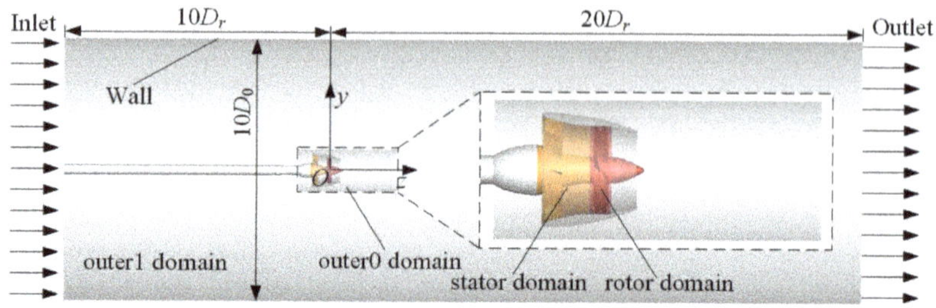

Figure 5. The computational domain and boundary conditions.

Table 4. Mesh count (milllions).

PJP Subdomains	S8R6	S9R7	S11R7	S11R9	S13R7	S13R9	S13R11
rotor	6.97	6.83	6.83	6.99	6.83	6.99	7.51
stator	5.73	6.14	6.76	6.76	6.96	6.96	6.96
outer0				8.15			
outer1				1.89			
outer0 (refined)				20.63			
outer1 (refined)				5.70			

Figure 5 presents the boundary conditions of calculation. The turbulence intensity and turbulent viscosity ratio of the inlet are 1% and 10%, respectively. These values are based on the experiments and a previous investigation [40]. The domain outlet is a pressure outlet where constant static pressure is placed. The lateral wall is specified as a zero shear stress wall boundary. The revolution speed of the rotor is $n = 20\ r/s$. Then, the inlet velocity is specified according to the advance coefficient J. Six advance coefficients from 0.2 to 1.2 with a step of 0.2 are considered in all IDDES calculations, while $J = 0.2$, 0.6, and 1.0 are mainly considered in the ELES calculations. According to previous investigations, the timestep Δt is the time of the rotor rotating one degree: $\Delta t = 1/(360n)$. For IDDES, the simulation time is eight times the period of the rotor. The discussion on the force fluctuation is based on the next ten times period calculation. The ELES simulation time is also eight times the period of the rotor, and the next five times period calculation is used to get the phase-averaged flow field. The pressure-based solver in all simulations is under the SIMPLEC algorithm. The bounded central differencing scheme and bounded second-order implicit scheme are used for the momentum and time terms, respectively.

3.3. Comparison with Experiment

The y+ distribution of PJP S8R6 at J = 1.0 condition, shown in Figure 6, suggests that the near-wall mesh satisfies the y+ requirement of IDDES and ELES. The instantaneous LES region in IDDES calculation identified by the ratio of l_{IDDES}/l_{RANS} indicates that the concerned regions are in LES resolution. In ELES calculations, the stator, rotor, and outer 0 regions are all in LES resolution as these regions are further refined and manually specified. The comparison between the IDDES and experimental results is given in Figure 7, where the numerical results are the time-averaged values of the last two-period. The deviation is mainly shown in torque coefficient and has the largest value of 6.8% at $J = 1.2$. The torque deviation is potentially caused by the insufficient prediction of the friction on rotor blades. The deviation shows a larger value owing to a larger contribution of friction to torque at a high advance coefficient. Overall, the IDDES prediction has an acceptable deviation with the experiment.

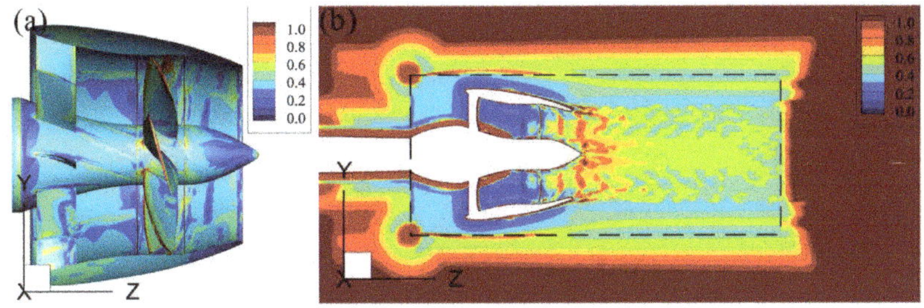

Figure 6. (**a**) the y+ distribution at $J = 1.0$, and (**b**) the LES region in corresponding IDDES calculation, identified by the ratio l_{IDDES}/l_{RANS}.

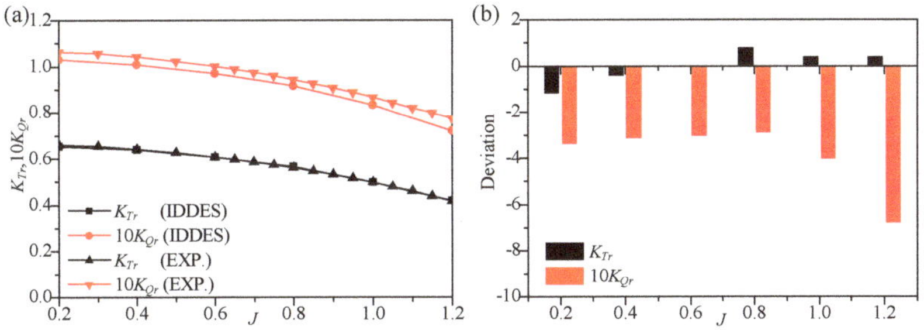

Figure 7. The comparison between the numerical results and the experiment: (**a**) hydrodynamic curves, (**b**) deviation.

4. Results and Discussion

4.1. Propulsion

Though the main purpose of changing the blade number is to inspect the thrust fluctuation and the flow characteristics, a check on the PJP loading is also needed. The total loading change of PJP is negligible when changing the blade numbers of the stator and rotor with constant solidities. As shown in Figures 8 and 9, the notable relative deviations are mainly shown in the thrust and torque of the stator. The stator loading is also not affected when only changing the rotor blade number. However, the rotor loading has a noticeable but very slight change when only changing the stator blade number. Comparatively, the relative deviation on the duct loading is very small when changing the blade numbers. According to Figure 9c, increasing the blade number of the stator or rotor, the zero torque balance point moves towards the high advance coefficient. However, this movement is almost negligible.

The rotor undergoes a non-uniform oncoming flow generated by the stator and shows noticeable fluctuations of thrust and torque, as presented in Figure 10. The stator also has an obvious torque fluctuation owing to the rotor revolution. Comparatively, the thrust fluctuation is dominated by the rotor in the axial direction, while in the side directions, it is dominated by the side forces of the duct, as depicted in Figure 11, where $K_{Tr}(x)$, $K_{Tr}(y)$, $K_{Ts}(x)$, $K_{Ts}(y)$, $K_{Td}(x)$, and $K_{Td}(y)$ indicate the side forces in the x and y directions of the rotor, stator, and duct. The duct has almost an equivalent fluctuation level of side forces to the axial thrust fluctuation of the rotor. However, the duct side force shows slight changes in fluctuant degree between different advance coefficients. Figures 12 and 13 give the RMSE (root mean squared error) values of all mentioned dimensionless coefficients of all PJP models. K_{Tr}, $K_{Td}(x)$, and $K_{Td}(y)$ have larger RMSE values under different blade numbers. The main difference is that the fluctuation of K_{Tr} considerably increases when PJP works at a low advance coefficient. K_{Qr} and K_{Qs} also have almost equivalent RMSE values, and they notably increase at heavy loading. Nevertheless, the fluctuation of K_{Qs}

and its change tendency with the PJP loading condition is not affected when changing the blade numbers.

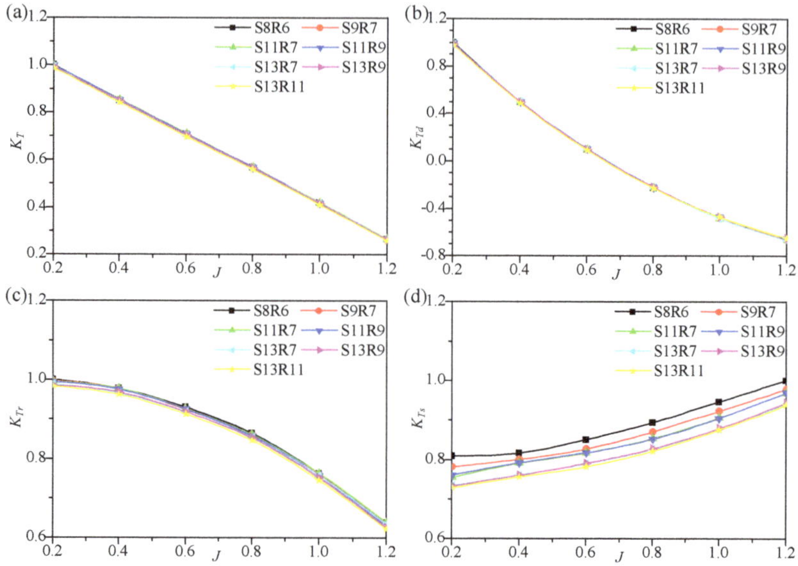

Figure 8. Thrust coefficients of the rotor, stator, duct, and the total thrust coefficient: (**a**) K_T, (**b**) K_{Td}, (**c**) K_{Tr}, (**d**) K_{Ts}, where K_T, K_{Tr}, and K_{Td} are normalized by the corresponding ones of PJP S8R6 at $J = 0.2$ while K_{Ts} is normalized by the corresponding one of PJP S8R6 at $J = 1.2$.

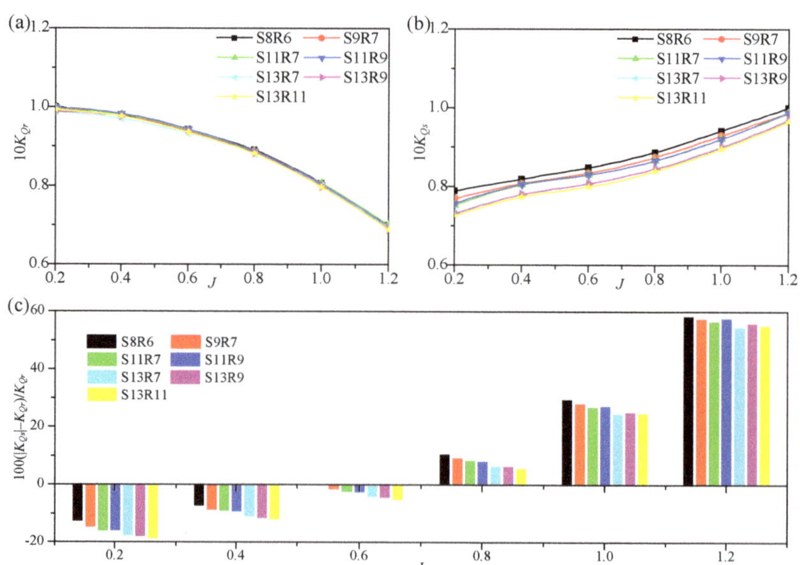

Figure 9. Torque coefficients of the rotor and stator, and the torque coefficient difference between the rotor and stator: (**a**) $10K_{Qr}$, (**b**) $10K_{Qs}$, (**c**) $100(|K_{Qs}| - K_{Qr})/K_{Qr}$, where K_{Qr} is normalized by the corresponding one of PJP S8R6 at $J = 0.2$ while K_{Qs} is normalized by the corresponding one of PJP S8R6 at $J = 1.2$.

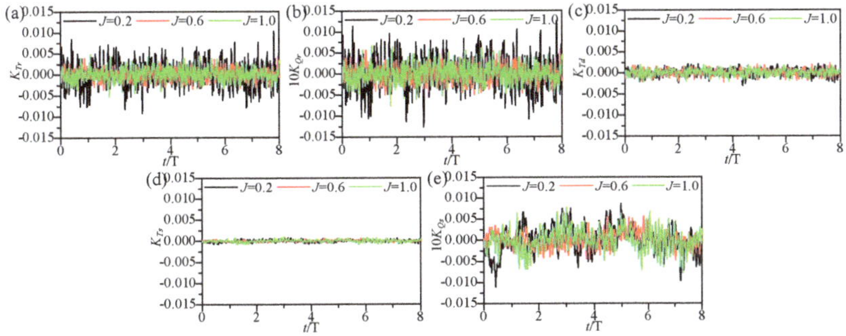

Figure 10. Thrust and torque fluctuations of the rotor, stator, and duct of the PJP model S8R6: (**a**) K_{Tr}, (**b**) $10K_{Qr}$, (**c**) K_{Td}, (**d**) K_{Ts}, (**e**) $10K_{Qs}$.

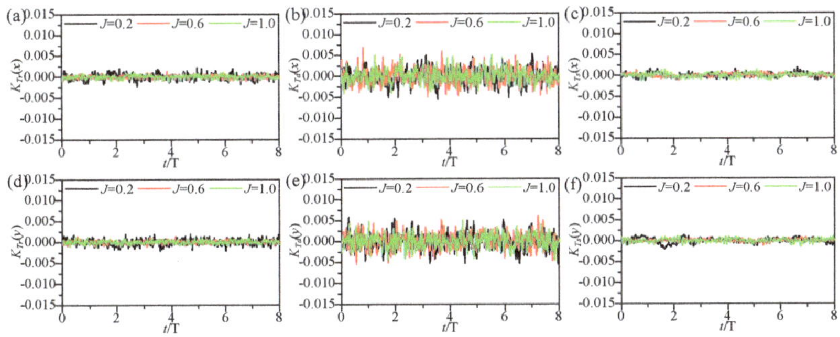

Figure 11. Side force fluctuations of the rotor, stator, and duct of the PJP model S8R6: (**a**) $K_{Tr}(x)$, (**b**) $K_{Td}(x)$, (**c**) $K_{Ts}(x)$, (**d**) $K_{Tr}(y)$, (**e**) $K_{Td}(y)$, (**f**) $K_{Ts}(y)$.

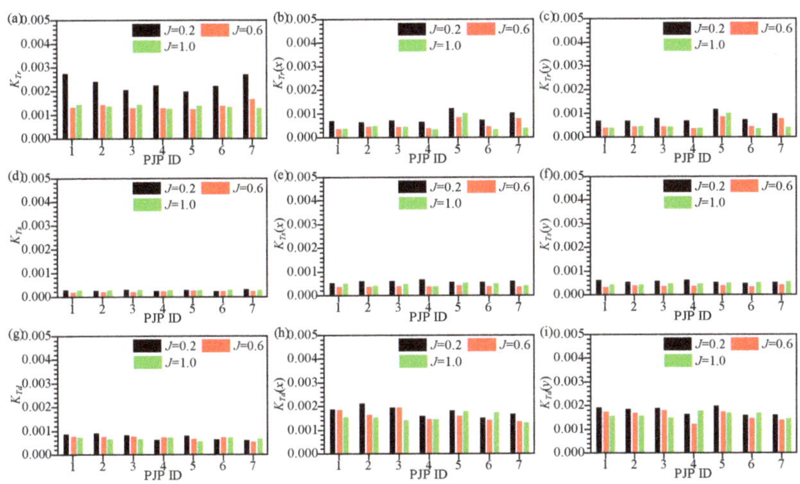

Figure 12. RMSE values of the fluctuation component of thrust and side forces: (**a**) K_{Tr}, (**b**) $K_{Tr}(x)$, (**c**) $K_{Tr}(y)$, (**d**) K_{Ts}, (**e**) $K_{Ts}(x)$, (**f**) $K_{Ts}(y)$, (**g**) K_{Td}, (**h**) $K_{Td}(x)$, (**i**) $K_{Td}(y)$.

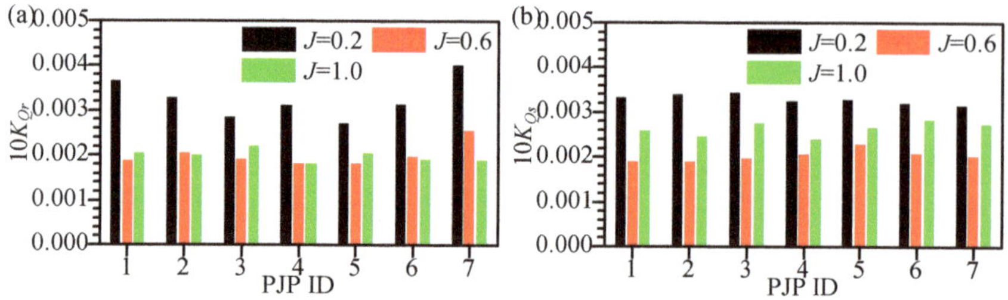

Figure 13. RMSE values of the fluctuation component of torque: (**a**) $10K_{Qr}$, (**b**) $10K_{Qs}$.

As shown in Figure 14, K_{Tr} has a dominant peak at the frequency of four times BPF (blade passing frequency) in the frequency domain while $K_{Td}(x)$ does not show a noticeable peak. It should be noted that the value of BPF is different for different PJP models. Unlike the fluctuation of K_{Td} in the side directions, the fluctuation of K_{Tr} notably decreases in the whole frequency band when the PJP loading decreases. The range of thrust fluctuation has a considerable increment with increasing the PJP loading when the blade numbers of the rotor and stator are close, as depicted in Figure 15, where the PJP models, S8R6, S9R7, S11R9, and S13R11 have large thrust fluctuations at the heavy loading ($J = 0.2$). It is not beneficial for the thrust fluctuation in the axial direction when the rotor blade number is close to the stator's. In the frequency domain, as shown in Figure 16, the number of dominant peaks decrease or disappear when the blade numbers between the rotor and stator are coprime. At the light loading ($J = 1.0$), a visible high peak occurs at BPF or 2BPF when the difference in blade number between the rotor and stator is large, while at the heavy loading ($J = 0.2$), the frequency curve has more peaks when the blade numbers of the rotor and stator are close. According to Figures 17 and 18, the fluctuant range of the duct side force is not affected when changing the blade number. However, unlike the original model, the fluctuation of the duct side force produces visible peaks in the frequency domain, particularly when the blade numbers between the rotor and stator have a large difference. Overall, focusing on weakening the characteristic peaks of thrust fluctuation in the frequency domain, adopting a rotor and a stator with coprime blade numbers and a large difference in blade number is better. However, this configuration will exaggerate the fluctuation of side force on the duct.

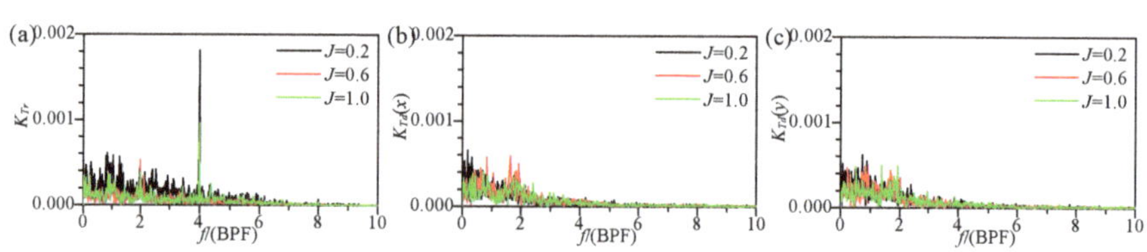

Figure 14. Frequency domain curves of K_{Tr}, $K_{Td}(x)$ and $K_{Td}(y)$ of the PJP model S8R6: (**a**) K_{Tr}, (**b**) $K_{Td}(x)$, (**c**) $K_{Td}(y)$.

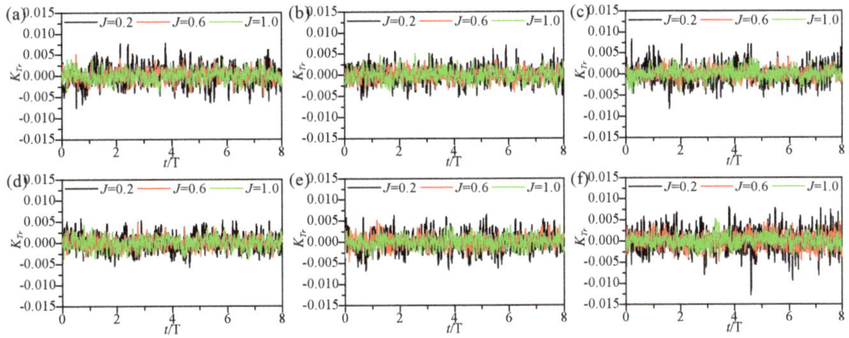

Figure 15. Time domain curves of fluctuation component of K_{Tr}. (from (**a–f**) corresponding to PJP model ID 2 to 7).

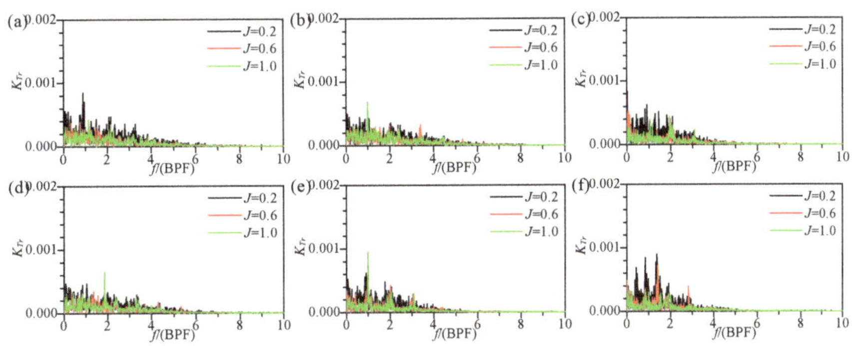

Figure 16. Frequency domain curves of fluctuation component of K_{Tr}. (from (**a–f**) corresponding to PJP model ID 2 to 7).

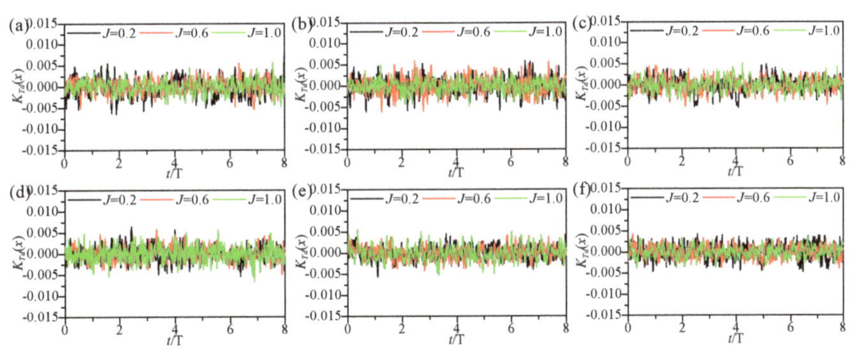

Figure 17. Time domain curves of fluctuation component of $K_{Td}(x)$. (from (**a–f**) corresponding to PJP model ID 2 to 7).

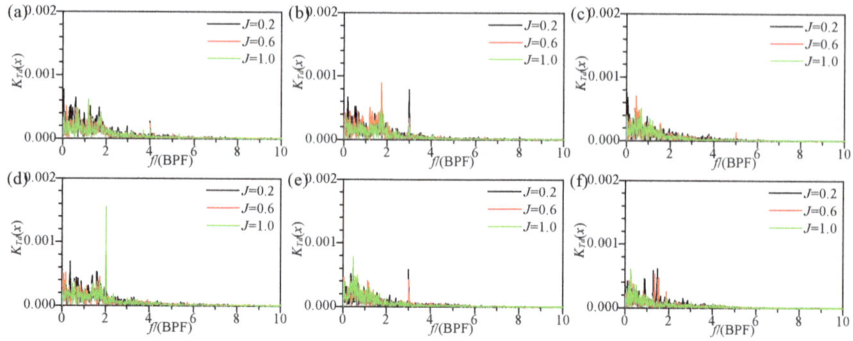

Figure 18. Frequency domain curves of fluctuation component of $K_{Td}(x)$. (from (**a–f**) corresponding to PJP model ID 2 to 7).

4.2. Flow Field

The pressure distribution on blades determines the rotor thrust, and its low value mainly distributes on the suction side and the region under the effects of the tip clearance leakage flow. Figure 19a shows the pressure on the suction side of the rotor blade of the original PJP model at different advance coefficients. The low-pressure area on the blade leading edge decreases as the loading decreases. At the lightest loading, the low pressure mainly exists on the inner radii part of the leading edge. The tip clearance leakage flow causes considerable low pressure on the suction side near the blade tip. This low-pressure area gradually decreases and disappears as the advance coefficient goes up. Changing the blade number under a constant solidity means the total loading on the rotor or the pre-swirl effect of the stator does not change. Hence, increasing the rotor blade numbers decreases the loading per blade. As shown in Figure 19b, the pattern of pressure distribution does not change. For the low-pressure region on the suction side caused by the tip clearance leakage flow, increasing the blade number decreases its size in the chord direction but does not change its radial size. The tip clearance leakage flow is driven by the pressure difference between the pressure side and the suction side. Consequently, its relative beginning place along the blade chord is not changed.

Figure 20 presents the C_p iso-surface lower than −0.25. The iso-surface of the low-pressure region on the leading edge of the rotor blade decreases as the loading decreases. Though the low-pressure area caused by the tip clearance leakage flow gradually thins as the loading decreases, it maintains a long distance downstream at the high advance coefficient. Besides, low pressure also occurs on the duct leading edge due to the PJP duct undergoing a large angle of attack in the open water condition. Like the pressure distribution on the blade surface, the main difference is the change of the low-pressure region in the tip clearance leakage flow under different blade numbers. Increasing the blade number weakens the effect of low-pressure in tip clearance leakage flow on the rotor suction side and decreases this low-pressure region size. However, this low-pressure region does not shorten its evolution distance downstream.

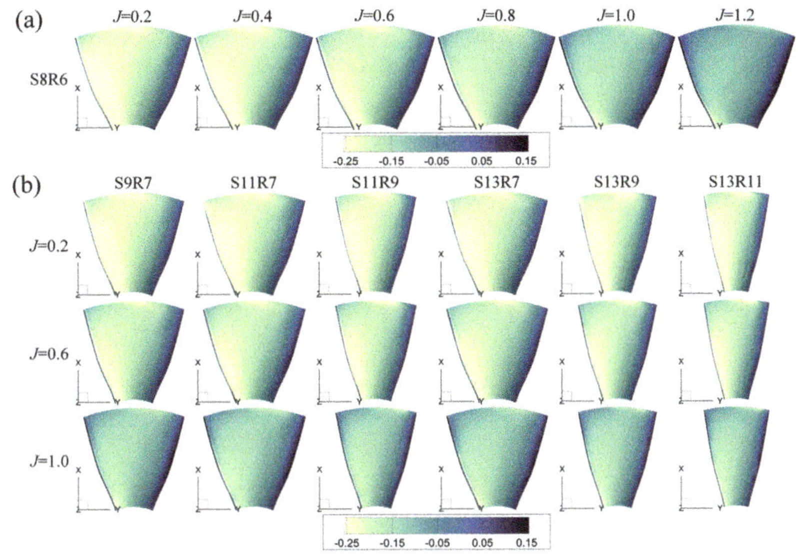

Figure 19. Pressure distribution on the suction side of a rotor blade. (**a**) PJP model S8R6, (**b**) derived PJP models ID 2-7.

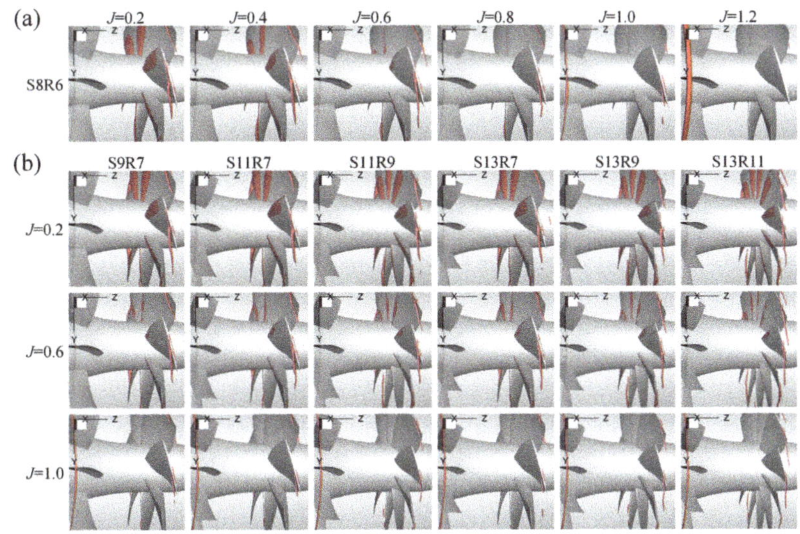

Figure 20. Pressure iso-surfaces ($C_p = -0.25$) in the PJP flow region. (**a**) PJP model S8R6, (**b**) derived PJP models ID 2-7.

The stator has a pre-swirl effect on the rotor oncoming flow. As shown in Figure 21, increasing the advance coefficient, the inner radii of the stator blades have a considerable circumferentially wide wake, whose radial length and circumferential width depend on the flow velocity through the stator blade cascade. Besides, at the heaviest loading ($J = 0.2$), the duct leading edge forms a flow separation on the inside, exaggerating the non-uniformity of the stator wake. This phenomenon further increases the thrust fluctuation of the rotor and causes more peaks in the frequency domain when the blade number of the rotor is close to the stator's, as discussed in Section 4.1. At the middle ($J = 0.6$) and light ($J = 1.0$)

loadings, the aforementioned wide wake of the stator blade mainly has a notable decrease in the circumferential direction when increasing the blade number of the stator. Moreover, the velocity gradient in this wake is also weakened.

After the duct, the PJP forms a high-velocity wake, as shown in Figure 22. In the wake inner region, there exists distinguishable wake topologies of the stator and rotor (hereafter, the two wake topologies are named stator wake and rotor wake, respectively). Different style circles are used to mark the radial areas of these distinguishable wake topologies. According to the previous investigations, the two wakes come from the trailing edge wake of the inner radii blade [11,12]. The rotor wake depends on the rotor loading, while the stator wake depends on the flow velocity through the stator. However, the two wakes are concurrently weakened as the advance coefficient increases owing to their interaction [12]. Unlike the stator wake between the stator and rotor, as shown in Figure 21, increasing the stator blade number decreases the radial size of the stator wake in the PJP wake. Increasing the rotor blade number has the same tendency, but the rotor wake's effect on the stator wake is weakened. As shown in Figure 22b, the stator wake is more distinguishable when increasing the rotor blade number and holding the stator blade number.

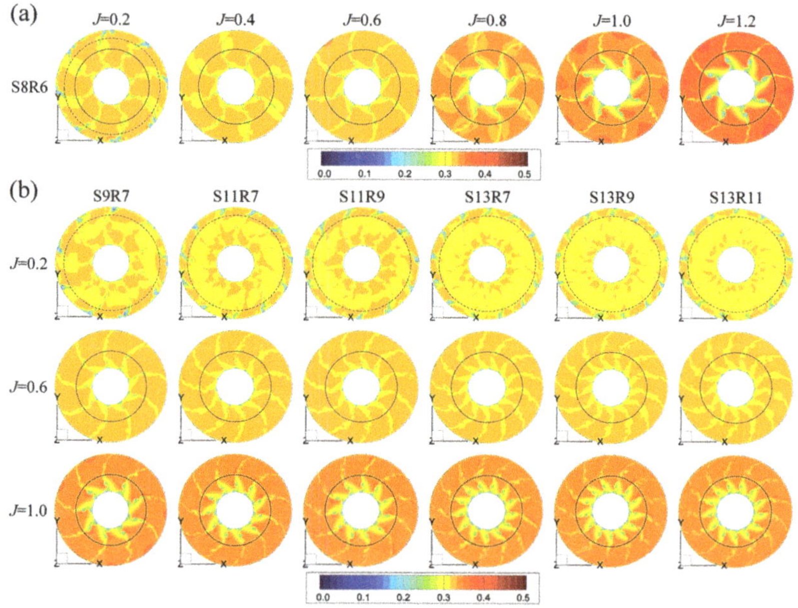

Figure 21. Velocity magnitude V_m at $z = -0.3D_r$ slice, where the radii of the dashed circle and the solid circle are $0.45D_r$ and $0.35D_r$, respectively. (**a**) PJP model S8R6, (**b**) derived PJP models ID 2-7.

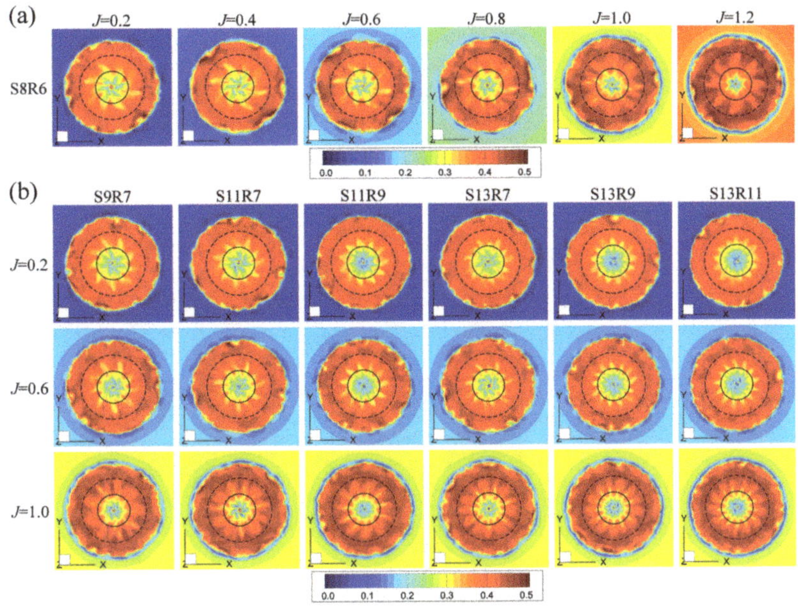

Figure 22. Velocity magnitude V_m at $z = 0.5D_r$ slice, where the radii of dashed circle and solid circle are $0.30D_r$ and $0.15D_r$, respectively. (**a**) PJP model S8R6, (**b**) derived PJP models ID 2-7.

4.3. Vortices

The vortices in the PJP flow field are identified by the Ω method [41,42]. Figure 23 presents the vortical structures of the original PJP model S8R6 at loading $J = 0.6$, and the size of iso-surface shows the vortex intensity. More discussion on these vortices under the effects of the mesh and turbulence model and the evolution mechanism is documented in Refs. [11,12], respectively. Here, the interaction between the duct shedding vortex (DSV) and tip clearance leakage vortex (TCLV) near the duct outlet, the interaction between the wake vortices of the stator and rotor, and hub, are mainly discussed under the effects of blade numbers, where the rotor wake vortices represent the vortices shedding from the trailing edge of the rotor blade, particularly the ones in the inner radii. More about the discrimination of these vortices are in the investigation [12].

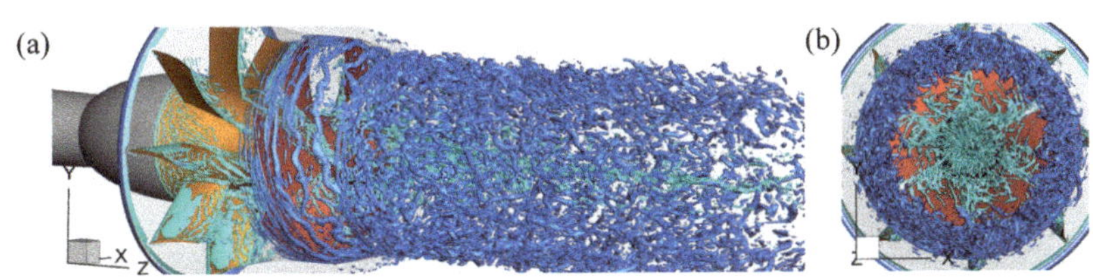

Figure 23. (**a**) vortices identified by the Ω method in the flow field of PJP model S8R6 at $J = 0.6$; (**b**) back view.

Increasing the rotor blade number results in a blade loading decrement. As shown in Figure 24, the TCLV intensity decreases when the rotor has more blades under the same constant. Hence, the TCLV is more unstable during the evolving downstream along the contracted duct. Near the duct shrinkage stage, TCLVs become twisted. They lost their helical topologies as their intensities are not enough to maintain their shape (as depicted in Figure 24 when the rotor blade numbers are nine and eleven). Then, they are broken into 'Ω' or 'C' shape segments. These segment vortices twine or bridge each other. Increasing the rotor loading exaggerates this phenomenon. The DSVs bridge to these upstream segment vortices at the duct outlet and many secondary vortices occur. According to the investigation on propeller vortical structure [12,43], these secondary vortices include the ones induced by primary vortices and the ones developing from the DSVs. The primary structure of TCLV is not distinguishable after the duct outlet. When the TCLV intensity is high enough to maintain its primary vortex shape, the essential break of the primary vortex comes from the DSV interaction and the development of the secondary vortices. So, the primary structure is distinguishable in the near field after the duct outlet owing to an incomplete break from secondary vortices at this time.

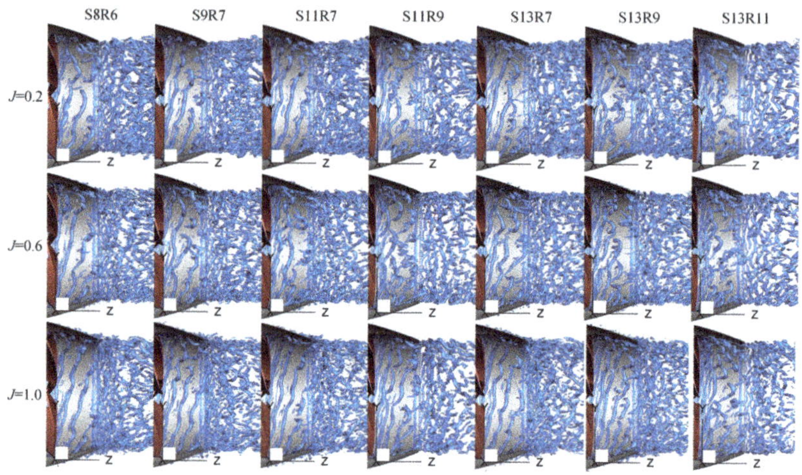

Figure 24. Interaction between the tip clearance leakage vortex and the duct shed vortex.

The stator wake includes many wake vortices interacting with the downstream rotor, as presented in Figure 25. The stator wake vortices depend on the flow velocity through them. Every stator blade forms a high-intensity vortex evolving downstream when the stator undergoes a high-velocity flow. The flow separation occurs near the duct leading edge at the heaviest loading condition and generates many vortices. As documented in the investigation [12], the aforementioned high-intensity wake vortex of the stator blade plays the main role in interacting with the rotor blade and rotor blade wake vortices.

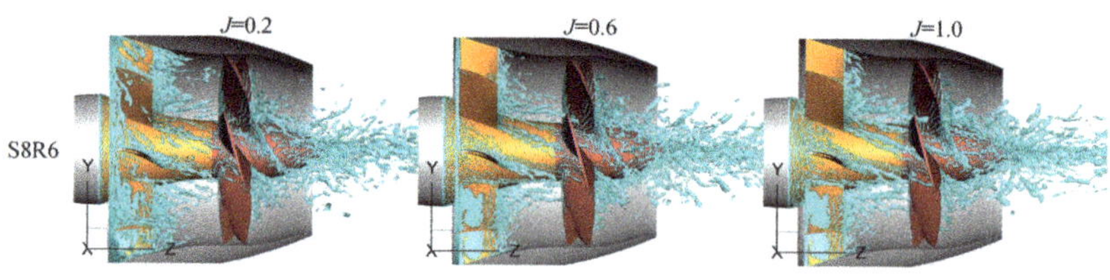

Figure 25. Interaction between the blade wake vortices of the rotor and stator of PJP model S8R6.

Figure 26 shows the wake vortices of the rotor and stator under different rotor blade numbers. The evolution of rotor blade wake vortices is considerably affected due to the decrement of the intensity of per blade wake vortex. Though the total loading on the rotor is not changed, the wake vortices quickly dissipate downstream owing to the lower intensity when increasing the rotor blade number. The downstream rotor does not show noticeable effects on the wake vortices of the stator. As shown in Figure 27, increasing the stator blade number means the wake vortices per blade are weakened. However, the stator blade also has considerable separating vortices on its suction side owing to the constant camber. Increasing the stator blade number mainly affects the high-intensity vortices in the blade trailing edge wake. As the flow velocity through the stator determines these high-intensity vortices, the rotor wake vortices show a noticeable difference caused by the stator wake vortices when the PJP is at a high advance coefficient. A high-intensity vortex occurs in the wake vortices of the stator blade at the high advance coefficient ($J = 1.0$) when the blade number is less ($Z_s = 9$). This high-intensity vortex has a longer evolution distance downstream and successfully interacts with the rotor blade and its wake vortices. After interacting with the rotor wake vortices, the high-intensity stator wake vortices are broken into segments but show distinguishable structures after the duct outlet.

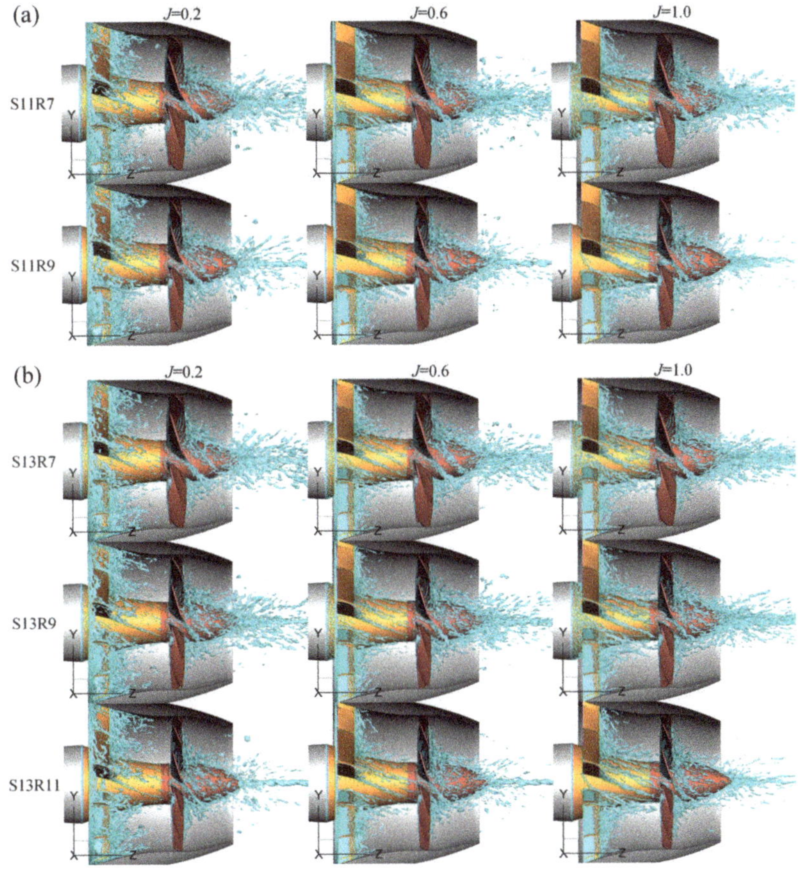

Figure 26. Interaction between the blade wake vortices of the rotor and stator of PJP models withholding the stator blade numbers: (**a**) different rotor blade numbers under 11 stator blades, (**b**) different rotor blade numbers under 13 stator blades.

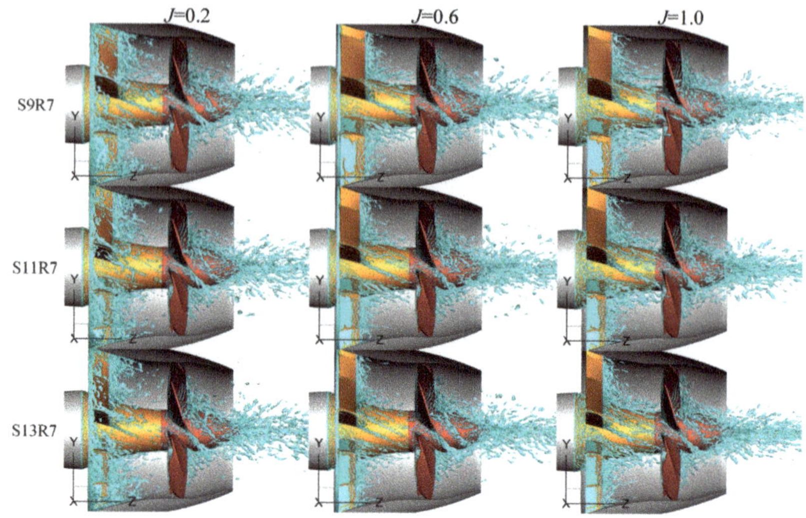

Figure 27. Interaction between the blade wake vortices of the rotor and stator of PJP models withholding the rotor blade numbers.

After the duct outlet, the wake vortices of the rotor and stator considerably affect the evolution of hub vortices, where the wake vortices of the rotor blade play the dominant role. The wake vortices can be weakened by increasing the blade number. Increasing the stator blade number results in a longer distance of blade wake vortices in PJP wake and a more unstable topology of hub vortices, which can be clearly evidenced from the Figure 28 where the rotor blade number is seven or nine. The effects of stator blade wake vortices on rotor wake vortices are decreased, and hence the rotor blade wake vortices can more affect the hub vortices. This phenomenon is noticeable at the low advance coefficient. However, it presents an inverse phenomenon at the high advance coefficient owing to the high-intensity vortices in the stator wake vortices when the stator blade number is less. As the rotor blade wake vortices play the main role in affecting the stability of hub vortices, increasing the rotor blade number results in fewer rotor wake vortices around the main hub vortex tube. Then, the hub vortex tube can maintain its stable primary vortex structure long distance.

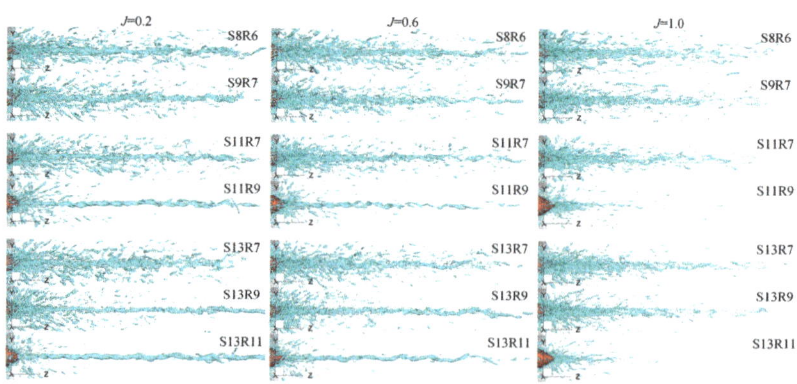

Figure 28. Hub wake vortices under different rotor and stator blade numbers.

5. Conclusions

In the present study, the effects of blade number on the propulsion, flow, and vortices of a pre-swirl stator pump-jet propulsor are concerned under constant solidities. The original has an eight-bladed stator and a six-bladed rotor. The stator and rotor with coprime blades numbers are derived for new PJP models. The thrust fluctuation, flow field, vortical structures of these PJP models are discussed and analyzed after discussing the derivation method of these PJP models, the numerical method, and the verification of the original PJP model. The conclusions can be drawn as follows:

(1) Changing the blade number does not affect the PJP performance when the solidities of the stator and rotor are not changed;
(2) Adopting coprime blade numbers of the stator and rotor decreases the peak amplitude of the rotor thrust fluctuation in the frequency domain. Nevertheless, the rotor thrust shows a large fluctuant range when the rotor blade number is close to the stators;
(3) The duct side force shows a noticeable fluctuation, and a large difference in blade numbers between the stator and rotor causes high peaks in the frequency domain;
(4) Increasing the rotor blade number decreases the effect range of tip clearance leakage flow on the suction side in the chord direction, while the effect range does not change in the radial direction. A rotor with more blades also results in a radially narrow core region in its wake. The intensity of the tip clearance leakage vortex is also decreased, and hence this vortex prematurely becomes more unstable before it is out of the duct. However, the total break of this vortex is completed after the interaction with the duct shedding vortex. The rotor wake vortices are notably weakened, and hence their effects on the stability of hub vortices decrease. Compared with the effects of stator wake vortices, the rotor wake vortices dominate the effects on the instability of hub vortices. Consequently, the hub vortices maintain a stable primary vortical structure with a long distance when increasing the rotor blade;
(5) Increasing the stator blade number decreases the velocity gradient in the inner radii of the stator wake and results in a radial narrow stator wake in the rotor downstream. The decreased loading per stator blade weakens the high-intensity vortex in the stator wake vortices. Hence, the effects of this high-intensity vortex on the rotor wake vortices decrease.

Author Contributions: This study is the result of collaborative teamwork. Conceptualization, methodology, investigation, writing—original draft, H.L.; project administration, writing—review and editing, Q.H.; supervision, G.P.; visualization, X.D.; Software, F.L. All authors have read and agreed to the published version of the manuscript.

Funding: This research was funded by the National Natural Science Foundation of China under Grant No. 51979226, and the Fundamental Research Funds for the Central Universities under [Grant No. 3102019HHZY030019 and 3102020HHZY030018]. The APC was funded by the National Natural Science Foundation of China under Grant No. 51979226.

Institutional Review Board Statement: Not applicable.

Informed Consent Statement: Not applicable.

Data Availability Statement: The data presented in this study are available on request from the corresponding author upon reasonable request.

Conflicts of Interest: The authors declare no conflict of interest.

References

1. Suryanarayana, C.; Satyanarayana, B.; Ramji, K.; Saiju, A. Experimental evaluation of pumpjet propulsor for an axi-symmetric body in wind tunnel. *Int. J. Nav. Archit. Ocean Eng.* **2010**, *2*, 24–33. [CrossRef]
2. Suryanarayana, C.; Satyanarayana, B.; Ramji, K. Performance evaluation of an underwater body and pumpjet by model testing in cavitation tunnel. *Int. J. Nav. Archit. Ocean Eng.* **2010**, *2*, 57–67. [CrossRef]

3. Suryanarayana, C.; Satyanarayana, B.; Ramji, K.; Rao, M.N. Cavitation studies on axi-symmetric underwater body with pumpjet propulsor in cavitation tunnel. *Int. J. Nav. Archit. Ocean Eng.* **2010**, *2*, 185–194. [CrossRef]
4. Ivanell, S. *Hydrodynamic Simulation of a Torpedo with Pumpjet Propulsion System;* Royal Institute of Technology: Stockholm, Sweden, 2001.
5. Shirazi, A.T.; Nazari, M.R.; Manshadi, M.D. Numerical and experimental investigation of the fluid flow on a full-scale pump jet thruster. *Ocean Eng.* **2019**, *182*, 527–539. [CrossRef]
6. Li, H.; Pan, G.; Huang, Q. Transient analysis of the fluid flow on a pumpjet propulsor. *Ocean Eng.* **2019**, *191*, 106520. [CrossRef]
7. Yu, H.; Zhang, Z.; Hua, H. Numerical investigation of tip clearance effects on propulsion performance and pressure fluctuation of a pump-jet propulsor. *Ocean Eng.* **2019**, *192*, 106500. [CrossRef]
8. Yu, H.; Duan, N.; Hua, H.; Zhang, Z. Propulsion performance and unsteady forces of a pump-jet propulsor with different pre-swirl stator parameters. *Appl. Ocean Res.* **2020**, *100*, 102184. [CrossRef]
9. Huang, Q.; Li, H.; Pan, G.; Dong, X. Effects of duct parameter on pump-jet propulsor unsteady hydrodynamic performance. *Ocean Eng.* **2021**, *221*, 108509. [CrossRef]
10. Qiu, C.; Pan, G.; Huang, Q.; Shi, Y. Numerical analysis of unsteady hydrodynamic performance of pump-jet propulsor in oblique flow. *Int. J. Nav. Archit. Ocean Eng.* **2020**, *12*, 102–115. [CrossRef]
11. Li, H.; Huang, Q.; Pan, G.; Dong, X. The transient prediction of a pre-swirl stator pump-jet propulsor and a comparative study of hybrid RANS/LES simulations on the wake vortices. *Ocean Eng.* **2020**, *203*, 107224. [CrossRef]
12. Li, H.; Huang, Q.; Pan, G.; Dong, X. Wake instabilities of a pre-swirl stator pump-jet propulsor. *Phys. Fluids* **2021**, *33*, 085119.
13. Shi, Y.; Pan, G.; Huang, Q.; Du, X. Numerical Simulation of Cavitation Characteristics for Pump-jet Propeller. *J. Phys.* **2015**, *640*, 012035. [CrossRef]
14. Pan, G.; Lu, L.; Sahoo, P.K. Numerical simulation of unsteady cavitating flows of pumpjet propulsor. *Ships Offshore Struct.* **2016**, *11*, 64–74. [CrossRef]
15. Sun, Y.; Liu, W.; Li, T.Y. Numerical investigation on noise reduction mechanism of serrated trailing edge installed on a pump-jet duct. *Ocean Eng.* **2019**, *191*, 106489. [CrossRef]
16. Wang, C.; Weng, K.; Guo, C.; Gu, L. Prediction of hydrodynamic performance of pump propeller considering the effect of tip vortex. *Ocean Eng.* **2019**, *171*, 259–272. [CrossRef]
17. Wang, C.; Weng, K.; Guo, C.; Chang, X.; Gu, L. Analysis of influence of duct geometrical parameters on pump jet propulsor hydrodynamic performance. *J. Mar. Sci. Technol.* **2020**, *25*, 640–657. [CrossRef]
18. Kumar, P.; Mahesh, K. Large eddy simulation of propeller wake instabilities. *J. Fluid Mech.* **2017**, *814*, 361–396. [CrossRef]
19. Ahmed, S.; Croaker, P.; Doolan, C.J. On the instability mechanisms of ship propeller wakes. *Ocean Eng.* **2020**, *213*, 107609. [CrossRef]
20. Posa, A.; Broglia, R.; Balaras, E. The wake structure of a propeller operating upstream of a hydrofoil. *J. Fluid Mech.* **2020**, *904*. [CrossRef]
21. Jing, Z.; Ducoin, A. Direct numerical simulation and stability analysis of the transitional boundary layer on a marine propeller blade. *Phys. Fluids* **2020**, *32*, 124102. [CrossRef]
22. Long, Y.; Han, C.; Long, X.; Ji, B.; Huang, H. Verification and validation of Delayed Detached Eddy Simulation for cavitating turbulent flow around a hydrofoil and a marine propeller behind the hull. *Appl. Math. Model.* **2021**, *96*, 382–401. [CrossRef]
23. Li, H.; Pan, G.; Huang, Q.; Shi, Y. Numerical Prediction of the Pumpjet Propulsor Tip Clearance Vortex Cavitation in Uniform Flow. *J. Shanghai Jiaotong Univ.* **2020**, *25*, 352–364. [CrossRef]
24. Yuan, J.; Chen, Y.; Wang, L.; Fu, Y.; Zhou, Y.; Xu, J.; Lu, R. Dynamic analysis of cavitation tip vortex of pump-jet propeller based on DES. *Appl. Sci.* **2020**, *10*, 5998. [CrossRef]
25. Su, Z.; Shi, S.; Huang, X.; Rao, Z.; Hua, H. Vibro-acoustic characteristics of a coupled pump-jet–Shafting system–SUBOFF model under distributed unsteady hydrodynamics by a pump-jet. *Ocean Eng.* **2021**, *235*, 109429. [CrossRef]
26. Ji, X.Q.; Dong, X.Q.; Yang, C.J. Attenuation of the Tip-Clearance Flow in a Pump-Jet Propulsor by Thickening and Raking the Tips of Rotor Blades: A Numerical Study. *Appl. Ocean Res.* **2021**, *113*, 102723. [CrossRef]
27. Gong, J.; Guo, C.Y.; Zhao, D.G.; Wu, T.C.; Song, K.W. A comparative DES study of wake vortex evolution for ducted and non-ducted propellers. *Ocean Eng.* **2018**, *160*, 78–93. [CrossRef]
28. Villa, D.; Gaggero, S.; Tani, G.; Viviani, M. Numerical and experimental comparison of ducted and non-ducted propellers. *J. Mar. Sci. Eng.* **2020**, *8*, 257. [CrossRef]
29. Stark, C.; Shi, W.; Troll, M. Cavitation funnel effect: Bio-inspired leading-edge tubercle application on ducted marine propeller blades. *Appl. Ocean Res.* **2021**, *116*, 102864. [CrossRef]
30. Felli, M.; Guj, G.; Camussi, R. Effect of the number of blades on propeller wake evolution. *Exp. Fluids* **2008**, *44*, 409–418. [CrossRef]
31. Felli, M.; Camussi, R.; Di Felice, F. Mechanisms of evolution of the propeller wake in the transition and far fields. *J. Fluid Mech.* **2011**, *682*, 5–53. [CrossRef]
32. Malmir, R. A CFD study on the correlation between the skew angle and blade number of hydrodynamic performance of a submarine propeller. *J. Braz. Soc. Mech. Sci. Eng.* **2019**, *41*, 1–14. [CrossRef]
33. Wang, L.Z.; Guo, C.Y.; Su, Y.M.; Wu, T.C. A numerical study on the correlation between the evolution of propeller trailing vortex wake and skew of propellers. *Int. J. Nav. Archit. Ocean Eng.* **2018**, *10*, 212–224. [CrossRef]

34. Spalart, P.R. Comments on the feasibility of LES for wings, and on a hybrid RANS/LES approach. In Proceedings of the First AFOSR International Conference on DNS/LES, Ruston, LA, USA, 4–8 August 1997; p. 1.
35. Strelets, M. Detached eddy simulation of massively separated flows. In Proceedings of the 39th Aerospace Sciences Meeting and Exhibit, Reno, NV, USA, 8–11 January 2001; p. 879.
36. Gritskevich, M.S.; Garbaruk, A.V.; Schütze, J.; Menter, F.R. Development of DDES and IDDES formulations for the k-ω shear stress transport model. *Flow Turbul. Combust.* **2012**, *88*, 431–449. [CrossRef]
37. Menter, F.R. Two-equation eddy-viscosity turbulence models for engineering applications. *AIAA J.* **1994**, *32*, 1598–1605. [CrossRef]
38. Menter, F.R.; Smirnov, P.E.; Liu, T.; Avancha, R. A one-equation local correlation-based transition model. *Flow Turbul. Combust.* **2015**, *95*, 583–619. [CrossRef]
39. Nicoud, F.; Ducros, F. Subgrid-scale stress modelling based on the square of the velocity gradient tensor. *Flow Turbul. Combust.* **1999**, *62*, 183–200. [CrossRef]
40. Li, H.; Huang, Q.; Pan, G.; Dong, X. Assessment of transition modeling for the unsteady performance of a pump-jet propulsor in model scale. *Appl. Ocean Res.* **2021**, *108*, 102537. [CrossRef]
41. Liu, C.; Wang, Y.; Yang, Y.; Duan, Z. New omega vortex identification method. *Sci. China Phys. Mech. Astron.* **2016**, *59*, 1–9. [CrossRef]
42. Liu, J.m.; Wang, Y.Q.; Gao, Y.s.; Liu, C. Galilean invariance of Omega vortex identification method. *J. Hydrodyn.* **2019**, *31*, 249–255. [CrossRef]
43. Di Mascio, A.; Muscari, R.; Dubbioso, G. On the wake dynamics of a propeller operating in drift. *J. Fluid Mech.* **2014**, *754*, 263–307. [CrossRef]

Journal of
Marine Science and Engineering

MDPI

Article

Influence of Various Stator Parameters on the Open-Water Performance of Pump-Jet Propulsion

Fuzheng Li [1,2], Qiaogao Huang [1,2,*], Guang Pan [1,2], Denghui Qin [1,2] and Han Li [1,2]

[1] School of Marine Science and Technology, Northwestern Polytechnical University, Xi'an 710072, China; jinglelfz@mail.nwpu.edu.cn (F.L.); panguang@nwpu.edu.cn (G.P.); qindenghui@mail.nwpu.edu.cn (D.Q.); Lihan_nwpu@mail.nwpu.edu.cn (H.L.)

[2] Key Laboratory of Unmanned Underwater Vehicle, Northwestern Polytechnical University, Xi'an 710072, China

* Correspondence: huangqiaogao@nwpu.edu.cn

Abstract: In order to improve the hydrodynamic performance of pump-jet propulsion (PJP) when matching stator with the rotor, the RANS method with SST k-ω turbulence model is employed to study the influence of six kinds of stator parameters, which are classified into three groups, i.e., stator solidity, stator angles and rotor–stator spacing (S). Results show that the stator solidity involves the blade number (Ns) and chord length (L), has an obvious acceleration effect at and after stator, and produces a higher thrust and torque with a slight efficiency change. Further comparing Ns and L results, we find greater distinctions between the two cases when stator solidity is greatly adjusted. Three stator angles, i.e., stagger angle (α), lean angle (γ), and sweep angle (β), are studied. The α has the biggest effect on the thrust, torque, and efficiency; meanwhile, it shifts the advance number that corresponds to maximum efficiency. The effect of γ is similar to α, but its influence is far less than α. However, there is little difference between various β cases except for off-design conditions, where the efficiency drops dramatically as β increases. The S has a slight effect on PJP performance. Even though S decreases 34% relative to the original PJP, the rotor thrust and torque increase by less than 1%. In addition, we compare torque balance locations under various parameters, and each component force is analyzed in detail to explain the reason for performance variation. The present work is conducive to future optimization in PJP design.

Keywords: pump-jet propulsor; SST; stator solidity; stator angle; rotor–stator spacing

Citation: Li, F.; Huang, Q.; Pan, G.; Qin, D.; Li, H. Influence of Various Stator Parameters on the Open-Water Performance of Pump-Jet Propulsion. *J. Mar. Sci. Eng.* **2021**, *9*, 1396. https://doi.org/10.3390/jmse9121396

Academic Editor: Alon Gany

Received: 27 October 2021
Accepted: 24 November 2021
Published: 7 December 2021

1. Introduction

The pump-jet propulsion (PJP), which consists of a stator (a stationary blade row), a rotor (a rotating blade row), and a duct, is a special type of underwater thruster. PJP has been widely used for various underwater vehicles. According to the position of the stator, there are two types of PJP: pre-swirl PJP and post-swirl PJP. The submarines and some offshore vessels generally adopt the former for noise reduction [1]. In contrast, underwater vehicles such as torpedoes typically employ post-swirl PJP, for the post-stator can recover part of the wake energy that is lost with conventional propellers. Hence, the stator generates more thrust and improves the open-water efficiency of PJP.

In earlier years, PJP research mainly focused on the design method. Mccormick et al. [2] are the first academics who discussed the design of a post-stator PJP. They performed the numerical evaluation of the efficiency and cavitation of PJP. The shroud and propeller were replaced by an equivalent system of ring source and sinks, ring vortices, and vortex filaments. Henderson et al. [3] introduced a method for the design of PJP. A quasi-one-dimensional method of blade design was described that used compressor data. Meanwhile, the arrangement of shroud and vanes is discussed. Furuya et al. [4] presented a report describing the blade-to-blade design with corrections due to the three-dimensional approach.

There is some literature about hydrodynamic research on PJP by experimental and numerical methods. In the experimental study, Zierke et al. [5] are earlier researchers who performed various test methods in the underwater turbomachine with stator. They conducted experimental measurements of a high Reynolds number axial pump in the Applied Research Laboratory (ARL) at Penn State. Flow visualization and velocity measurement by laser doppler velocimeter (LDV) were carried out. In addition, both slow-response and fast-response pressure probes were used to test pressure fluctuation. Suryanarayana et al. [6–8] conducted a series of experiments of PJP on an axisymmetric underwater body. The investigation of PJP was tested at a wind tunnel, which focuses on the axial velocity and tangential velocity, the self-propulsion point, thrust magnitude, and torque balance. In addition, experiments at the cavitation tunnel showed that the inception position took place at the rotor tip at a high advance ratio and the leading edge of the suction side at a low advance ratio. However, the duct and stator were free from cavitation. Shirazi et al. [9] also investigated a post-swirl PJP on a full-scale underwater vehicle experimentally and numerically. The experiments include the bollard pull, self-propulsion, and bare hull resistances were conducted to validate the design procedure and verify the numerical results. Further conditions in which PJP works were numerically studied.

In numerical simulation, many academics utilized the RANS method with various turbulence to study PJP; Lee et al. [10] investigated an IGV/rotor propulsion pump using a standard k-ε turbulence model with a low-Reynolds-number approach near the wall. Results such as blade loadings and wake vortex patterns are compared with Zierke's [5] experimental results. The authors of [11–13] also used the k-ε model to investigate the hydrodynamic performance of PJP in torpedo propulsion. Park et al. [14,15] presented a numerical simulation of the waterjet propulsion system and single-stage ducted marine propeller, respectively; both of which solved RANS equations with the k-ε model. Huyer [16] also employed the k-ε model to study the lateral force of a post-swirl propulsor under a controllable pitch stator; both ducted configuration and open case were researched. Ahn [17] studied a pump-jet with and without a ring by solving the Spalart–Allmaras equation. The ring cases showed lower vortex strength at the tip. The tip clearance is a key factor for the shrouded rotor and has been a research hotspot for PJP. The authors of [18–21] numerically investigated the influence of tip clearance on post-stator PJP. Results showed that efficiency is closely related to the size of clearance due to the energy loss generated by tip leakage vortex (TLV). The larger the tip clearance size, the lower efficiency of PJP, while the efficiency remains unchanged when it increases to a certain value. The tip vortices, including tip separation vortex (TSV) caused by separation at the leading edge of the rotor and TLV, are also influenced by clearance. As the tip clearance size increases, the TSV spreads toward the suction side. Decreasing the advance ratio (increasing rotation speed or decreasing the inlet velocity) will reinforce the vortices and make the cavitation phenomenon more possible to happen.

Reviewing the previous literature, most researchers focus on the hydrodynamics of the rotor and flow details around the rotor, especially the tip clearance between the rotor and duct inner surface. Little research has been conducted on pre-swirl stator parameters except for [22], which mainly investigated the rotor fluctuation but ignored some important results of PJP components, such as the stator force, stator torque, and duct force. From Yu's work, it is obvious that the pre-whirl stator has a significant impact on PJP global performance. The fundamental reason is that the hydrofoil performance [23] of rotor blade is influenced by the inflow condition caused by various stator parameters. The comprehensive study of stator parameters is worthwhile for future PJP design. In this paper, six kinds of stator parameters, organized into three groups, are thoroughly studied based on the RANS method. Apart from the hydrodynamics of all components of PJP, we emphasize the flow field between stator and rotor. The layout is as follows:

Section 2 presents the governing equations and the details of numerical discretization method. In Section 3, the geometry of prototype PJP and the definition of stator parameters are given. The computational domain and numerical setup are described in detail. Then, we perform the validation of the numerical method. Section 4 presents the detailed hydrodynamic results of PJP with various parameters. Finally, Section 5 gives the summary of key results.

2. Numerical Simulation Methods

PJP works in a uniform flow, and the fluid is water at a constant temperature, and thus heat transfer effect is neglected. Due to an extremely low Mach number, the water is treated as an incompressible fluid. In the present work, the steady simulation is adopted to investigate the open-water performance of PJP under the different stator parameters. Reynolds Averaged Navier–Stokes (RANS) equations include mass conservation equation and momentum conservation equation are:

$$\frac{\partial U_i}{\partial x_i} = 0 \tag{1}$$

$$\rho \overline{u}_j \frac{\partial \overline{u}_i}{\partial x_j} = \rho \overline{F}_i - \frac{\partial \overline{p}}{\partial x_i} + \frac{\partial}{\partial x_j}\left(\mu \frac{\partial \overline{u}_i}{\partial x_j} - \rho \overline{u_i' u_j'}\right) \tag{2}$$

In the equations, ρ is water density, x_i and x_j are cartesian coordinate components (i = 1, 2, 3; j = 1, 2, 3), and F represents the mass force of an element, such as gravity. For the present work, the F equals to zero. μ and p are the dynamic viscosity and pressure, respectively. u_i and u_j are the absolute velocity component. $\rho \overline{u_i' u_j'}$ is the Reynold stress. $\overline{u}_i$ and $\overline{u}_j$ represent the time-averaged item, the turbulence model is needed for the closure of Equation (2).

The SST k-ω turbulence model is a two-equation eddy-viscosity model proposed by Menter [24], which has become widely used in flow simulation of underwater vehicles and PJP [25–31]. The shear stress transport (SST) formulation combines the best of k-ω model and k-ε model, is able to investigate a model of low Reynolds number without using the wall function method. In the free-stream far away from the wall, the SST can freely switch to the k-ε formulation and thus avoids the common problem exists in k-ω formulation that the model is too sensitive to the inlet free-stream turbulence conditions. Meanwhile, the SST has a better behavior in adverse pressure gradients and separating flow. Qin [19] carried out numerical simulations of PJP with three different turbulence models, namely standard k-ε, realizable k-ε, and SST k-ω turbulence models. The results showed that the SST configuration is much more accurate. Therefore, the SST k-ω formulation is chosen to study the performance and flow characteristics of PJP. The turbulent kinetic energy k and specific dissipation rate ω formulations can be written as follows:

$$U_j \frac{\partial k}{\partial x_j} = P_k - \beta^* k\omega + \frac{\partial}{\partial x_j}\left[(\nu + \sigma_k \nu_T)\frac{\partial k}{\partial x_j}\right] \tag{3}$$

$$U_j \frac{\partial \omega}{\partial x_j} = \alpha {S_0}^2 - \beta\omega^2 + \frac{\partial}{\partial x_j}\left[(\nu + \sigma_\omega \nu_T)\frac{\partial \omega}{\partial x_j}\right] + 2(1 - F_1)\sigma_{\omega 2}\frac{1}{\omega}\frac{\partial k}{\partial x_i}\frac{\partial \omega}{\partial x_i} \tag{4}$$

where the P_k is the production rate of turbulence used to prevent the built-up of stagnation regions:

$$P_k = \min\left(\tau_{ij}\frac{\partial U_i}{\partial x_j},\ 10\beta^* k\omega\right) \text{ with } \tau_{ij} = \nu_T\left(\frac{\partial U_i}{\partial x_j} + \frac{\partial U_j}{\partial x_i}\right)$$

To properly predict the onset and amount of flow separation from the surface, the eddy-viscosity ν_T should be limited, where $v_T = a_1 k / \max(a_1\omega,\ SF_2)$. Notice that $v_T = \mu_T/\rho$. S_0 is an invariant measure of strain rate and F_2 is a blending function similar to F_1. Specific formulations and constant values can be found in reference [32].

In the present work, the Ansys/CFX solver is employed to numerically investigate the global performance of PJP. The solver uses the element-based finite volume method. The governing equations are reorganized in the volume integral and surface integral forms and then discretized with each element. The advection scheme for four equations from (1) to (4) all adopt a high-resolution scheme to ensure a more accurate result and good robustness.

3. Numerical Simulation of PJP

3.1. Model Geometry

Figure 1 shows the typical model-scaled PJP. The pre-swirl stator with eight blades is fixed on the conical hub, which is the extension of the submarine tail and has the same conical degree. The six-post rotor blades are installed on a cylindrical hub, with a tip clearance equal to 1 mm. In this paper, the rotor has a diameter of D_r = 146 mm. The area ratio and hub diameter ratio of the rotor are 0.8 and 0.3, respectively. Both the profile curves of stator hub and shroud are not parallel to the flow direction, thus resulting in unequal diameters of stator at upstream and downstream, which are D_{sin} = 101.8 mm and D_{sout} = 91.6 mm, respectively. The duct length is Ld = 177.8 mm, and the diameters of inlet and outlet are D_{in} = 210.4 mm and D_{out} = 146.0 mm, respectively.

Figure 1. The pump-jet propeller model.

At present, we classify six stator parameters into three groups, i.e., stator solidity, stator angle, and rotor stator spacing (S). The first includes two parameters, the blade numbers (N_s) and chord length (L). The second contains three angles, as shown in Figure 2.

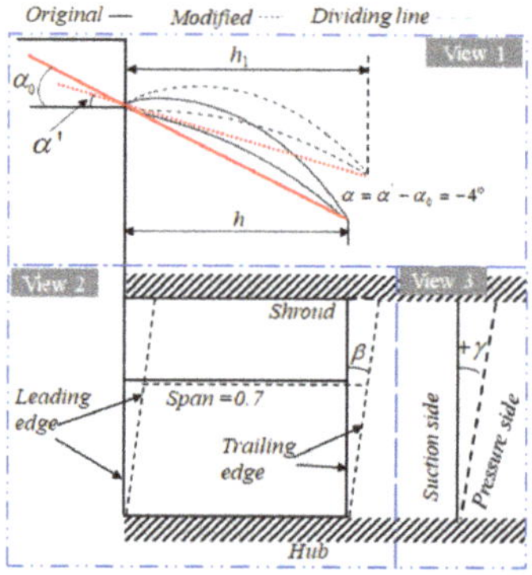

Figure 2. Stator angles in three views.

In Figure 2 view1, i.e., the circumferential view, the original foil and the modified foil at 0.7 span are presented. We adjust the stagger angle (α) of the original model with a step of 4°. However, the blade height (h) increases to h_1 when decreasing α. In order to exclude the influence of blade height on performance of PJP, the chord length (C) is scaled in a proper ratio. The scaled ratio in Figure 2 should be h_1/h.

In Figure 2 view2, i.e., the meridional view, a swept blade with a sweep angle β = 4° in dashed line is shown.

In Figure 2 view3, i.e., the axial view, the solid line, and dashed line are the leading edge of the stator. The lean angle (γ) with a positive value means that the blade leans to the pressure side. Accordingly, negative γ denotes blade leans to the suction side.

The third parameter S is the axial distance between the trailing edge of the stator and the leading edge of the rotor at the root of the blade. It will be modified by translating the stator blade along the axial direction with a step 5 mm and treated as a dimensionless value normalized by D_r. The S for the original PJP is 0.35 D_r.

In the present work, five sets of cases for each parameter will be studied. The computational cases are listed in Table 1.

Table 1. Computational cases for each stator parameters.

Case ID	Stator Parameters					
	Solidity		Angle			Rotor–Stator Spacing
	N_s	L	α	β	γ	S
1	6	6/8C	−8°	−8°	−4°	$0.35D_r$ (origin)
2	7	7/8C	−4°	−4°	0° (origin)	$0.32D_r$
3	8 (origin)	1.0C (origin)	0° (origin)	0° (origin)	4°	$0.29D_r$
4	9	9/8C	4°	4°	8°	$0.26D_r$
5	10	10/8C	8°	8°	12°	$0.23D_r$

3.2. Mesh and Numerical Setup

The entire computational domain is divided into three subdomains: external domain, stator domain, and rotor domain. The external domain is shown in Figure 3. PJP works in a cylindrical tunnel with a length of $15D_r$, and a radius of $5D_r$. The inlet is located $5D_r$ upstream of the rotor, and the outlet is located $10D_r$ downstream of the rotor. Both the stator and rotor domains are generated by a single periodic domain that contains one blade.

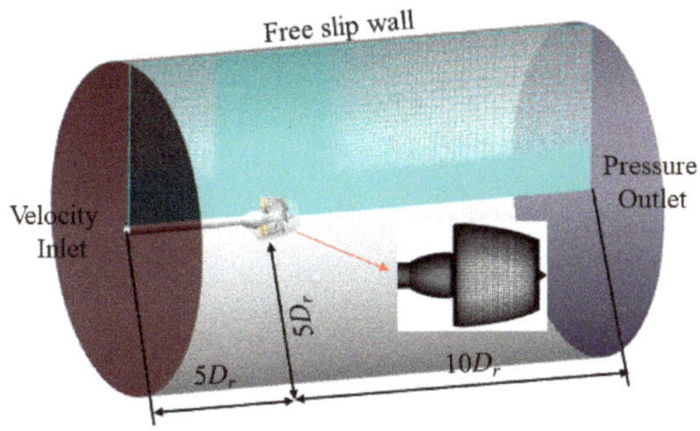

Figure 3. The mesh of external domain and boundary conditions.

Figure 4 shows the coarse grid, the medium grid, and the fine grid of the single domain of rotor and stator. Table 2 lists the number of meshes. The relation between y^+ and the width of the first cell to the wall y is driven by the Blasius equation, expressed as follows:

$$y = 6\left(\frac{V_{ref}}{\nu}\right)^{-\frac{7}{8}}\left(\frac{L_{ref}}{2}\right)^{\frac{1}{8}}y^+ \quad (5)$$

where the V_{ref} is reference velocity, defined as $\sqrt{V_A^2 + (\pi n D_r)^2}$ (V_A denotes the inflow velocity, n denotes the rotation speed). The L_{ref} is reference length, corresponding to the chord length of rotor blade at $r/R = 0.75$. According to Equation (5), the height of first layers, under the typical operating condition of $V_A = 2.6$ m/s and $n = 20$ r/s, is 2×10^{-6} m for all solid surface except the tunnel wall.

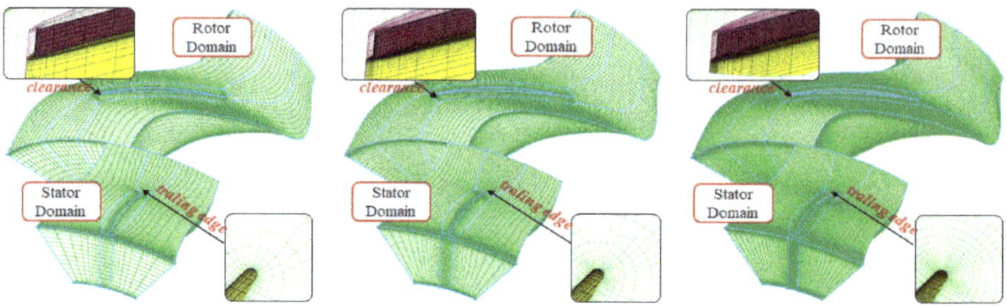

Figure 4. Structure mesh of stator and rotor domain (from **left** to **right**: coarse, medium, fine).

Table 2. Mesh number.

Mesh Groups	ID	Stator Domain (M)	Rotor Domain (M)	External Domain (M)	Total (M)
fine	1	5.13	6.07	3.73	14.92
medium	2	1.35	1.45	2.69	5.49
coarse	3	0.52	0.61	0.87	2.01

The boundary conditions can be seen in Figure 3. A uniform inflow velocity V_A is added at the inlet boundary, and the turbulence intensity is set to 1%. For outlet boundary, an average static pressure of zero is imposed to simulate the far-field. The solid surface of all PJP components is consider to be the no-slip wall. The outer tunnel wall is set to a free-slip wall. In addition, three subdomains need to be connected with the interface boundary. There are three types of interfaces provided by CFX for frame change models, the mixing plane, frozen rotor, and transient rotor–stator methods. Among them, the frozen rotor is the most useful way to produce a steady-state solution for the multi-frame problem, and is adopted in this work. The external domain and stator domain are set to stationary. The rotor domain is set to rotate around the negative z-axis with a constant speed of 20 r/s. The alternate rotation model, in which the absolute frame velocity is advected instead of relative frame velocity, is chosen for a significant reduction in numerical error.

3.3. Verification of Grid

In order to assess the performance of PJP, the dimensionless hydrodynamic coefficients are defined as follows.

$$J = \frac{V_A}{nD_r},\ K_{Tr} = \frac{T_r}{\rho n^2 D_r^4},\ K_{Ts} = \frac{T_s}{\rho n^2 D_r^4},\ K_{Td} = \frac{T_d}{\rho n^2 D_r^4},\ K_T = \frac{T_r + T_s + T_d}{\rho n^2 D_r^4} = K_{Tr} + K_{Ts} + K_{Td}$$

$$K_{Qr} = \frac{Q_r}{\rho n^2 D_r^5},\ K_{Qs} = \frac{Q_s}{\rho n^2 D_r^5},\ \eta = \frac{TV_A}{2\pi n Q_r} = \frac{(T_r + T_s + T_d)V_A}{2\pi n Q_r} = \frac{J}{2\pi}\frac{K_T}{K_{Q_r}} \tag{6}$$

where J is the advance ratio. T is the thrust, with the subscript of r, s, d denote stator, rotor and duct. Q represents torque, Q_r and Q_s are the torque of rotor and stator, respectively. K_{Tr}, K_{Ts} and K_{Td} are the thrust coefficient of the rotor, stator, and duct, respectively. K_{Qr} and K_{Qs} are the torque coefficient of the rotor and stator, respectively. η is the open-water efficiency. During calculation, the rotor speed is fixed at 20 r/s. By adjusting the V_A, we can acquire the PJP's performance at various J.

This section mainly discusses the results of the K_T and K_{Qr} to verify the grid. The formulations will be illustrated by taking the K_T as example. The convergency ratio of thrust coefficient $R(K_T)$ is:

$$R(K_T) = \frac{S_2(K_T) - S_1(K_T)}{S_3(K_T) - S_2(K_T)} \tag{7}$$

where the 'S' with subscript '1', '2', '3' represent three different configurations with fine, medium, and coarse meshes. According to the results of $R(K_T)$, three conditions are obtained as follows:

(1) $0 < R(K_T) < 1$: Monotonic convergence.
(2) $-1 < R(K_T) < 0$: Oscillatory convergence.
(3) $|R(K_T)| > 1$: Divergence.

The uncertainty can be calculated by a correction of factor [33,34] or factor of safety [35]. The present work uses the factor of safety, i.e., value of 1.25 to calculate the uncertainty, which has an alternate definition: the Grid Convergency Index (GCI). The GCI for a fine mesh is defined as:

$$GCI^{21}_{Fine}(K_T) = \frac{1.25\delta^a_{21}(K_T)}{r^{p_l(K_T)} - 1} = U(K_T) \tag{8}$$

The following formulations are used to calculated the $GCI^{21}_{Fine}(K_T)$,

$$\begin{aligned} p(K_T) &= \frac{1}{\ln(r_i)}\left[\ln\left(\frac{S_3(K_T) - S_2(K_T)}{S_2(K_T) - S_1(K_T)}\right) + q(p(K_T))\right] \\ q(p(K_T)) &= \ln\left(\frac{r_{21}^{p_l(K_T)} - s(K_T)}{r_{32}^{p_l(K_T)} - s(K_T)}\right) \\ s(K_T) &= 1 * \mathrm{sign}\left(\frac{S_3(K_T) - S_2(K_T)}{S_2(K_T) - S_1(K_T)}\right) \\ \delta^a_{21} &= \left|\frac{S_2(K_T) - S_1(K_T)}{S_1(K_T)}\right| \end{aligned} \tag{9}$$

where the order of accuracy $p(K_T)$ is generally calculated by the fixed-point iteration method, but the refinement ratio r_i is set to a constant value of 1.4 in this work, which means the $q(p(K_T))$ is zero. Meanwhile, we can use the K_T and r_i to solve $p(K_T)$. Following the above steps, we can also obtain the CGI of K_{Qr}.

Table 3 lists the verification results under different J. The convergency ratio of K_T and K_{Qr} are all located between 0 to 1, illustrating that the mesh groups are monotonic convergence. Although the GCI of K_T and K_{Qr} are discrepant at different J, all of them are below 1%, demonstrating that the refinement of grid does not cause a significant change of hydrodynamic coefficients. In order to get a more detailed flow field, the fine mesh is adopted for a further study of the effect of the stator on the PJP's performance.

Figure 5 presents PJP performance curves with fine mesh. The curves of K_T, K_{Tr} and η are plotted in solid line. The dash lines are applied to other thrust coefficients and torque coefficients of PJP components. In addition, the experiment data of K_{Tr} and K_{Qr} [36] are given in the style of the symbol. The results show that the changing trend of K_T, K_{Qr}, and η is similar to conventional propellers. The difference is that η for PJP has a relatively large value in a wide range of J. The maximum η location is approximately 0.9 of J, where the η reach up to approximately 0.6. Furthermore, the duct generates additional effective thrust at low J, but produces resistance at high J; the turning point of J after which the duct never produce effective thrust is approximately 0.7. In contrast, the stator always provides drag force.

Table 3. Uncertainty analysis under different advance coefficients.

	J = 0.2		J = 0.4		J = 0.6	
	$\xi = K_T$	$\xi = 10K_{Qr}$	$\xi = K_T$	$\xi = 10K_{Qr}$	$\xi = K_T$	$\xi = 10K_{Qr}$
$S_1(\xi)$	0.6306	0.9818	0.6306	0.9818	0.6306	0.9818
$S_2(\xi)$	0.6284	0.9655	0.6284	0.9655	0.6284	0.9655
$S_3(\xi)$	0.6218	0.9403	0.5975	0.9152	0.6218	0.9403
$R(\xi)$	0.3447	0.6445	0.6579	0.8444	0.3493	0.7900
$p(\xi)$	3.1656	1.3054	1.2443	0.5025	3.1259	0.7005
$\delta^{a}_{21}(\xi)$	0.0036	0.0166	0.0058	0.0258	0.0013	0.0182
$GCI^{21}_{Fine}(\xi)$	0.0024	0.0376	0.0140	0.1751	0.0009	0.0857
	J = 0.8		J = 1.0		J = 1.2	
	$\xi = K_T$	$\xi = 10K_{QR}$	$\xi = K_T$	$\xi = 10K_{QR}$	$\xi = K_T$	$\xi = 10K_{QR}$
$S_1(\xi)$	0.5274	0.8669	0.4638	0.8062	0.3755	0.6867
$S_2(\xi)$	0.5263	0.8513	0.4632	0.7867	0.3751	0.6723
$S_3(\xi)$	0.5246	0.8288	0.4625	0.7622	0.3745	0.6512
$R(\xi)$	0.6683	0.6906	0.7390	0.8008	0.5705	0.6818
$p(\xi)$	1.1976	1.1003	0.8990	0.6603	1.6678	1.1386
$\delta^{a}_{21}(\xi)$	0.0021	0.0180	0.0012	0.0243	0.0010	0.0209
$GCI^{21}_{Fine}(\xi)$	0.0054	0.0501	0.0041	0.1220	0.0017	0.0561

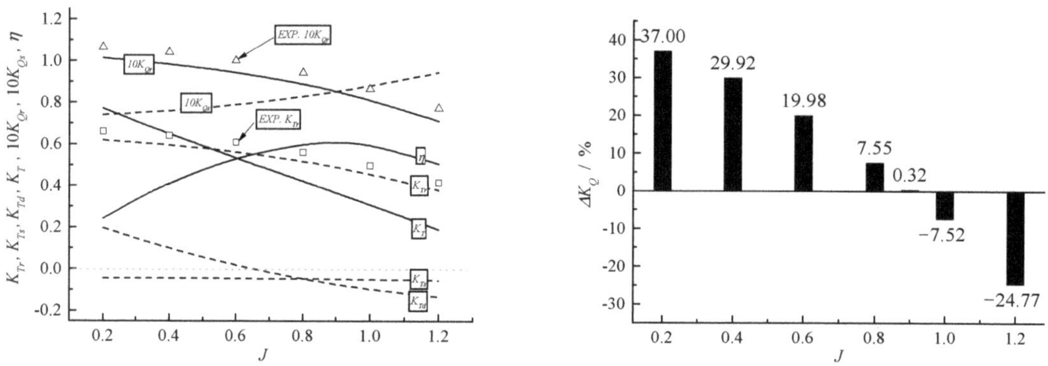

(**a**) Hydrodynamic coefficient curves (**b**) Torque difference between stator and rotor

Figure 5. PJP performance with fine mesh.

Comparing the K_{Tr} and K_{Qr} obtained by numerical simulation with the experimental data, it is evident that the numerical results at the whole range of J are relatively smaller than experiment data, especially at large J. The maximum error of K_{TR} and K_{QR} appears at J = 1.2 is 8%; other errors range 4%~7%. We think this relatively high error should be attributed to inherent numerical analysis, plus an unconformity with experimental environments, such as attachment used for fixing PJP components or relatively small flow tunnel may have an undesirable influence. However, the trend of numerical results corresponds well to experimental data. The numerical method is acceptable to get a further study on PJP. In addition, the torque directions of the stator and rotor are opposite to each other. The larger difference in rotor torque and stator torque may generate a high overturning moment for the submarine, affecting its maneuverability. Therefore, we define the torque difference ΔK_Q as $(K_{QR} - K_{QS})/K_{QS} \times 100$ for a quantitative evaluation. Figure 5b exhibits the ΔK_Q at different J. It is clear that the absolute value of ΔK_Q reduces first and then increase with the increase in J. The optimum condition is where the ΔK_Q equals zero, which means that the submarine does not need to produce extra force to balance the overturning moment. The balance condition of torque for the original PJP locates at slightly higher than 0.9 of J.

Figure 6 shows the contour of wall y^+ at $J = 0.8$. The rotor exhibits a high y^+ compared with the stator and duct. Meanwhile, the y^+ at the leading edges of the stator, rotor, and duct is relatively higher than at other regions. The mean y^+ at the stator and duct are 0.3 but 0.8 at the rotor, satisfying the turbulence model demand and is appropriate for high Reynolds number flow.

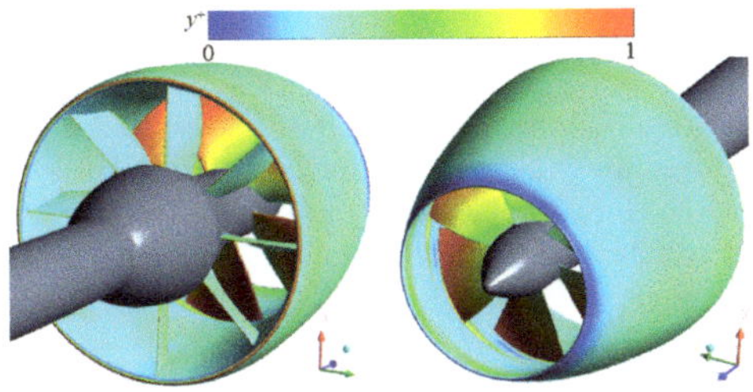

Figure 6. The y^+ distribution at wall under $J = 0.8$.

4. Flow Results of Different Stator Parameters

4.1. Influence of Stator Solidity

The stator solidity, determined by N_s and L, is generally one of the first key factors in the design procedure. The stator solidity increases proportionally to the increase in N_s or L. In order to compare the effect of these two parameters on the PJP's performance, the range and interval of stator solidity of N_s cases and L cases are kept the same.

A. Stator number

The hydrodynamic performance of PJP at $J = 0.2$~1.2 with five sets of stator numbers is presented in Figure 7, where the case of $N_s = 8$ represents the original case. As shown in Figure 7a, the thrust and torque of PJP increase linearly with the increase in N_s. The difference in total thrust and torque at low J between two adjacent N_s cases ranges 4% ~ 5%. The efficiency of PJP slightly changes or even presents a negligible change with the increase in N_s, and the maximum efficiency locates at $J = 0.9$. However, after the 0.9 of J, the efficiency reduces dramatically with the decrease in N_s, especially at $J = 1.2$. For instance, the efficiency of $N_s = 6$ at $J = 1.2$ is reduced by 16% relative to the original case. The thrust coefficients of PJP, which consist of K_{Tr}, K_{Ts}, K_{Td}, are plotted in histogram form in Figure 7b. It is clearly seen that the larger the N_s, the more resistance the stator creates. For each additional stator blade, the resistance increases by 2.5 N at $J = 0.8$, or causes an increase in K_{Ts} by 15%~20%. For K_{Tr} and K_{Td}, there is an approximate 5% and 2% increase, respectively. Figure 7c exhibits torque difference at the whole operating conditions. Generally, the change rules of torque and thrust with J are similar. That is, both K_{Qr} and K_{Qs} increase with the increase in N_s. However, the different increased rate of K_{Qr} and K_{Qs} with N_s eventually shifts the torque balance position. As N_s increase, the balance condition moves toward lower J. Specifically, the torque balance positions for $N_s = 6$~10 are located at 1.1, 1.0, 0.9, 0.8, and 0.7, respectively.

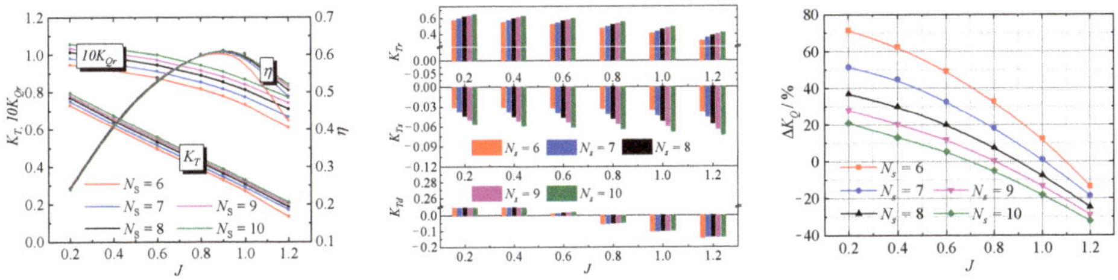

(a) Open-water performance (b) Thrust coefficient of different parts (c) Torque difference in stator and rotor

Figure 7. PJP performance with different stator number.

Figure 8 presents pressure contour at the plane of x = 0 and velocity contour at four different planes under J = 0.8 (principal analysis condition in later). The four planes are separately located upstream of the stator, in the middle of the stator, downstream of the stator, and upstream of the rotor. Note that the plane of A_4 located in the rotor domain, the variable 'velocity in stn frame', which denotes the absolute velocity, is plotted to avoid all-one-color at A_4 or other planes due to excessive velocity difference between stator domain and rotor domain. In addition, both the pressure and velocity are dimensionless quantities normalized by $0.5\rho {V_A}^2$ and V_A, respectively. The contour of C_P shows some local-low pressure regions which correspond to free vortices in the wake of the rotor. Results indicate that the local low-pressure position and magnitude are essentially unchanged as N_s varies. The velocity contours at A_3 and A_4 show some low-speed zones, and the shape of which get narrow from the hub to the shroud. The radial-going trail of the low-speed zone at A_3 resembles the form of the trailing edge of the stator.

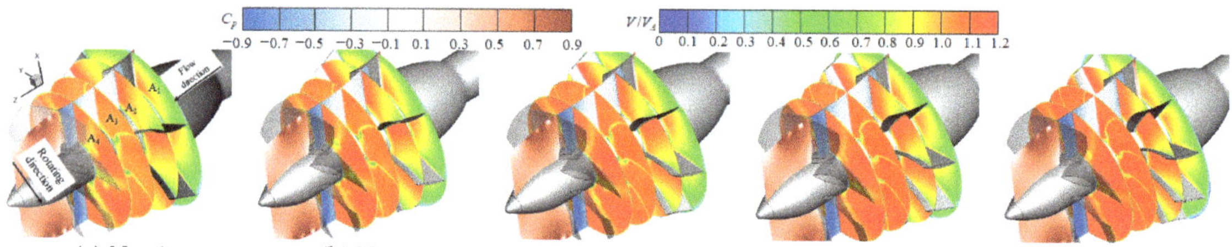

(**a**) N_s = 6 (**b**) N_s = 7 (**c**) N_s = 8 (**d**) N_s = 9 (**e**) N_s = 10

Figure 8. The contours of velocity and pressure.

Comparing the velocity contour at the surface of A_1~A_4, there is a significant acceleration effect for flow velocity. It is worth noting that flow velocity gradually becomes more uniform between two adjacent stator blades with N_s increase. Meanwhile, the velocity magnitude at A_3 is gradually increased from V_A to 1.2 V_A as N_s increases from 6 to 10. Since the larger N_s, the narrower flow passage between two adjacent stator blades and the more obvious acceleration effect it performs. The increased velocity of incidence flow for the rotor eventually causes an increase in rotor thrust. Since the increased amplitude of rotor thrust is greater than the resistance of stator, the total thrust of PJP keeps growing even though more drag is produced by added stator blade.

B. Chord length

It is necessary to investigate the effect of chord length on the PJP's performance before studying other paraments. Because when changing other paraments, such as stagger angle or sweep angle, the chord at each stator profile will vary inevitably. In this section, the chord is modified on the basis of N_s studied before. The chord of the original PJP is taken as a reference to other cases with extended foil or shrunken foil, namely L = 6/8C, 7/8C, 9/8C, and 10/8C. Figure 9 presents the performance of PJP with different chord length. In order to get a distinct contrast with the N_s effect, the ordinate scales are the same as that of Figure 7. According to the figures, the trend of all coefficients with L and with N_s is similar. The increase in K_T and K_{QR} range 3%~5% as L increase. The efficiency basically stays the same value at lower J but drops quickly at high J. Further, both K_{Ts} and K_{Tr} grow with the increase in L as rapidly as with the increase in N_s, and the K_{Td} also become larger as L increases when the duct produces thrust but become smaller when it generates drag. As for the torque difference, there are distinct differences of balance position between N_s configurations and L configurations under the same stator solidity. For L = 6/8C and 10/8C, the balance condition of J is slightly smaller than 1.1 and slightly higher than 0.7, respectively. However, for L = 7/8C and 9/8C, the balance conditions are the same as with the corresponding N_s cases (N_s = 7 and N_s = 9), i.e., J = 0.8 and J = 1.0, respectively.

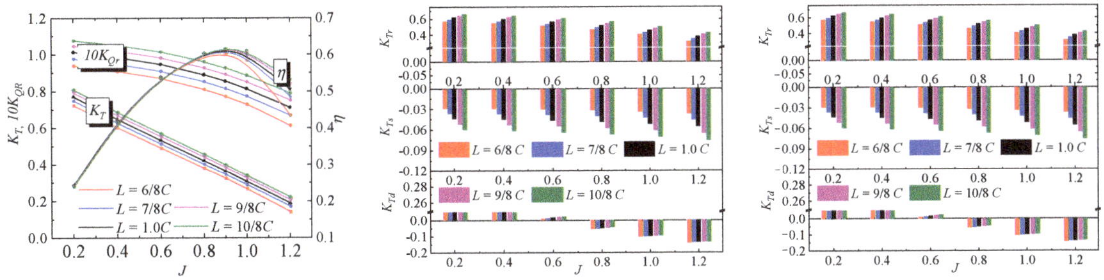

(**a**) Open-water performance (**b**) Thrust coefficient of different parts (**c**) Torque difference in stator and rotor

Figure 9. PJP performance with different chord length.

To further figure out the difference in the effect of N_s and L on the PJP's open-water performance, each hydro-coefficient difference between two sets under the same stator solidity is plotted in Figure 10. We define the difference as $\Delta\varphi = 100 \cdot (\varphi_1 - \varphi_2)/\varphi_2$, among which the φ represents hydrodynamic coefficients such as K_{Tr} or K_{Ts}, with the subscript '1' and '2' separately denote the coefficients of L and N_s configuration. The labels from 'conf.1' to 'conf.4' at the horizontal axis represent the cases of $N_s = 6, 7, 9, 10$, and each of them corresponds to $L = 6/8C, 7/8C, 9/8C, 10/8C$. Note that the conf.2 and conf.3 are reduced and increased stator solidity, respectively, based on the original PJP cases with a same magnitude, so are the conf.1 and conf.4.

As is shown in the figure, the difference in all coefficients under conf.1 is distinctly larger than conf.2. Take the K_{Tr} as an example, the $\Delta\varphi$ for K_{Tr} is 3.16% under conf.1, but reduces to 1.03% under conf.2; similarly, the $\Delta\varphi$ under the conf.4 is larger than that under the conf.3 except for the K_{Qs}, which illustrates the higher extent of stator solidity modified, the more distinct difference between N_s effect and L effect on the PJP's performance even under the same stator solidity. Furthermore, all the signs of coefficients for increased stator solidity are opposite to those for reduced except for K_{Qs}. For the increased stator solidity cases, i.e., the conf.3 and conf.4, the $\Delta\varphi$ are positive except for K_{Td}, which indicates the effect of L is more significant than N_s for increasing the PJP's thrust or efficiency, and the higher of stator solidity, the more obvious this effect. However, for the reduced stator solidity cases, i.e., the conf.1 and conf.2, the $\Delta\varphi$ is negative except for K_{Td} and K_{Qs}, which indicates the reduction in the N_s is more likely to deteriorate the PJP's performance compared with the decrease in L. Based on the above analyses, we can draw a conclusion that it is more obvious to modify L for the improvement of PJP's performance rather than adjust N_s.

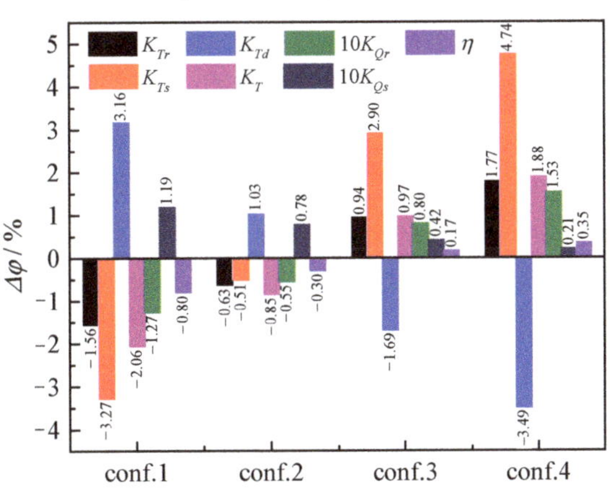

Figure 10. Hydro-coefficients difference under same stator solidity.

To further investigate the reason for the L effect on the PJP's performance, it is essential to study the flow field at and downstream of the stator. Figure 11 presents the C_P contour at the suction side and pressure side of the stator blade. The limiting streamlines are also exhibited. The label 'H' at ordinate represents the maximum span length in x-axis.

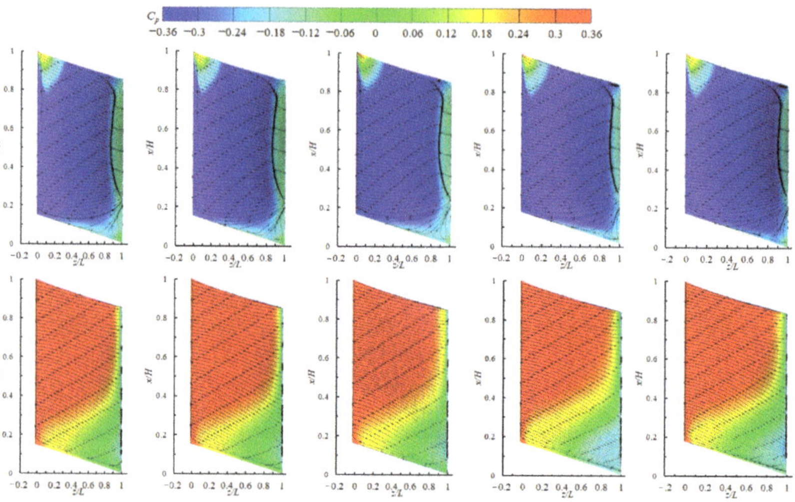

Figure 11. Limiting streamlines at stator (top: suction side; bottom: pressure side; the columns from left to right: L = 6/8C, 7/8C, 1.0C, 9/8C, 10/8C).

The contour of the pressure side presents a decrease in pressure from top left to bottom right, with streamlines parallel to the conic surface of the hub and shroud. It seems that the chord length L has little impact on streamline direction. The corner between the trailing edge and hub exhibits a low-pressure region that gradually enlarged with the increase in L. The results of the suction side surface show a local high-pressure region (axial direction: 0~0.3L, radial direction: 0.8~1H) in the corner between the shroud and leading-edge, which may be influenced by high stagnation pressure of duct inlet. The separation of the boundary layer occurs near the trailing edge. As L increases, the separation position moves downstream, namely from 0.8L to 0.9L. At the same time, the streamlines on the root of the suction side present radial-going flow and concentrate near the trailing edge. Furthermore, a small vortex is obviously presented close to the shroud and trailing edge. Due to the enormous pressure difference between the pressure side and suction side, together with a low-energy at the stator boundary layer, a secondary crossflow occurs, and the flow direction points to the hub.

Figure 12 presents the velocity distribution at the slice of A_3 and A_4. The figure shows narrow low-velocity regions along circumferential direction periodically. The periodicity of the low-speed zones at A_3 along circumference direction is consistent with N_s, i.e., eight; this is because of the occurrence of separation at stator trailing edge and its prolonged effect on downstream. However, the influence on location at A_4 by stator wake is not as strong as at A_3. Thus, the periodicity turns into six (the number of rotor blades), with some small strip-shaped and round-shaped low-speed regions existing due to the slight impact of stator wake. The results also show that a slight grow of velocity magnitude with the increase in L.

In order to investigate the flow details at A_3 and A_4, the velocity V and velocity components: axial velocity V_z, circumferential velocity V_c, radial velocity V_r along the circumferential direction at three different spans are presented in Figure 13. The circumferential angle at the horizontal axis is described in Figure 12. For the V_r, the direction from hub to shroud denotes positive; the opposite direction is negative.

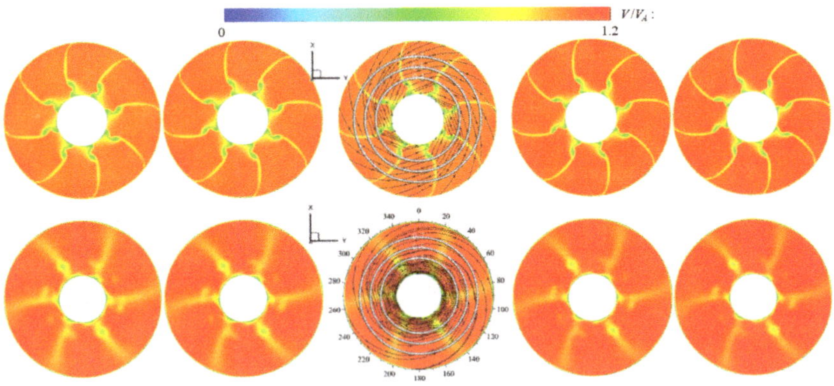

Figure 12. Velocity contour at the A_3 plane and the A_4 plane (top: A_3; bottom: A_4; the columns from left to right: $L = 6/8C$, $7/8C$, $1.0C$, $9/8C$, $10/8C$).

The velocity components change periodically except at 0.3span of A_4. The periodicity at A_3 is eight but six at A_4 except for V_r. However, the V_r at A_4 has the same periodicity as at A_3, which indicates V_r is deeply influenced by the stator. The stator effect is also observed at 0.3span of A_4, since there are induced peaks by stator blade for V_z and V_c, thus causing an uncertain periodicity. The proportions of the three velocity components vary greatly. Take the 0.3span of A_3 as an example, the V_z curves are especially close to V curves, meaning that the highest velocity proportion is V_z, the maximum value is above $0.9V_A$. Followed by the V_c, the maximum value is approximately 0.4 V_A.

Comparing the velocity components with various L, the large L seems to accelerate flow. Both V_z and V_c get larger as L increases. That explains why the larger L, the closer the boundary layer separation position to the trailing edge of the stator. In contrast, the L has little effect on V_r. The V_r curves at A_3 and A_4 under for various L cases almost coincide with each other. The resultant velocity finally has a 1.5%~2% increase with the increase in chord length.

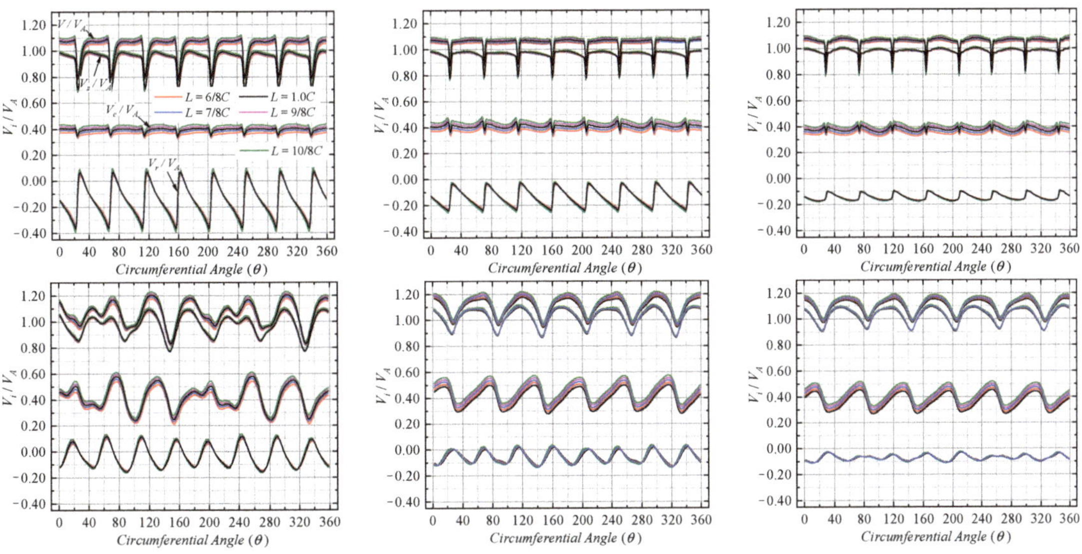

Figure 13. Detailed Velocity component (the rows from top to bottom: A_3, A_4; the columns from left to right: span = 0.3, 0.5, 0.7).

4.2. Influence of Stator Angles

The stator angles can be summarized as three parameters in cylindrical coordinate views as described previously: the stagger angle for the circumferential view, the lean angle for the axial view, and the sweep angle for the meridional view. Each parameter is modified with a same step (4°) to investigate and compare their effect.

A. Stagger angle

The stator stagger angle α is an essential parameter for PJP performance, since it greatly influences the effective incidence angle of rotor, while the blade height will be changed when adjusting the angle directly. Therefore, the chord length of hydrofoil at each section needs to be scaled to maintain a constant blade height. Table 4 lists the specific scaled factor of various angle cases of each stator profile.

Table 4. The scaled factor of various stator blade.

Cases	#1	#2	#3	#4	#5	#6	#7	#8	#9	#10
$\alpha = -4°$	0.9922	0.9903	0.9866	0.9833	0.9814	0.9804	0.9801	0.9808	0.9825	0.985
$\alpha = -8°$	0.9894	0.9856	0.9782	0.9719	0.968	0.9661	0.9655	0.967	0.9702	0.9751
$\alpha = 4°$	1.0128	1.0149	1.0188	1.0223	1.0245	1.0255	1.0259	1.025	1.0232	1.0205
$\alpha = 8°$	1.0311	1.0353	1.0436	1.0508	1.0554	1.0577	1.0584	1.0566	1.0528	1.0471

Figure 14 shows the PJP's performance with different α. In Figure 14a, as α increases, the K_T and K_{Qr} increase considerably. The α has a greatly obvious effect on the PJP's performance compared with the previous parameters. The growth rate of K_T from $-8°$ to $-4°$ is 13.96% relative to the former. The K_T increases by 13.78% from $-8°$ to $0°$, followed by 12.68% and 12.96% for $0°$ to $4°$ and $4°$ to $8°$, respectively. The growth rates of K_{Qr} are 11.58%, 12.16%, 12.09%, and 13.7%, respectively, for $-8°$ to $-4°, \ldots,$ and $4°$ to $8°$. The different growth rates of K_T and K_{Qr} eventually lead to a slight efficiency difference at $J = 0.8$, which is a turning point of the efficiency difference between various α cases. For $J < 0.8$, the η stays almost the same, but when J is greater or equal than 0.8, the η increase considerably as α increase. Further, the large α will cause the highest efficiency to move slightly toward high J.

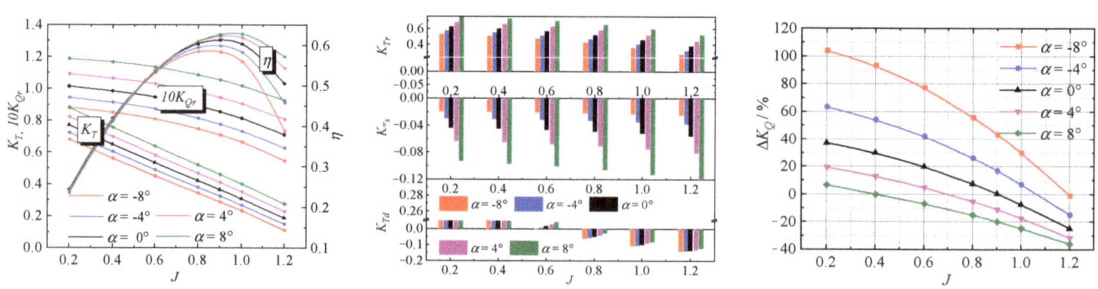

(**a**) Open-water performance (**b**) Thrust coefficient of different parts (**c**) Torque difference in stator and rotor

Figure 14. PJP performance with different chord length.

As shown in Figure 14b, the K_{Tr} and K_{Ts} positively correlate with α. However, for the K_{Td}, it increases with the increase in α when the duct provides thrust but decreases when the duct generates drag. In general, K_{Ts} and K_{Tr} are the most affected coefficients with the rise of α. The average growth rates for *KTs* and K_{Tr} are approximately 50% and 13% at $J = 0.8$.

Figure 14c exhibits a relatively larger torque difference value compared with the configurations of the previous two parameters. The maximum ΔKQ locates at $J = 0.2$

exceeds 100% when $\alpha = -8°$. As α increases, the torque balance position moves toward low J. For $\alpha = -8°\sim 4°$. The balance points locate separately at J = 1.2, 1.05, 0.9, 0.7, 0.4.

Figure 15 shows the C_p contour for various cases at three different spanwise surfaces. Each subfigure exhibits only half of the computational domain for saving space. Obviously, the α has a dramatic influence on the pressure distribution in both stator and rotor domains. The pressure at the suction side of the stator and rotor reduces with the increase in α, but there is little change at the pressure side. It can be inferred that the rise of α will produce a larger stagnation pressure at the leading edge of the stator and cause the separation on the suction side of the stator to occur in advance. Meanwhile, the velocity flow direction is changed along the stator section as α increases and eventually leads to a larger attack angle for hydrofoil at each rotor blade section, which is similar to the function of increasing the pitch angle of the rotor. The large pressure-reduced area near the suction side of the rotor not only produces thrust and torque but may affect the contiguous region around the stator. The drag force caused by the stator is much higher at the same time when α increases, but the increased resistance of the stator is insufficient to offset the increased thrust of the rotor. Further, the increased rate of stator torque is also far less than the rotor. Consequently, the large α contributes to the increase in PJP total thrust and causes a higher torque difference at usual working conditions, further making torque balance point move to higher J. In addition, a strong change of pressure at 0.95 span near the tip of the rotor blade is observed, which results in tip leakage flow and rotor wake.

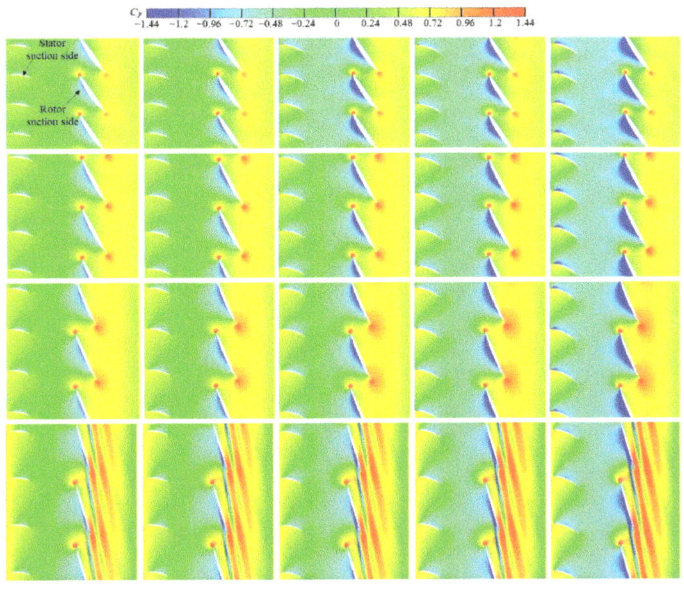

Figure 15. Pressure around stator and rotor blade sections under condition of $J = 1$ (the rows from top to bottom: span = 0.3, 0.5, 0.7, 0.95; the column from left to right: $\alpha = -8°, -4°, 0°, 4°, 8°$).

B. Lean angle

In this section, the lean angle γ, as one of the secondary parameters in PJP design, is modified with the same step of 4° as α does. A series of PJP performances with various γ are presented in Figure 16. Similar to other previous parameters, as γ increases, K_T, K_{Qr}, and η all increase in varying degrees. The increased rate of K_T and K_{Qr} between two adjacent γ cases distributes in the range of 2%~4% at J = 0.8. In contrast, the η has a slight change but no more than 1%. For the component thrust, the K_{Tr} and K_{Ts} also increase linearly at 3% and 11% with the increase in γ, respectively. However, for K_{Td}, the relative variation between two adjacent configurations does not exceed 3%. As for torque difference, the change in ΔK_Q with γ is opposite to K_T or K_{Qr}. The maximum variation of ΔK_Q is located at $J = 0.2$ for $\gamma = -8°$, and it is smaller than that for previous parameters that have been analyzed. Consequently, the torque balance condition of J for $\gamma = -8°$ turns small, i.e., $J = 1.0$; For $\gamma = 8°$, it locates at $J = 0.8$. However, in other cases, the balance positions distribute between $J = 0.8$ and 1.0.

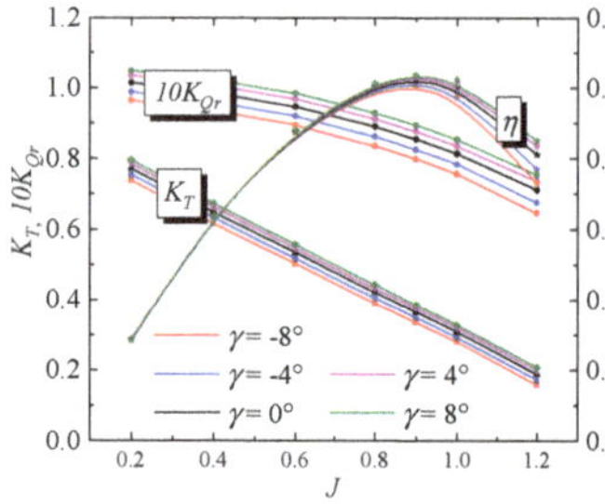

(**a**) Open-water performance

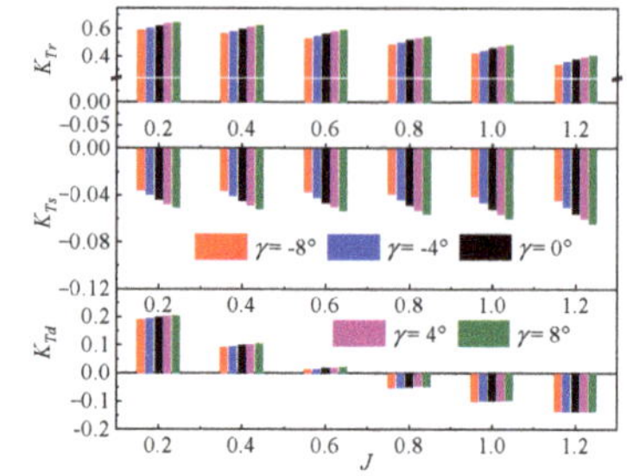

(**b**) Thrust coefficient of different parts

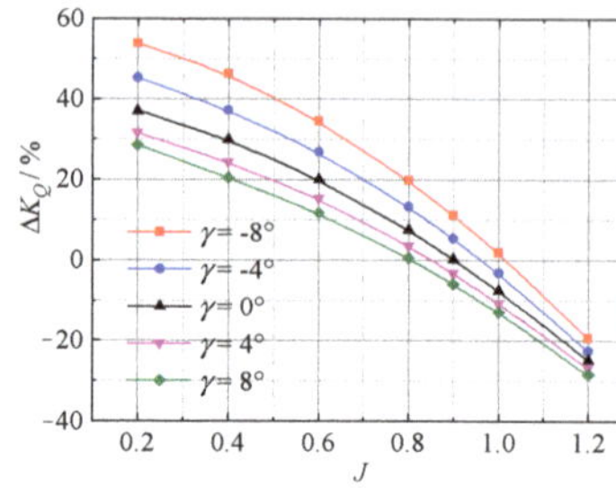

(**c**) Torque difference

Figure 16. PJP performance with different lean angle.

Further analysis on velocity distribution, as shown in Figure 17 between stator and rotor, is needed to better understand the reason for performance variation. As has been shown in Figure 12, there are six radial-going low-velocity-belts distributed evenly at constant Z plane due to stator wakes, with high-velocity between two adjacent belts. When γ increases, the velocity at the high-velocity region becomes slightly larger. The same phenomenon is obviously seen at 0.5 span, especially at the stator wake region. In Figure 18, the flow angle φ, defined as *arctan* (V_z/V_c), is presented at 0.5 span. It is found that φ decreases slightly with the increase in γ, especially at the region between stator and rotor. However, the decreased rate of flow angle is far less than the increased rate of velocity, thus causing an increase in K_T and K_{Qr}.

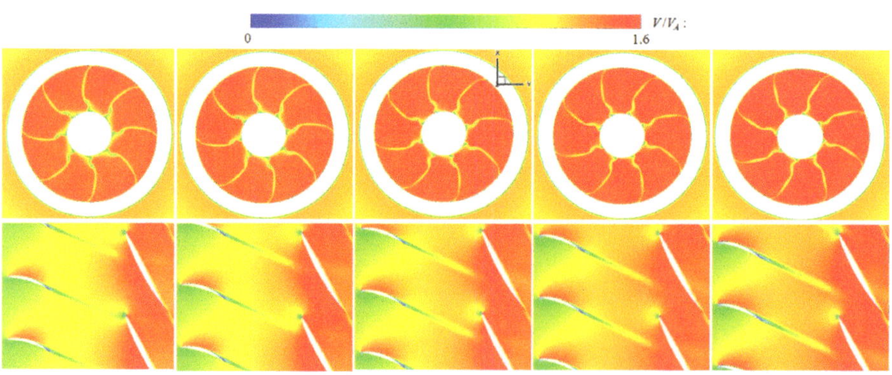

Figure 17. Velocity distribution at surface of A_4 and 0.5 span (top: A_3, bottom: 0.5 span; from left to right: $\gamma = -8°$~$8°$).

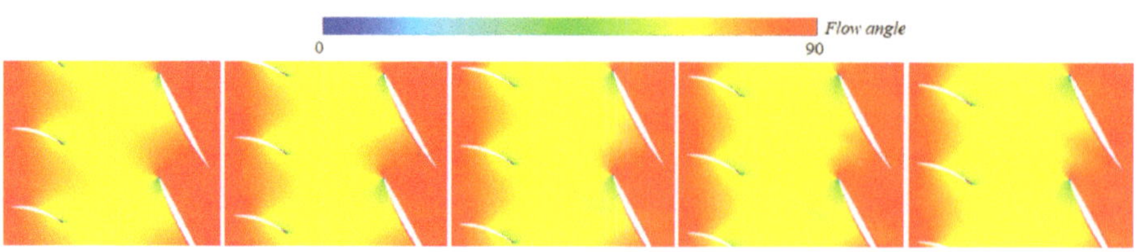

Figure 18. Flow angle at 0.5 span.

C. Sweep angle

In previous parameters, the stator always maintains a constant value of blade height to avoid the possible influence on the PJP's performance. In this section, we will investigate the global performance of PJP with various sweep angle β. Considering the mate dimensions of duct and stator, the lower bound of β is -4°, with a step length and upper bound are 4° and 12°, respectively. The open-water performance and thrust coefficient of all components are plotted in Figure 19a. The results show that the β has little effect on the performance of the rotor, with no more than a 1% difference in K_T, K_{Qr}, η between two different β situations without considering the off-design condition. The thrust of all components, including the stator, rotor, and duct, are plotted in Figure 19b. The K_{Tr} changes little. However, there are slight changes in K_{Ts} and K_{Td}. Take J = 0.8 as an instance, the K_{Ts} and K_{Td} separately decrease and increase with an average rate of 2% as a 4° step increase in β. Since both of K_{Ts} and K_{Td} are in the same order of magnitude, the resistance decreased on stator just offset the increased drag of duct, hence the thrust coefficient of PJP is essentially unchanged. For torque difference, a slight difference yields at low and high J compared with the original PJP, whereas the ΔK_Q stays almost the same at middle J, which results in a fixed torque balance position.

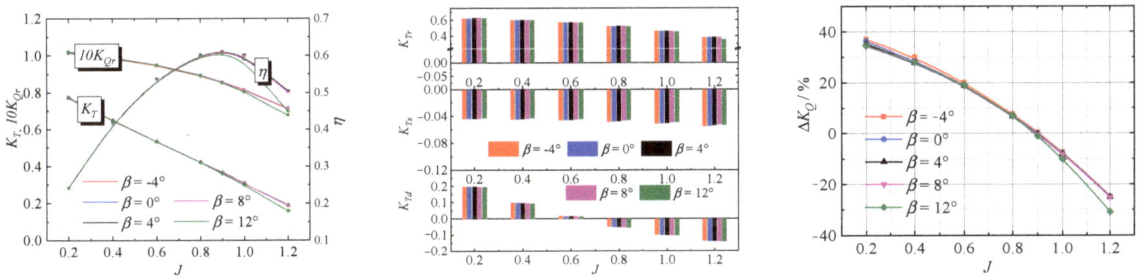

(a) Open-water performance (b) Thrust coefficient of different parts (c) Torque difference

Figure 19. PJP performance with different sweep angle.

Figure 20 presents the C_P at a constant span of 0.5. There are intense pressure changes at the leading edge and trailing edge. However, the curves of various β situations coincide with each other well. The same phenomena exist at other profile foils, thus demonstrating that changing β has essentially no effect on the pressure of the rotor.

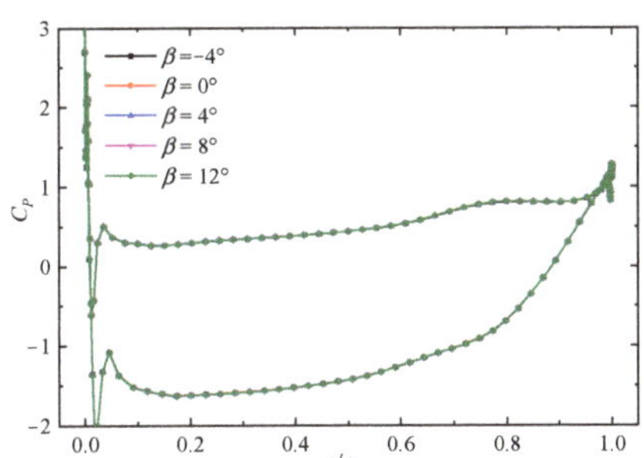

Figure 20. The static pressure coefficient of rotor at 0.5 span (s: the distance from leading edge; c: chord length).

The pressure coefficient of the stator blade is presented in Figure 21. The local high pressure in the upstream shroud-corner region on the suction surface for $\beta = -4^\circ$ is much higher than for other cases. With the increase in β, namely the increase in the distance from the duct inlet to the high-span leading edge of the stator, the local high pressure gradually fades away. Finally, it forms a low-pressure belt at three sides, enclosing a much lower pressure region, which is obvious for high β. Thus, we can deduce that the pressure at the upstream shroud corner is deeply affected by the distance from the duct inlet to the leading edge of the stator. Compared with the negative β or lower β case, the higher β may benefit to structural performance due to a more uniform pressure distribution on the surface rather than a high local pressure that may result in stress concentration. Another noticeable phenomenon is that the concentrated point of streamlines near the hub and the trailing edge shifts slightly toward upstream when β increases. The vortex at the shroud corner of downstream is observed when β equals 8° and 12°, which will give rise to a large shear action on fluid tightly close to this area. Similarly, the *CP* of pressure surface at the upstream shroud-corner position is higher in the β equals $-4°$ case than others, with the increase in β, the high-pressure area gradually reduced, thus causing the decrease in stator resistance.

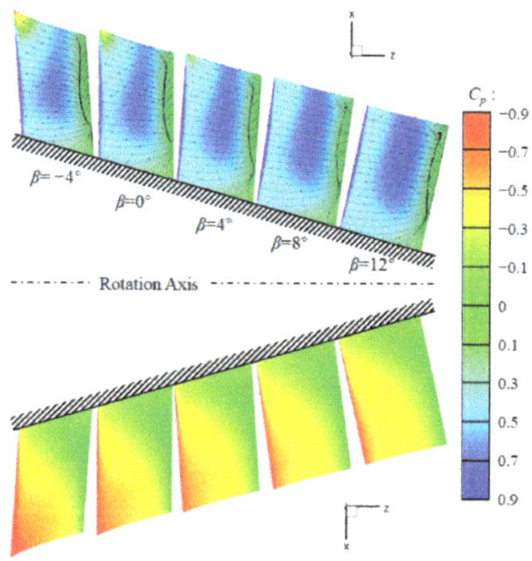

Figure 21. The C_P contour of stator blade (top: suction side covered with limiting streamlines; bottom: pressure side).

Figure 22 shows the velocity components *Vi* obtained from 0.5 span at the cut plane of A_3 and A_4. The influence of β on *Vi* at A_3 is more obvious than at A_4. A tiny distinction between various β cases is observed, while at A_4, the curves almost coincide. As for the flow angle, the maximum value is generally smaller at A_3 than at A_4. Specifically, the value of the flat part does not exceed 68° at A_3, but it is larger than 70° when the flow enters the rotor domain. Since the absolute increase in V_z is larger than V_c, the flow gradually deflects towards the axis direction after passing through the stator. It is also found that φ curves are much more identical at A_4 than at A_3 except for some special flat parts at A_4, thus explaining the identical pressure distribution of the rotor blade section for various β cases, as shown in Figure 20.

4.3. *Influence of Rotor–Stator Spacing*

The stator-rotor spacing *S* is an important parament in the design of PJP. Since the rotor–stator interaction become intense when the *S* becomes small and causes a disastrous hydro-oscillation force on the rotor. In contrast, a big *S* will cause hydraulic loss that may have an adverse impact on the performance of PJP. According to some design manuals, the values of *S* typically shall not be less than $0.05D_r$ in axial-flow pump design, and the smallest *S* modified in this paper far satisfy the devising standard. In this section, the interface stays the same location. Only the stator blades are moved toward the rotor. Numerical simulation settings remain the same except that the number of nodes in the stator domain is properly increased and decreased before and after the stator.

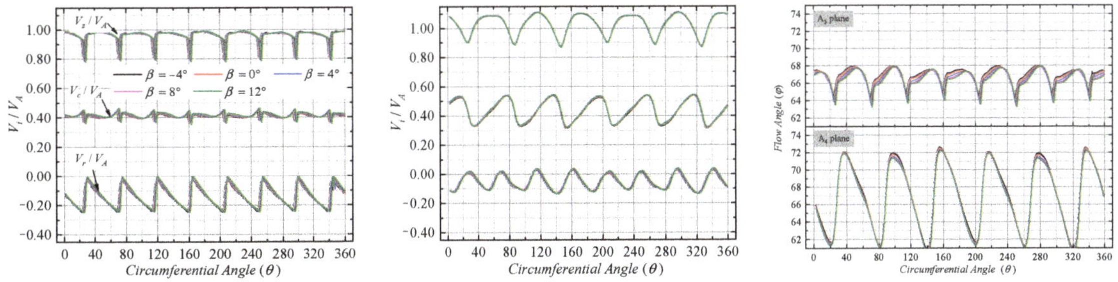

Figure 22. Velocity component and flow angle at 0.5 span (the first two columns: V_i at A_3 and A_4, respectively; the third column: flow angle at A_3 and A_4).

Figure 23 presents the performance of different S situations, the K_{Qr} and η of PJP are slightly affected by S, while the K_T gives the same value for all cases. By further studying the force of each component, it is found that the variation of force for duct and stator between all cases is highly varied than that of rotor force. For example, when J = 0.8, although the increased rates of K_{Ts} and K_{Td} are separately approximately 2.5% and 1.5%, the K_{Tr} increases less than 1%. Eventually, the K_T is almost unchanged because of one magnitude order larger than K_{Ts} and K_{Td} that K_{Tr} owns. For torque difference, as J reduces, the variation of ΔK_Q gradually becomes larger. The maximum difference between the case of maximum S and minimum S (i.e., 0.35Dr and 0.23Dr) is less than 10%. The balance conditions for all S cases regularly distribute at approximately J = 0.9.

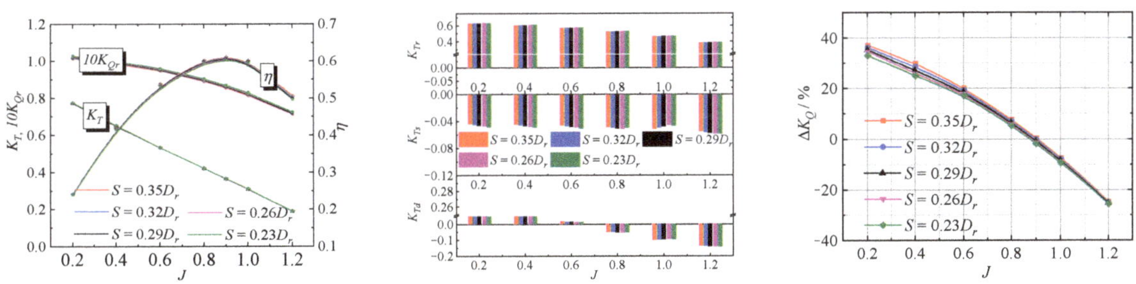

(**a**) Open-water performance (**b**) Thrust coefficient of different parts (**c**) Torque difference

Figure 23. PJP performance with different rotor–stator spacing.

Due to the slight change of K_{Tr}, the pressure curves at 0.5 span rather than pressure contour are presented in Figure 24 to investigate the main reason for slight change of the performance coefficients. The figure shows that the C_p at the pressure side has no change with variation of S. However, the C_p at the suction side decreases slightly with the reduction in S, especially within the range of 0~0.5 chord of the rotor. Thus, it causes a slight increase in KTR when S becomes smaller.

Figure 25 presents the velocity component at 0.5 span. Due to the frozen interface between stator and rotor, the flow is interrupted when entering into the rotor domain, thus causing the length of wake to decrease accordingly with the decrease in S, which is non-physical. The rotor acceleration is obviously seen in the figure of V_z contour. The V_z increases to approximately 1.4 V_A at the channel and wake of the rotor. However, the V_c, plotted in the absolute coordinate reference system, is close to zero on the pressure side, so are the regions after the rotor. However, the high V_c exists on the suction side of the rotor and stator. The difference is that the high V_c locates at the trailing edge of the stator, but it locates at the leading edge of the rotor. As for V_r, there are negative values in a wide area, especially near the pressure side of the stator, which demonstrates that the flow has a trend towards the hub. Additionally, the velocities distribution becomes more complicated as S decreases due to the more intense interaction between stator and rotor. Further transient simulation is worthwhile for the study of the fluctuant pressure and the PJP's oscillation force.

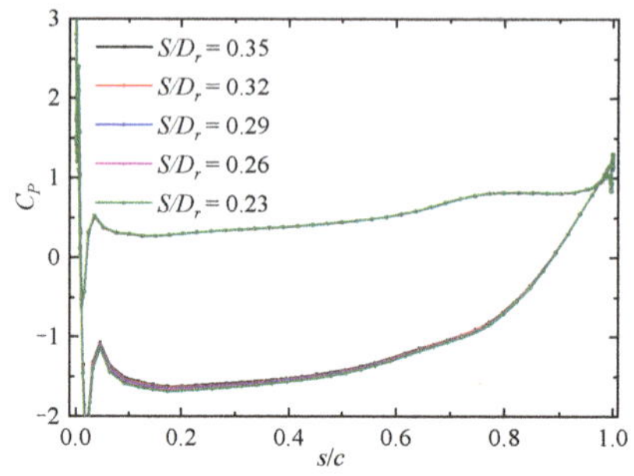

Figure 24. The static pressure coefficient of rotor at 0.5 span (s: the distance from leading edge; c: chord length).

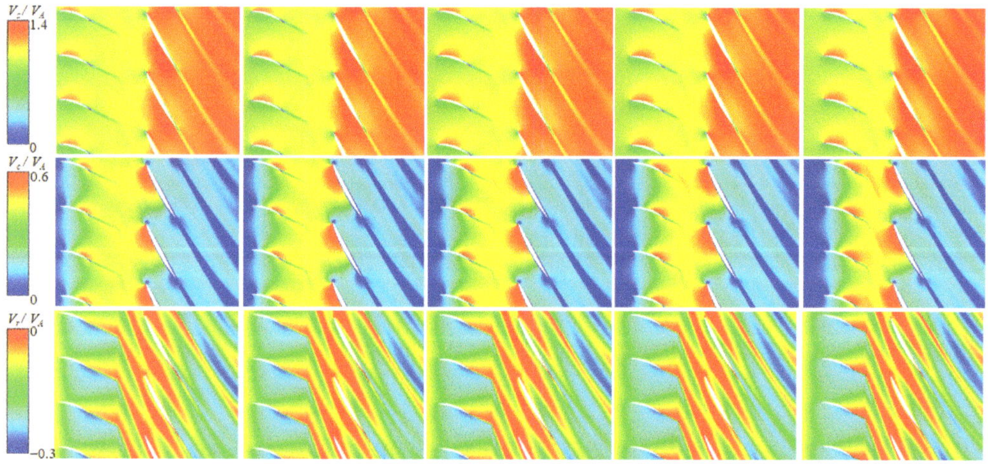

Figure 25. Contour of velocity component contour at 0.5 span (s: the distance from leading edge; c: chord length).

More specific velocity components along the circumferential direction of 0.5 span at A_3 and A_4 are plotted in Figure 26. It is evident that the lower S leads to a higher velocity at A_3. The maximum value of V_c increases tremendously with the decrease in S compared with V_z and V_r; however, the effect of S on the velocity at A_4 seems not as obvious as at A_3. The flow angle varies erratically at A_3, yet in most regions, it presents an increasing trend with the decrease in S; whereas at A_4, the change in flow angle becomes regular with S, i.e., increase slightly with S decreases, thus causing a slight increase in K_{Tr}.

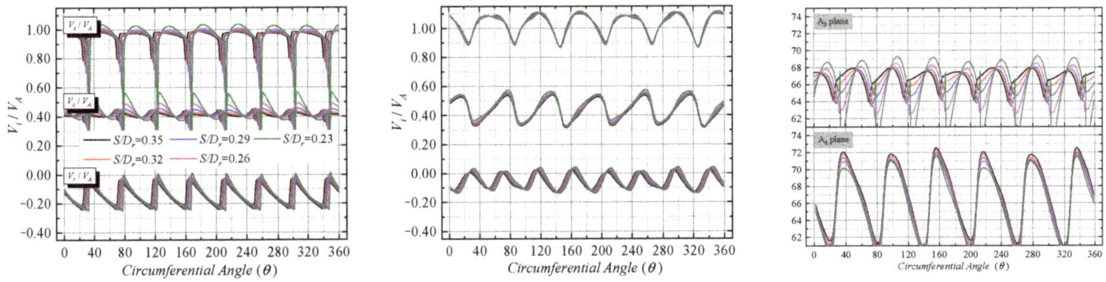

Figure 26. Velocity component and flow angle at 0.5 span (the first two columns: V_i at A_3 and A_4, respectively; the third column: flow angle at A_3 and A_4).

5. Conclusions

We have numerically investigated the hydro-performance of PJP with different stator parameters and classified these parameters into three groups:

(A) The blade number *Ns*, chord length *L*;
(B) The stagger angle α, lean angle γ, sweep angle β;
(C) The rotor–stator spacing *S*.

The SST *k*-ω turbulence model and frozen interface technique are employed to predict performance at various conditions. Based on RANS results, the verifications of mesh groups are performed to confirm appropriate mesh for further study. The results of the selected mesh give good accuracy with experiment results, which demonstrates the computation model is reasonable to predict the PJP's performance. To better understand why PJP global performance varies, we mainly analyze the global performance and flow results under $J = 0.8$, where the efficiency of the original PJP is a little lower than maximum efficiency. The numerical results are summarized as follows:

The parameters related to stator solidity, i.e., *Ns* and *L*, can accelerate flow when increasing them, yield higher incidence velocity for the rotor, and increase thrust and torque. At the same time, the efficiencies at design conditions stay almost the same. In addition, there are some distinctions between the two parameters for effect on PJP performance. The higher degree of stator solidity varies from the original PJP, the greater variation of results for two parameters.

The stator angles defined in various views are modified with the same step value (4°) to investigate their influence. The results indicate that *KT*, *KQr*, and η are positively associated with α and γ. However, no significant change occurs when adjusting β except for the off-design condition. The α has a greatly higher influence on the PJP's performance than the other two angles. The operating condition of *J* corresponding to maximum efficiency is shifted to a higher value when α increases. However, the γ only slightly affects the maximum efficiency, which increases with the increase in γ. In addition, we noted that a higher γ case gives a larger velocity and lower flow angle upstream of the rotor and eventually results in a rise of rotor thrust owing to a greatly higher change rate of velocity than flow angle. From the steady simulations, increasing α produces a large area of low-pressure regions on the suction side of the stator or rotor, which may lead to a higher fluctuation pressure and deteriorate the noise performance. In addition, the higher β, the lower chance of the local high-pressure region at shroud corner close to the leading edge of the stator, which can relieve the stress concentration without changing the PJP's performance and merits a further assessment.

The rotor–stator spacing has only a slight or limited influence on PJP performance. Even the stator has been translated towards the rotor 20 mm, or *S* decreases 34% relative to the original PJP, the rotor thrust at $J = 0.8$ rise only 1%, so does the K_{Qr} or η. Further investigation on the pressure coefficient of profile foil illustrates that *Cp* within 50% chord length from leading-edge becomes lower as *S* decrease.

Overall, among these stator parameters, the α mostly affects the PJP's global performance, followed by N_s and L, then is γ, the last two are S and β. In general, the torque and thrust for the blade (including stator and rotor) increase or decrease synchronously. The share of rotor thrust is extremely higher than the stator at design conditions. The increased K_{Ts}, caused by the increased α, stator solidity, or γ, does not significantly change K_T. In comparison, the duct provides thrust at heavy-loading conditions, which occupies 25% and 14% of the total thrust at J = 0.2 and J = 0.4 for the original PJP. A significant change of K_{Td} hence leads to an obvious change of K_T. For normal operating conditions, i.e., J = 0.8, the proportion of K_{Td} is as small as that of K_{Ts} and less than 10%, which is also not enough to change the total thrust. The ΔK_Q mainly is related to the change of K_{Qr}. The change rates of K_{Qr} for studied stator parameters are higher than K_{Qs}. The higher of K_{Qr}, the ΔK_Q becomes lower. Consequently, the J corresponding to torque balance location becomes lower.

Author Contributions: Conceptualization, F.L. and Q.H.; data curation, F.L.; investigation, F.L.; methodology, F.L.; visualization, F.L.; writing—original draft, F.L.; supervision, Q.H. and G.P.; writing—review and editing, Q.H., D.Q. and H.L. All authors have read and agreed to the published version of the manuscript.

Funding: This work is supported by the National Natural Science Foundation of China under Project Nos. 51979226 and by the Fundamental Research Funds for the Central Universities under Project Nos. 3102019HHZY030019 and 3102020HHZY030018.

Institutional Review Board Statement: Not applicable.

Informed Consent Statement: Not applicable.

Data Availability Statement: Not applicable.

Conflicts of Interest: The authors declare no conflict of interest.

References

1. Qin, D.; Huang, Q.; Shi, Y.; Pan, G.; Shi, Y.; Dong, X. Comparison of hydrodynamic performance and wake vortices of two typical types of pumpjet propulsor. *Ocean. Eng.* **2021**, *224*, 108700. [CrossRef]
2. McCormick, B.W.; Elsenhuth, J.J. Design and performance of propellers and pumpjets for underwater propulsion. *AIAA J.* **1963**, *1*, 2348–2354. [CrossRef]
3. Henderson, R.E.; McMahon, J.F.; Wislicenus, G.F. *A Method for the Design of Pumpjets*; ORL Report, No. 63-0209-0-7; Pennsylvania State Univ State College Ordnance Research Lab.: Centre County, PA, USA, 1964.
4. Furuya, O.; Chiang, W.L. *A New Pumpjet Design Theory*; Honeywell Inc.: Hopkins MN, USA, 1988.
5. Zierke, W.C.; Straka, W.A.; Taylor, P.D. An experimental investigation of the flow through an axial-flow pump. *J. Fluids Eng.* **1995**, *117*, 485–490. [CrossRef]
6. Suryanarayana, C.; Satyanarayana, B.; Ramji, K.; Saiju, A. Experimental evaluation of pumpjet propulsor for an axi-symmetric body in wind tunnel. *Int. J. Nav. Archit. Ocean. Eng.* **2010**, *2*, 24–33. [CrossRef]
7. Suryanarayana, C.; Satyanarayana, B.; Ramji, K. Performance evaluation of an underwater body and pumpjet by model testing in cavitation tunnel. *Int. J. Nav. Archit. Ocean. Eng.* **2010**, *2*, 57–67. [CrossRef]
8. Suryanarayana, C.; Satyanarayana, B.; Ramji, K.; Rao, M.N. Cavitation studies on axi-symmetric underwater body with pumpjet propulsor in cavitation tunnel. *Int. J. Nav. Archit. Ocean. Eng.* **2010**, *2*, 185–194. [CrossRef]
9. Shirazi, A.T.; Nazari, M.R.; Manshadi, M.D. Numerical and experimental investigation of the fluid flow on a full-scale pump jet thruster. *Ocean. Eng.* **2019**, *182*, 527–539. [CrossRef]
10. Lee, Y.T.; Hah, C.; Loellbach, J. Flow analyses in a single-stage propulsion pump. *J. Turbomach.* **1996**, *118*, 240–248. [CrossRef]
11. Ivanell, S. *Hydrodynamic Simulation of a Torpedo with Pumpjet Propulsion System*; Royal Institute of Technology: Stockholm, Sweden, 2001.
12. Das, H.N.; Jayakumar, P.; Saji, V.F.; Yerram, R. CFD examination of interaction of flow on high-speed submerged body with pumpjet propulsor. In Proceedings of the HIPER 06: 5th International Conference on High-performance Marine Vehicles, Launceston, Australia, 8–10 November 2006; Australian Maritime College: Newnham, Australia, 2006; p. 466.
13. Dong, Y.; Duan, X.; Feng, S.; Shao, Z. Numerical simulation of the overall flow field for underwater vehicle with pump jet thruster. *Procedia Eng.* **2012**, *31*, 769–774. [CrossRef]
14. Park, W.G.; Jang, J.H.; Chun, H.H.; Kim, M.C. Numerical flow and performance analysis of waterjet propulsion system. *Ocean. Eng.* **2005**, *32*, 1740–1761. [CrossRef]
15. Warn-Gyu, P.; Jung, Y.-R.; Kim, C.-K. Numerical flow analysis of single-stage ducted marine propulsor. *Ocean. Eng.* **2005**, *32*, 1260–1277.

16. Huyer, S.A. Postswirl maneuvering propulsor. *J. Fluids Eng.* **2015**, *137*, 9225. [CrossRef]
17. Ahn, S.J.; Kwon, O.J. Numerical investigation of a pump-jet with ring rotor using an unstructured mesh technique. *J. Mech. Sci. Technol.* **2015**, *29*, 2897–2904. [CrossRef]
18. Lu, L.; Pan, G.; Wei, J.; Pan, Y. Numerical simulation of tip clearance impact on a pumpjet propulsor. *Int. J. Nav. Archit. Ocean. Eng.* **2016**, *8*, 219–227. [CrossRef]
19. Lu, L.; Gao, Y.; Li, Q.; Du, L. Numerical investigations of tip clearance flow characteristics of a pumpjet propulsor. *Int. J. Nav. Archit. Ocean. Eng.* **2018**, *10*, 307–317. [CrossRef]
20. Qin, D.; Pan, G.; Huang, Q.; Zhang, Z.; Ke, J. Numerical investigation of different tip clearances effect on the hydrodynamic performance of pumpjet propulsor. *Int. J. Comput. Methods* **2018**, *15*, 1850037. [CrossRef]
21. Li, H.; Pan, G.; Huang, Q. Transient analysis of the fluid flow on a pumpjet propulsor. *Ocean. Eng.* **2019**, *191*, 106520. [CrossRef]
22. Yu, H.; Duan, N.; Hua, H.; Zhang, Z. Propulsion performance and unsteady forces of a pump-jet propulsor with different pre-swirl stator parameters. *Appl. Ocean. Res.* **2020**, *100*, 102184. [CrossRef]
23. Amini, Y.; Kianmehr, B.; Emdad, H. Dynamic stall simulation of a pitching hydrofoil near free surface by using the volume of fluid method. *Ocean. Eng.* **2019**, *192*, 106553. [CrossRef]
24. Menter, F. Zonal two equation kw turbulence models for aerodynamic flows. In Proceedings of the 23rd Fluid Dynamics, Plasmadynamics, and Lasers Conference, Orlando, FL, USA, 6–9 July 1993; p. 2906.
25. Rao, Z.Q.; Li, W.; Yang, C.J. Simulation of unsteady interaction forces on a ducted propeller with pre-swirl stators. In Proceedings of the Third International Symposium on Marine Propulsors, SMP, Launceston, Tasmania, Australia, 5–8 May 2013; Volume 13, pp. 149–155.
26. Lü, X.J.; Zhou, Q.D.; Bin, F.A.N.G. Hydrodynamic performance of distributed pump-jet propulsion system for underwater vehicle. *J. Hydrodyn.* **2014**, *26*, 523–530. [CrossRef]
27. Motallebi-Nejad, M.; Bakhtiari, M.; Ghassemi, H.; Fadavie, M. Numerical analysis of ducted propeller and pumpjet propulsion system using periodic computational domain. *J. Mar. Sci. Technol.* **2017**, *22*, 559–573. [CrossRef]
28. Qiu, J.T.; Yang, C.J.; Dong, X.Q.; Wang, Z.L.; Li, W.; Noblesse, F. Numerical simulation and uncertainty analysis of an axial-flow waterjet pump. *J. Mar. Sci. Eng.* **2018**, *6*, 71. [CrossRef]
29. Qiu, C.; Pan, G.; Huang, Q.; Shi, Y. Numerical analysis of unsteady hydrodynamic performance of pump-jet propulsor in oblique flow. *Int. J. Nav. Archit. Ocean. Eng.* **2020**, *12*, 102–115. [CrossRef]
30. Lu, L.; Wang, C.; Qin, D. Numerical investigations of flow characteristics of a pumpjet propulsor in oblique inflow. *Appl. Ocean. Res.* **2020**, *103*, 102343. [CrossRef]
31. Yuan, J.; Chen, Y.; Wang, L.; Fu, Y.; Zhou, Y.; Xu, J.; Lu, R. Dynamic analysis of cavitation tip vortex of pump-jet propeller based on DES. *Appl. Sci.* **2020**, *10*, 5998. [CrossRef]
32. Menter, F.R.; Kuntz, M.; Langtry, R. Ten years of industrial experience with the SST turbulence model. *Turbul. Heat Mass Transf.* **2003**, *4*, 625–632.
33. Wilson, R.; Stern, F. Verification and validation for RANS simulation of a naval surface combatant. In Proceedings of the 40th AIAA Aerospace Sciences Meeting & Exhibit, Reno, NV, USA, 14–17 January 2002; p. 904.
34. Wilson, R.; Shao, J.; Stern, F. Discussion: Criticisms of the "correction factor" verification method. *J. Fluids Eng.* **2004**, *126*, 704–706. [CrossRef]
35. Celik, I.B.; Ghia, U.; Roache, P.J.; Freitas, C.J. Procedure for estimation and reporting of uncertainty due to discretization in CFD applications. *J. Fluids Eng.-Trans. ASME* **2008**, *130*, 953. [CrossRef]
36. Li, H.; Huang, Q.; Pan, G.; Dong, X. The transient prediction of a pre-swirl stator pump-jet propulsor and a comparative study of hybrid RANS/LES simulations on the wake vortices. *Ocean. Eng.* **2020**, *203*, 107224. [CrossRef]

Journal of
Marine Science and Engineering

MDPI

Article

Numerical Prediction of Convective Heat Flux on the Flight Deck of Naval Vessel Subjected to a High-Speed Jet Flame from VTOL Aircraft

Ho-Sang Jang [1], Se-Yun Hwang [2] and Jang-Hyun Lee [1,*]

[1] Department of Naval Architecture and Ocean Engineering, Inha University, 100, Inha-ro, Michuhol-gu, Incheon 22212, Korea; hsjang@inha.edu

[2] Extreme Technology Research Center for Ship and Offshore Platform, Inha University, 100, Inha-ro, Michuhol-gu, Incheon 22212, Korea; syhwang@inha.ac.kr

* Correspondence: jh_lee@inha.ac.kr

Abstract: This study examines the heat flux and convective heat transfer generated when a vertical take-off and landing (VTOL) aircraft takes off and lands on the flight deck of a naval vessel. A procedure for analyzing the convective heat transfer imposed on the deck by the high-temperature and high-velocity impingement of a VTOL jet is described. For the analysis, the jet velocity and the deck arrival temperature were calculated by applying computational fluid dynamics (CFD), assuming that the heat flow is an impingement jet. The relationships between the diameter of the jet, the speed of impingement, and the exhaust temperature of VTOL are introduced to assess the inlet condition. Heat flow was analyzed using CFD techniques, and Reynolds-averaged Navier–Stokes (RANS) and k-ε models were applied to model the turbulent motion. A procedure for evaluating the convection coefficient and convective heat flux from the calculated local velocity and temperature is presented. Simultaneously, a method for compensating the convection coefficient considering the singular velocity at the stagnation point is proposed. Furthermore, the accuracy was verified by comparing the convective heat flux and deck temperature predicted using CFD with the existing experimental studies. Finally, by applying finite element analysis (FEA) based on the thermal-structural interaction, the magnitude of thermal deformation due to conductive temperature and heat flux was presented as a design application of the flight deck.

Keywords: convective heat transfer; impinging jet; VTOL (vertical take-off and landing); thermal flow; stagnation point; CFD (computational fluid dynamics)

Citation: Jang, H.-S.; Hwang, S.-Y.; Lee, J.-H. Numerical Prediction of Convective Heat Flux on the Flight Deck of Naval Vessel Subjected to a High-Speed Jet Flame from VTOL Aircraft. *J. Mar. Sci. Eng.* **2022**, *10*, 260. https://doi.org/10.3390/jmse10020260

Academic Editor: Joško Parunov

Received: 24 December 2021
Accepted: 5 February 2022
Published: 14 February 2022

1. Introduction

Flame jets from vertical take-off and landing (VTOL) aircraft harm the flight decks of naval vessels, generating excessive thermal deformation and thermal stress. The purpose of this study is to predict the magnitude of the convective heat transfer of jet flames that generate thermal stresses on the flight deck. For predicting the distribution of local convective heat transfer, it is necessary to analyze the thermal flow and convective heat flux simultaneously. The turbulent thermal fluid flow of the VTOL represents an impinging jet. For predicting convective heat transfer, several variables such as the velocity and temperature of the flow and the temperature-dependent material properties must be considered. The complexity of the jet impingement makes it challenging to predict the convective heat transfer to surfaces that receive these violent flows.

Various studies have been conducted on the heat transfer of jet impingement for the Reynolds number, the shape and array of the nozzle, and the standoff distance of the nozzle and impingement wall. Most of them use both experimental measurement and numerical analysis to predict convective heat transfer. Jambunathan et al. [1] reviewed experimental data on heat transfer rates for circular jets impinging on a flat surface. The effects of the

Reynolds number, jet-to-surface distance, nozzle geometry, and jet orientation and shape have been most widely studied. Katti and Prabhu [2] experimentally investigated the effect of the nozzle-to-plate spacing and Reynolds number on the heat transfer distribution to normally impinging circular air-jet on a flat surface. VTOL heat flow can be expressed as a turbulent thermal flow of exhaust gas on the thermal deck. Crosser [3] and Naval Surface Warfare Center (NSWC) research [4] performed real-scale experimental results for a VTOL. Crosser [3] experimentally presented the convective heat transfer coefficient and temperature distribution of VTOL exhaust gas. NSWC [4] measured the heat effect of a land-based experiment of a VTOL for the flight deck. This experiment presented the history of maximum temperature on the thermal deck according to the operating conditions of the VTOL

Zuckerman and Lior [5] presented a series of different turbulence models used to examine their performance in simulating jet impingement cooling of a flat target under a round jet. Pattamatta et al. [6] focused on the theoretical treatment of the problem by numerical model or experiments with laboratory-controlled systems using axisymmetric jets impinging on flat plates. Barata [7] performed laser doppler anemometry (LDA) measurements and numerical simulations for impinging jets and studied the effect of flow on the impinging jets below a VTOL aircraft in the ground. Matsumoto et al. [8] examined the heat transfer and the flow pattern from an impinging jet. However, the Reynolds number and scale frame jet VTOL are distinct problems from those in the investigated studies. Furthermore, several studies use a Nusselt number or convection coefficient with interest in heat convection at the stagnation point. Gauntner et al. [9] suggest that the local radial velocity gradient parameter at the stagnation influences stagnation point heat transfer coefficients. Because the horizontal velocity (parallel to the impingement wall) is zero in the stagnation region, Martin and Boyd [10] considered the vertical velocity of the arrival stream. Although several studies on jet impingement have proposed convection coefficient equations for various Reynolds numbers and nozzle sizes, it is difficult to use the proposed convection coefficient equations because the Reynolds number and nozzle size of VTOL jets are very large compared to the studies investigated.

Considering relatively little has been published about the computational model where the thermal deck is heated by hot impingement, this study aims to examine the heat transfer characteristics of a VTOL jet impinging onto the deck of a naval vessel. The primary goal of the present study is to numerically predict the distribution of convective heat flux delivered to the flight deck by the nozzle flame. The literature to date agrees on the use of computational fluid dynamics (CFD) using Reynolds-averaged Navier-Stokes (RANS) equations in a turbulence model to study the effect of the heat transfer rate between the jet gas and the heated surface. CFD enables simulating convective heat flux and heat energy transfer on the impingement wall. This study analyzes the heat flow behavior using a three-dimensional (3D) finite volume method and a k-ε turbulence model based on the RANS equation. Transient heat conduction and thermal-structural analysis are then introduced.

2. Problem Description of VTOL with an Impinging Jet

2.1. Characteristic of Heat Transfer by VTOL

VTOL aircrafts operate on small-to-medium aircraft carrier naval vessels. VTOL can hover, take off, and land vertically on the flight deck. This study is concerned with the exhaust flame of the MV-22 Osprey, a tilt-rotor type VTOL aircraft designed to operate on naval vessels. The Osprey is unique because it uses two engines positioned on fixed wingtips housed in nacelles that rotate to enable the MV-22 to land and take off vertically. For take-off and landing, it typically hovers using vertical nacelles and horizontal propeller. Exhaust heat from the turboprop engines can potentially damage the flight decks of a ship [11]. A schematic view of an impinging jet issuing from a nozzle of the VTOL on a flat deck is depicted in Figure 1, which was adapted from Annaswamy (2003) [12] and Choi (2005) [13]. The exhaust gas of VTOL is vertically sprayed from two turboprops to the flight deck during landing and take-off operation. A ground effect occurs because there are two adjacent jets, and each impinging downward jet creates an upward reingestion of exhaust gases.

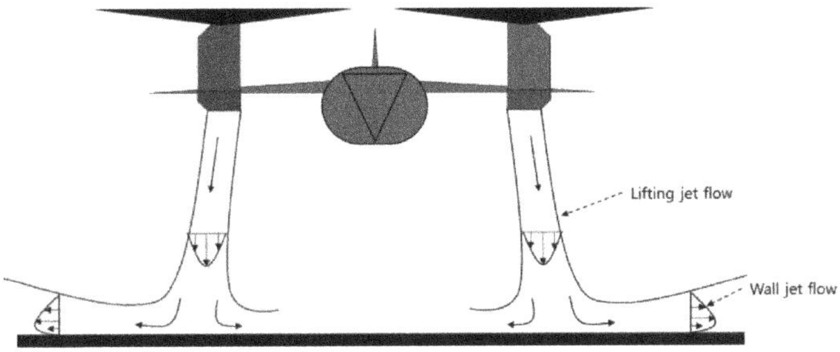

Figure 1. Schematic view of VTOL operation on naval vessel deck.

Considering that the impingement of each jet onto a surface leads to a highly localized heat flow rate at the centroid, the modeling of one jet can lead to a more conservative prediction of the amount of heat flow. Therefore, the present study focuses on the flow of a single jet rather than the wake of the two jets. Two assumptions are made to adopt the one impingement in the analysis: (1) The distance between the two jets is far enough so that the thermal flow of the jet is not amplified. (2) The distance between the jet nozzle and the deck was assumed to be that of the landing status. Therefore, the distance between the jet flame and the deck when the VTOL landed, was reflected in the analysis. This is because the impinging flame transfers the largest heat to the deck at the moment of landing. The distance between the nozzle and the deck is fixed for the computational model. The gas flow rate, nozzle diameter and shape, standoff distance from the nozzle to the deck, and operation condition of VTOL should be considered in the CFD model. In all cases, the temperature and velocity distribution of the jet stream over the surface are required to approximate the magnitude of the heat flow. This heat flow has two consequences: (1) the high-temperature gradient generated by the heat flux and (2) the development of thermal stress and deformation of the flight deck.

Beltaos [14,15] and Rajaratnam [16] also divided the flow region into three regions (Figure 2): free jet, impingement, and wall jet. Figure 2 was modified from Beltaos [14,15] and Rajaratnam [16]. Katti and Prabhu [6] also divided the three regions of the impingement wall that extend to a distance from the center by the spread of the fluid; three regions can be listed, shown in Table 1 as suggested by Katti and Prabhu [2]. The hydrodynamic pattern of an impinging jet is crucial in studying the thermal effects of exhaust gas acting orthogonally on the target surface. The potential core can be observed until $H/D = 4$–6 (H: standoff distance from nozzle to deck, D: diameter of nozzle) and exists within the free jet region where the jet exit velocity is conserved and the turbulence intensity level is relatively low. A shear layer exists between the potential core and the ambient fluid where the turbulence is relatively high, and the mean velocity is lower than the jet exit velocity. The potential core diminishes in width as the shear layer around the jet grows. After the jet is fully developed, the axial velocity profile can be represented by a Gaussian distribution. The shear layer entrains ambient fluid and causes the jet to spread radially. Beyond the potential core, the shear layer spreads to the point where it penetrates the centerline of the jet. At this stage, the centerline velocity decreases, and the turbulence intensity increases. The stagnation region spans approximately 1.00 times the nozzle diameter for a laminar flow and varies in size for a turbulent flow [17–20]. The stagnation region includes the stagnation point where the mean velocity is zero. Within this region, the free jet is deflected into the wall jet flow and the flow is affected by the presence of the impingement surface.

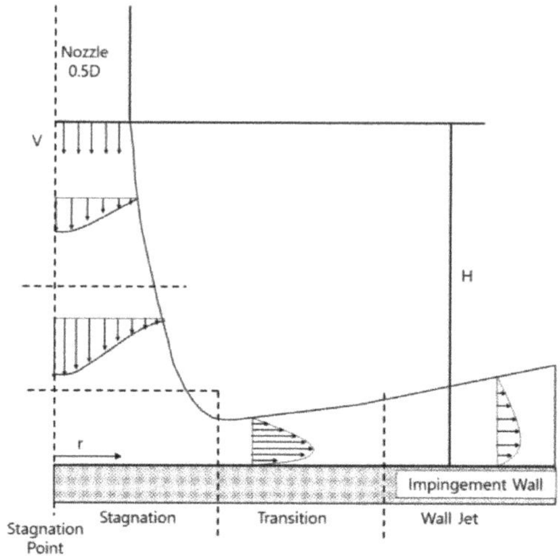

Figure 2. Schematic diagram of the flow region of a circular impinging jet.

Table 1. Regions of the impingement wall.

Region	Stagnation Region	Transition Region	Wall Jet Region
Section	$0 < r/D \leq 1$	$1 < r/D \leq 2.5$	$2.5 < r/D$

The flow accelerates as it advances through the stagnation region due to the difference in static pressure between stagnation and an outer region. The flow velocity becomes higher than the velocity at the nozzle exit to maintain continuity of the flow. This region is called the acceleration region. The viscosity effects and loss of momentum cause the flow velocity to decrease gradually as the fluid progresses through the acceleration region parallel to the surface. The behavior of the wall jet region is characterized by a flow in the outward radial or spanwise direction. The wall jet region has a lower velocity than the acceleration region due to the loss of momentum. In this study, the characteristic of heat transfer developed in the stagnant region was analyzed in Section 4.4.

2.2. Schematic Formulation of Convective Heat Transfer by the Impinging Jet

Jet impingement heat transfer and flow features depend on parameters such as the jet's Reynolds number, Prandtl number, nozzle geometry, spacing between nozzle exit and impingement plate, and distance from the stagnation point. The heat flow rate of impingement can be expressed as the convective heat transfer of a turbulent flow, where the convective heat transfer applies to heat transfer through a fluid to a solid. This mode propagates heat via the mixing of fluid regions on both molecular and large scales. The heat flux from the jet fluid to the target surface can be described by Newton's law as:

$$q'' = h\left(T_s - T_{jet}\right), \tag{1}$$

where q'' is the convective heat flux and h, T_s, and T_{jet} are the convection coefficient, the temperature on the target surface exposed to the jet flow, and the temperature of a jet fluid, respectively. The convection coefficient is derived from the Nusselt number (Nu), as defined by Equation (2).

$$h = Nu\frac{k}{\text{Characteristic lengh}} = Nu\frac{k}{D}, \tag{2}$$

where k represents the thermal conductivity. The Nusselt number (Nu) on the impinging wall is represented by the functions of the jet Reynolds number (Re), Prandtl number (Pr), and H/D, where H is the stand-off distance from the jet nozzle to the wall, respectively [2,17,21]. Because D is constant in this study, the Nusselt number can be expressed by the function of both the Reynolds number and the Prandtl number. Therefore, the numerical result of the jet flow is used to find both the Reynolds number and the Prandtl number.

The next section presents the results of the Nusselt number predicted based on the resultant motion of the jet flow. For the Nusselt number, the Reynolds number (Re) is determined according to the nozzle diameter (D) and jet velocity (V) as $\rho_f V \cdot D / \mu_f$, where μ_f is the viscosity of the jet fluid. In this study, Re is adjusted by a calibrated velocity at the stagnation point, as explained in Section 4.4 because ν is a temperature-dependent variable used in calculating the Reynolds number. The Prandtl number (Pr) means the ratio of fluid thermal diffusivity to viscosity and is obtained by ν/α, where ν and α means the kinematic viscosity and thermal diffusivity of the jet fluid, respectively. Based on the definition of thermal diffusivity, the value of α can be calculated by $k/\rho c$, where ρ and c denote the density of exhaust gas and the specific heat, respectively. The value of the Prandtl number is approximately 0.7~0.8 in the gas state [22]. Therefore, the Nusselt number is expressed by a function of Re and Pr in Equation (3).

$$Nu = f(Re,\ Pr,\ \text{with constant } H/D). \tag{3}$$

Finite element analysis (FEA) modeling of heat conduction and thermal deformation inside the impinging wall can then be implemented based on the heat flux models, heating and cooling boundary conditions, and temperature-dependent material properties. Based on the above equations, this paper divides the heat transfer behavior of exhaust gas into three steps, as depicted in Figure 3.

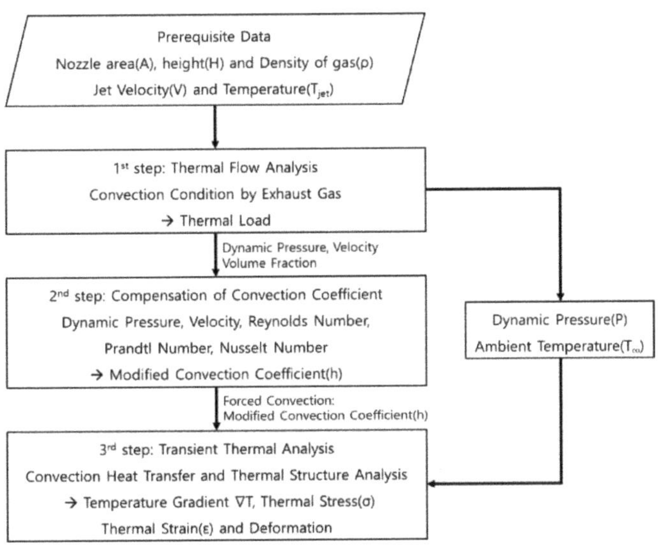

Figure 3. Analysis procedure of VTOL on a naval vessel deck.

First, the heat convection coefficient is calculated from the thermal flow behavior predicted by the analysis. The temperature distribution and velocity and pressure of the jet flow on the surface are then used to predict the Nusselt number and calculate the convection coefficient. Furthermore, the Reynolds number and Prandtl number are predicted. Second, the convective heat transfer coefficient is compensated by considering the vertical velocity and the dynamic pressure in the centroid stagnation zone. Third, the convective heat transfer coefficient predicted in the second step is applied to the coupled thermal-structural FEA. The coupled thermal-structural FEA is used to calculate the temperature distribution, thermal deformation, and thermal stress field inside the flight deck. The exhaust gas temperature distribution and dynamic pressure of impacting flow on the deck were applied as the load conditions.

2.3. Governing Equation

Many studies [5,23,24] have shown that CFD can be used to solve the convective heat transfer of impinging jets. In this study, a numerical procedure that simulates the thermal flow of VTOL impingement was performed using ANSYS™ CFX Solver®under Workbench™ version 2021R1 (ANSYS Inc., Canonsburg, PA, USA). The equations describing the thermal flow of the impinging jet are represented by transport equations of momentum (Navier-Stokes equations) and energy (heat diffusion equation for the flow), developed from the mass conservation law (the continuity equation), momentum, and energy conservation (Olsson et al. [23]). The thermal flow is governed by the incompressible form of the RANS equations and by the equations describing the transport of energy, momentum, and energy [23].

$$\frac{\partial U_j}{\partial x_j} = 0, \tag{4}$$

$$\rho \frac{\partial U_i}{\partial t} + \rho \frac{\partial (U_i U_j)}{\partial x_j} = -\frac{\partial P}{\partial x_i} + \frac{\partial}{\partial x_j}\left(\tau_{ij} + \tau_{ij}^{turb}\right), \tag{5}$$

$$\rho c_p \frac{\partial T}{\partial t} + \rho c_p \frac{\partial (U_j T)}{\partial x_j} = \frac{\partial}{\partial x_j}\left(q_j + q_j^{turb}\right), \tag{6}$$

where U_i is the velocity of the fluid, ρ is the density of the fluid, τ_{ij} is the Reynolds shear stress, c_p is the specific heat, T is the temperature, and q_i is the heat flux. P is the pressure, and superscript *turb* is the turbulence. Those variables are defined by: $\tau_{ij} = \mu\left(\frac{\partial U_i}{\partial x_j} + \frac{\partial U_j}{\partial x_i}\right)$, $\tau_{ij}^{turb} = -\rho \overline{u\prime_i u\prime_j}$, $q_j = \frac{\mu c_p}{Pr}\frac{\partial T}{\partial x_j}$, $q_j^{turb} = -\rho c_p \overline{u\prime_j T\prime}$, where μ is the dynamic viscosity of fluid, $u\prime_i$ and $T\prime$ are the fluctuating velocity and temperature, and Pr is the Prandtl number. It would be impossible to solve these equations analytically because of non-linearity and the stochastic nature of turbulence. The additional terms that appear due to averaging the velocity and temperature are the Reynolds stresses and the turbulent heat flux. These models are the closure problem of turbulence.

Angioletti et al. [24,25] extensively investigated the flow field behavior in the vicinity of the stagnation region. They found that the *k-w* SST(Shear Stress Transport) model produces suitable results for a lower Re, while *k-ε* performed better for a high Re. Furthermore, in the study by Achari and Das [26] and in the studies by Coussirat [27,28], the standard *k-ε* model was adopted. The standard *k-ε* model is widely used and has relatively high accuracy for fluid flow analysis. With the standard two-equation *k-ε* model, the turbulent viscosity is evaluated from $\mu_t = \rho C_\mu \frac{k^2}{\varepsilon}$. where μ_t is the turbulent viscosity, k is the turbulent kinetic energy, ε is the dissipation rate of the turbulent kinetic energy, C_μ is a turbulence constant, and ρ is the density of the fluid. The turbulent kinetic energy, k, and the dissipation rate of turbulent kinetic energy, ε, are obtained by solving a conservation equation for each of these two quantities. Those equations are defined by Equations (7) and (8).

$$\frac{\partial \rho k}{\partial t} + \frac{\partial (U_j T)}{\partial x_j} = \frac{\partial}{\partial x_j}\left[\left(\mu + \frac{\mu_t}{\sigma_k}\right)\frac{\partial k}{\partial x_i}\right] + P_k - \rho\varepsilon + P_{kb}, \tag{7}$$

$$\frac{\partial \rho \varepsilon}{\partial t} + \frac{\partial (\rho U_j \varepsilon)}{\partial x_i} = \frac{\partial}{\partial x_i}\left[\left(\mu + \frac{\mu_t}{\sigma_\varepsilon}\right)\frac{\partial \varepsilon}{\partial x_i}\right] + \frac{\varepsilon}{k}(C_{\varepsilon 1} P - C_{\varepsilon 2}\rho\varepsilon), \tag{8}$$

where P_k is the production rate of the turbulent kinetic energy defined as $P_k = -\overline{\rho u_i u_j}\frac{\partial U_j}{\partial x_j}$. C_μ, σ_ε, σ_k, $C_{\varepsilon 1}$, and $C_{\varepsilon 2}$ are empirical constants as shown in Table 2. These default values are defined from experiments with air or water for fundamental turbulent shear flows including homogeneous shear flows and decaying isotropic grid turbulence. [29–31].

Table 2. Empirical constants for the standard k-ε model.

Parameter	Value
C_μ	0.09
σ_k	1.0
σ_ε	1.3
$C_{\varepsilon 1}$	1.44
$C_{\varepsilon 2}$	1.92

3. VTOL Simulation Model

3.1. Estimation of Nozzle Velocity

In this study, it is assumed that the VTOL engine is a turboprop type, and the schematic view of that engine's propulsion is illustrated in Figure 4. The turboprop simplifies the VTOL thrust power so that the exhaust velocity can be estimated based on the two types of thrust with the propeller (T_{prop}) and turbojet (T_{jet}).

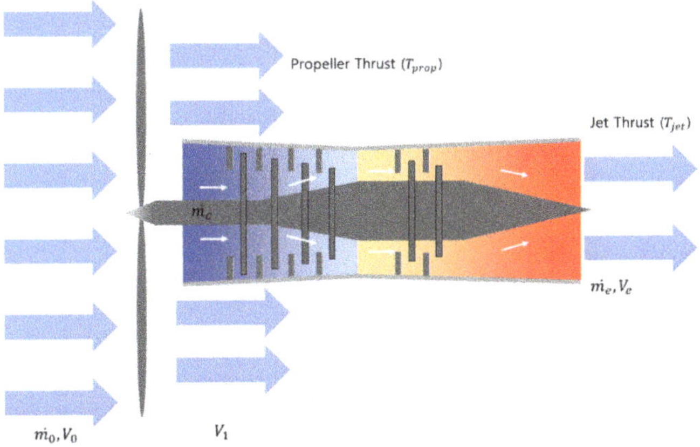

Figure 4. Schematic turboprop engine operation.

Bolkcom [32] suggests that the maximum weight of the take-off (W) was about 25 tons (55,000 lbs). The sum of thrust must be equal or greater than the weight of the VTOL to hover or take off. Benson [33] expressed the thrust force of turboprop as in Equation (9). Rotaru and Todorov [34] expressed the propeller thrust as (10) for the VTOL to maintain the hovering state.

$$W = \sum \text{Trust} = 2T_{prop} + 2T_{jet} = 2\left(\dot{m}_0(V_0 - V_1) + \dot{m}_e(V_e - V_1)\right) = 244.8[kN], \quad (9)$$

$$T_{prop} = \dot{m}_0(V_0 - V_1) = 2\rho_a A_p V_1^2, \quad (10)$$

$$T_{jet} = \dot{m}_e(V_e - V_1) = \rho_e A_e V_e(V_e - V_1), \quad (11)$$

where T_{porp} is the thrust of a propeller, ρ_a is the density of air, A_p is the area of a propeller, V_1 is the induced velocity, T_{jet} is the thrust of a jet, ρ_e is the density of exhaust gas, A_e is the area of a nozzle, and V_e is the exhaust gas velocity. The predicted VTOL velocity is

21.2 m/s [35]. Assuming that the motion of the VTOL is hovering, and the turboprop thrust force is the same as the weight of the VOTL, the exhaust gas velocity is assumed to be approximately 105 m/s.

3.2. Description of Simulation Model Set Up

The steady-state analysis in three dimensions of the heat transfer associated with the local Nusselt numbers from an exhaust jet impinging on a solid were performed in CFX. The Nusselt number describes the dimensionless heat transfer represented by convective heat flux, while the Reynolds number is based on the jet velocity and the width of the jet. This study was assumed to be completely turbulent when exiting the nozzle and adiabatic at the solid wall. The simulation model was defined by simplifying the geometry and the domain (Figure 5).

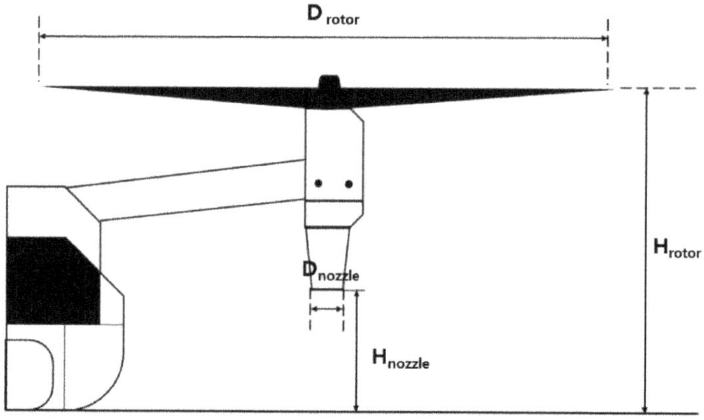

Figure 5. Schematic diagram of VTOL on the deck.

The length of an impingement wall (L) is defined as r/D = 11(L = 10,000 mm) from the center. The jet nozzle and propeller diameters are 900 mm and 11,580 mm, and their heights from the impingement wall are 1320 mm and 6100 mm. The geometry specifications of the analysis model are summarized in Table 3, and the simulation model is depicted in Figure 6.

Table 3. Simulation model geometry.

Specification	Value
Weight	24,950 (kg)
Nozzle Diameter (D_{nozzle})	0.9 (m)
Nozzle − to − Plate (H_{nozzle})	1.32 (m)
Propeller Diameter (D_{rortor})	11.6 (m)
Propeller − to − Plate (H_{rortor})	6.1 (m)
Length of simulation mode	10 (m)

The parameters for the computational domain and the boundary conditions for the simulations are summarized in Table 4 and Figure 7. CFD methods based on unstructured grids have the advantage of efficiently handling complex geometries and can improve solution accuracy by refining cells locally as required. In this study, unstructured volume meshes were created using ANSYS 2021 R1 Meshing tool (ANSYS Inc., Canonsburg, PA, USA). Although an unstructured mesh consisting of tetrahedral elements was generated, the finer mesh was generated in the nozzle and impinging wall to compute the local velocity and temperature in the region.

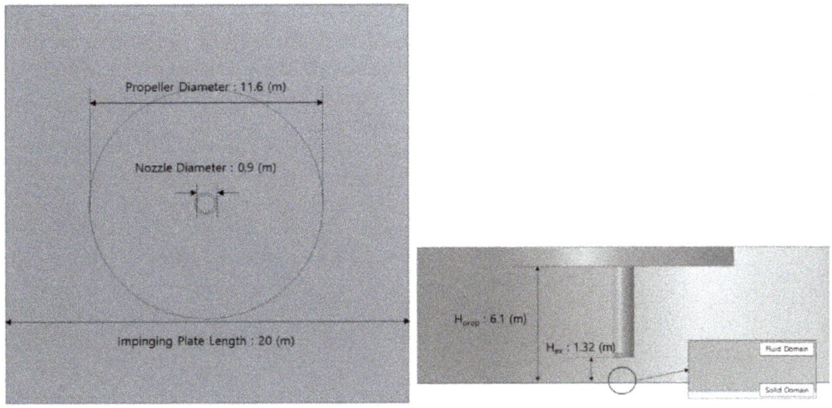

Figure 6. Simulation model geometry in the computational domain. (**left**) Top view (**right**) Side view.

Table 4. Simulation model summary.

Analysis Model	Description
Analysis Type	Steady state
Domain	Exhaust gas and air
Multiphase Model	Homogeneous model
Turbulence Model	Standard k-ε
Initial Condition	Temperature: 25 (°C)
	VF (volume fraction): Air (1.0), exhaust gas (0.0)
Boundary Condition (Inlet and Outlet)	Inlet condition: Exhaust nozzle - Velocity: 105 (m/s) - Temperature: 260 (°C) - VF (Volume fraction): Exhaust gas (1.0), Air (0.0)
	Inlet condition: Propeller - Velocity: 21.2 (m/s) - Temperature: 25 (°C) - VF (volume fraction): Exhaust gas (0.0), Air (1.0)
	Outlet: Opening condition

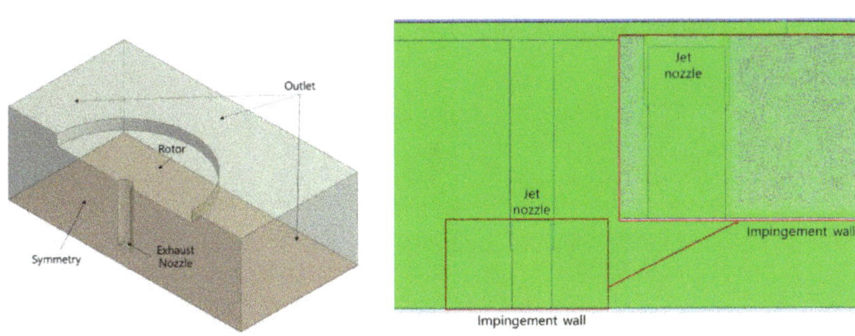

Figure 7. The computational domain Summary. (**left**) Boundary condition (**right**) Mesh.

For analysis efficiency, a half-model with a symmetrical condition applied at the center was used. The k-ε model was adopted for the turbulence model, and the analysis domain is defined as air and exhaust gas. The inlet condition of the propeller is air at 21.2 m/s and

25 °C; the inlet condition of a jet nozzle is exhaust gas at 100 m/s and 260 °C. An outlet with opening condition (Zero relative pressure) was given to the side walls of domain where the exhaust gas exits. The inlet and outlet boundary conditions used in the analysis are described in Figure 7. The values of the initial conditions and boundary conditions are also described in Table 4.

In order to check the grid dependence on the convergency of temperature and velocity, the results at the stagnation region according to the grid sizes are summarized in Table 5. Our values of interest are the velocity and temperature of flow, we make sure that these have converged to a steady value when the grid size is less than 80 mm. Ensuring that these values have reached an almost steady solution, the grid density of 21,791,040 elements for 50 mm was chosen for the CFD simulation.

Table 5. Convergence of temperature and velocity using grid refinement.

Grid Size (mm)	T [K]		V (m/s)	
	Value	Difference	Value	Difference
200	537.796	-	81.451	
180	538.585	0.15%	86.207	5.84%
150	538.892	0.06%	88.707	2.90%
120	539.643	0.14%	92.717	4.52%
100	539.874	0.04%	94.259	1.66%
80	539.974	0.02%	95.166	0.96%
50	540.007	0.01%	95.504	0.36%

4. Numerical Results and Discussion

4.1. *Velocity and Streamline of the Exhaust Gas*

The fully developed turbulent 3D, steady, incompressible, single confined impinging jet flow is numerically simulated. The radial-velocity profiles of exhaust gas in the impingement wall are also depicted in Figure 8, revealing that the velocity in the stagnation region is much lower than in the other locations. Furthermore, the velocity vector and streamline distributions are depicted in Figure 9. The stagnation region is characterized by the dramatic curvature caused by the flow obstructing the impingement wall. The local radial velocity gradient is a parameter influencing the heat transfer coefficients.

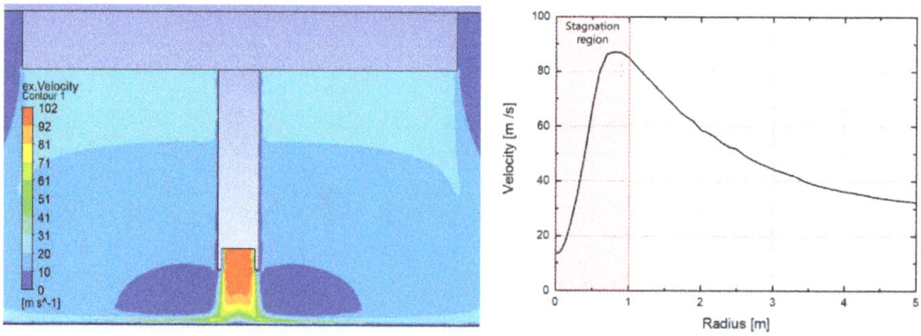

Figure 8. Velocity section contour (**left**) and distribution (**right**) of the exhaust gas.

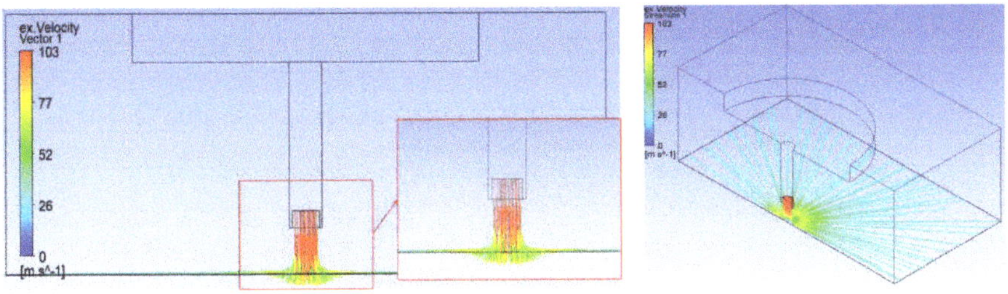

Figure 9. Velocity vector (**left**) and streamline (**right**) of the exhaust gas.

4.2. Volume Fraction of the Exhaust Gas

Volume fraction represents the ratio of exhaust gas to air. CFX calculates the convection coefficient based on momentum without considering the volume fraction of exhaust [36]. That is, despite the low volume fraction of exhaust gas, CFX yields an excessive convective heat transfer coefficient. As shown in Figure 10, a large convective heat transfer coefficient is calculated despite the very low volume fraction of the exhaust gas on the upper surface. Therefore, it is necessary to correct the convection coefficient by multiplying the volume fraction of the injection gas. The distribution of the volume fraction of exhaust gases at the impact wall is shown in Figure 11.

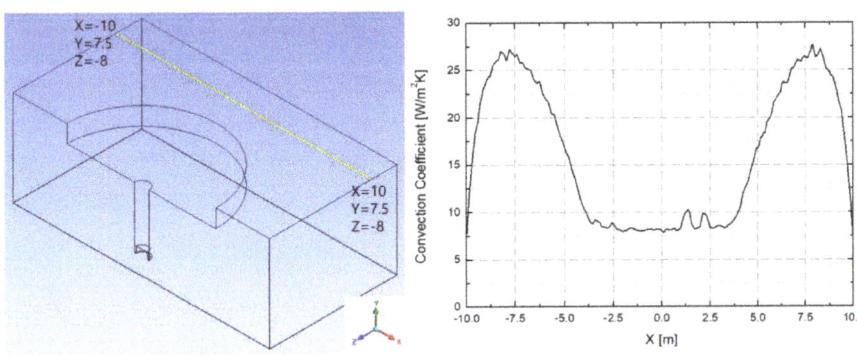

Figure 10. Convection coefficient location (**left**) and distribution (**right**) on the upper surface.

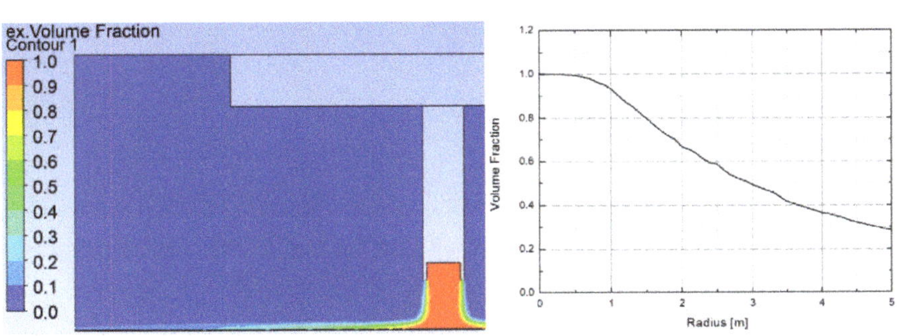

Figure 11. Volume fraction contour (**left**) and distribution (**right**) of exhaust gas.

4.3. Temperature and Pressure Distribution of Exhaust Gas

This section analyzes the temperature and dynamic pressure distribution as factors considered for thermal-structural analysis of the deck and temperature. The dynamic pressure distribution as r/D is depicted in Figure 12. The temperature distribution and the contour of the impingement gas are depicted in Figure 13. The maximum temperature reaches 545 K (276.6 °C), and the average temperature is 542.7 K (271.8 °C) in the stagnation region. As the radial distance increases, the temperature decreases on the wall jet region. Not only the temperature and the convection coefficient but also the dynamic pressure of exhaust gas is acting on the deck. The maximum pressure of exhaust gas is 3.8 kPa; the pressure profile is depicted in Figure 14.

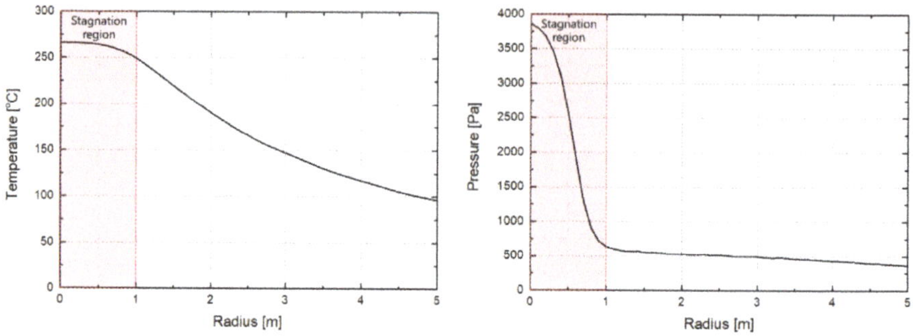

Figure 12. Temperature (**left**) and pressure (**right**) distribution by radius.

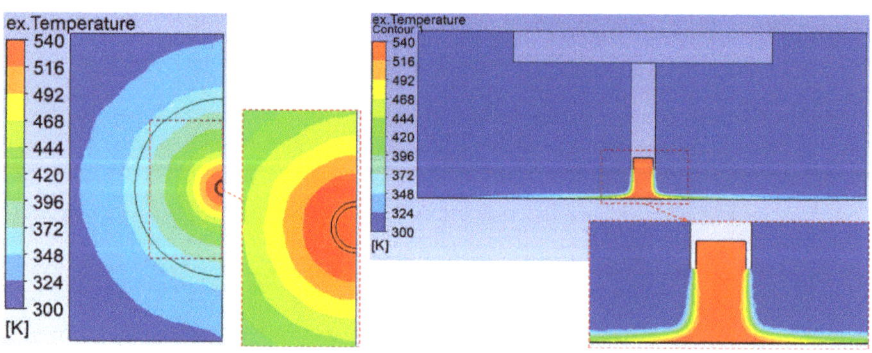

Figure 13. Temperature contour distribution of the exhaust gas. Top view (**left**) and section view (**right**).

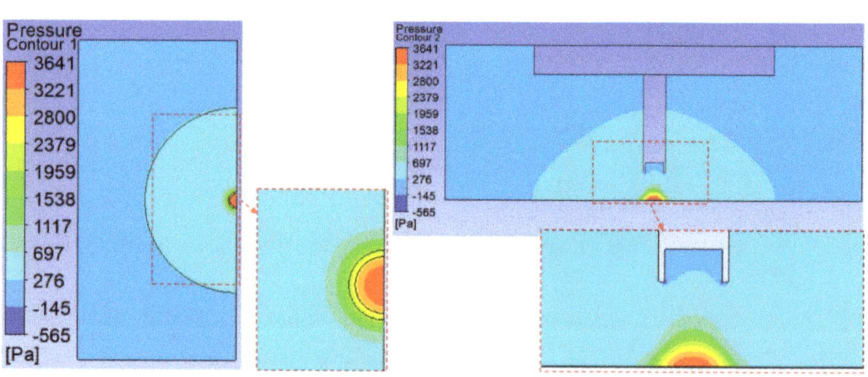

Figure 14. Pressure contour distribution of the exhaust gas. Top view (**left**) and section view (**right**).

4.4. Calibration of Convection Coefficient at the Stagnation Point

The convection coefficient is the major parameter of thermal load to evaluate the thermal effect on the deck. Although similar experiments [3,4] showed that the convection coefficient at the stagnant point was higher than that in the transition region, the convective coefficient predicted by CFX yields a maximum value in the transition region and a lower value in the stagnant region as shown in Figure 15. This discrepancy comes from the calculation procedure of CFX to compute the convection coefficient. CFX uses the difference between the temperature of the exhaust gas and the temperature of the impinging wall in steady state to calculate the convection coefficient as it calculates the convection coefficient through the equation $h = heat\ flux/(T_s - T_\infty)$, where T_s and T_∞ represent the temperature of the impinging wall and the gas, respectively [36]. The higher the temperature of the impinging walls, the lower the convection coefficient, whereas the opposite is observed in the experiment. Hannat and Morency [37], Park et al. [38], and Heyrichs et al. [39] also observed the heat flux profile falling at the stagnation point in the CFD analysis of the jet. The unreasonable profile of heat flux at the stagnation point is also due to the zero velocity at the stagnation point above the heat conduction media. Several numerical studies suggested methods to correct the convection coefficient and the Nusselt number at the stagnation point. In particular, they suggested that it is reasonable to consider the vertical velocity just before the jet impacts the impinging wall in order to calculate a reasonable convection coefficient at the stagnation point. Katti and Prabhu [2] and Vlachopoulos and Tomich [22] successfully presented an interpolation method for the Nusselt number at stagnation using the velocity field of fluid above the impinging wall. Therefore, it is reasonable to obtain the convection coefficient directly from the flow velocity and temperature at the moment of reaching the wall. That is, the Prandtl number, Reynolds number, Nusselt number, and convection coefficient should be assessed from the flow field of exhaust gas just above the wall. Therefore, in this study, the convection coefficient at the stagnation point was assessed based on the temperature and velocity above the impinging wall.

According to Gauntner et al. [9], the local radial velocity gradient at stagnation affects the stagnation point heat transfer coefficient. The local radial velocity gradient at the stagnation point is also considered as the static pressure distribution at the stagnation point, assuming incompressible flow. Therefore, in this study, the local convective heat transfer coefficient of the stagnation point was corrected by reflecting the flow pressure and velocity immediately before the impinging at the stagnation point, as defined by Equation (12).

$$V_{eq}(r) = \sqrt{2\left(\frac{P(r)}{\rho(r)} + \frac{V^2(r)}{2}\right)}, \quad (12)$$

where $V_{eq}(r)$, $P_1(r)$, and $\rho(r)$ are the equivalent velocity, pressure, and density of the exhaust gas at local point r, respectively. Furthermore, the Reynolds number in Equation (3) was calculated using the equivalent velocity with the dynamic viscosity and the diameter of a nozzle as Equation (13) in the impingement jet problem [1,2,5,6,22]. The maximum value is approximately 2.2 million at the stagnation point. The Prandtl number is the ratio of dynamic viscosity and thermal diffusion, as defined by Equation (14). In this study, *Pr(r)* is obtained at the local point *r* and its range is calculated as 0.7–0.71.

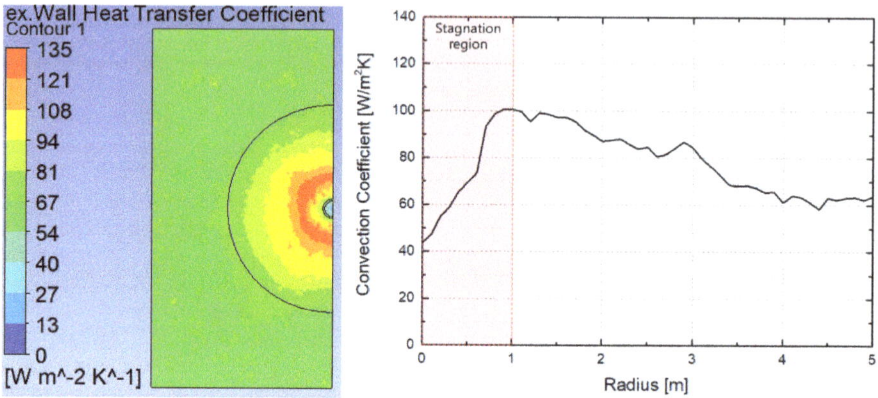

Figure 15. Convection coefficient contour (**left**) and distribution (**right**) on the impinging wall.

The local Nusselt number ($Nu(r)$) can be estimated with the local Prandtl ($Pr(r)$) and local Reynolds ($Re(r)$) numbers defined by Equation (15) [40]. The distribution of the local Nusselt and Reynolds numbers is depicted in Figure 16. The convection coefficient was then calculated from the equivalent velocity, and an appropriate thermal load distribution was presented.

$$Re(r) = \frac{\rho_f(r)\ V_{eq}(r)\cdot D_{nozzle}}{\mu_f(r)}, \tag{13}$$

$$Pr(r) = \frac{c_f(r)\cdot\mu_f(r)}{k_f(r)}, \tag{14}$$

$$Nu(r) = 0.0308\ Re(r)^{0.8}\ Pr(r)^{\frac{1}{3}}. \tag{15}$$

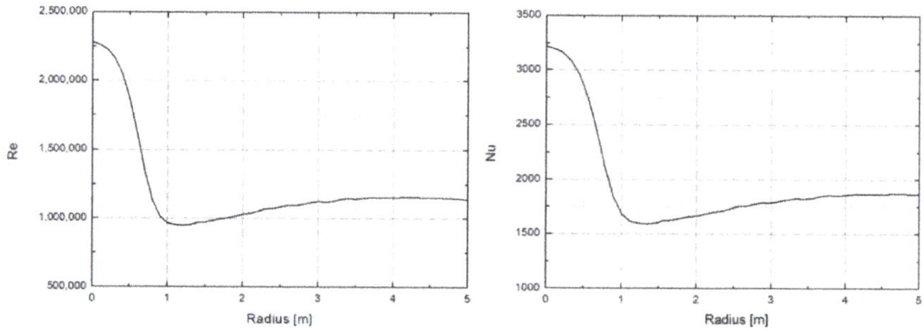

Figure 16. Local Reynolds number (**left**) and Nusselt number (**right**) distribution.

The heat flux by convection of exhaust gas analysis neglects the effect of a volume fraction of exhaust gas in the heat transfer. The efficiency of heat transfer where the VF is 1 and close to 0 are considered the same. Therefore, considering the heat transfer effect by the volume fraction, the convection coefficient was multiplied by the volume fraction. The maximum value of the convection coefficient is approximately 144 W/m^2K at the stagnation point.

To validate the accuracy of the convection coefficient and temperature results, their distributions are compared with the experimental data presented by Crosser [3]. The local convective heat transfer coefficient acting on the plate was calculated using the local Nusselt number relational Equation (16), with the distribution of the convection coefficient depicted in Figure 17. The distribution of the convection coefficient and the temperature were estimated from the contour obtained through experiments on VTOL. Compared with the results of Crosser's study, these results by distance (r) illustrate that the temperature decreases slowly, and the convection coefficient decreases rapidly, as shown in Figure 18.

$$h_{ex}(r) = \frac{Nu(r) \cdot k_f(r)}{L} * VF(r). \quad (16)$$

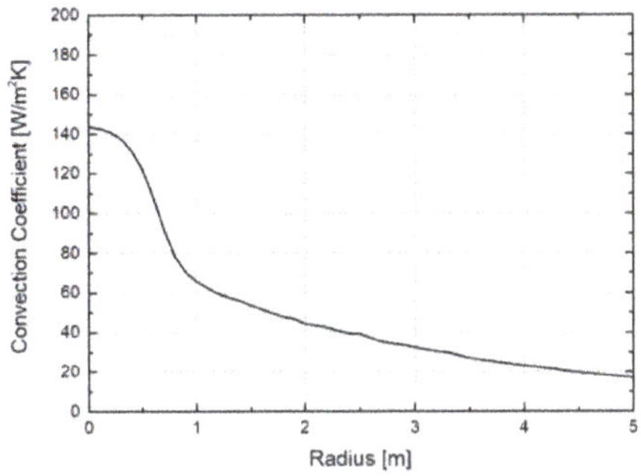

Figure 17. Calibrated convection coefficient distribution by radius (deck).

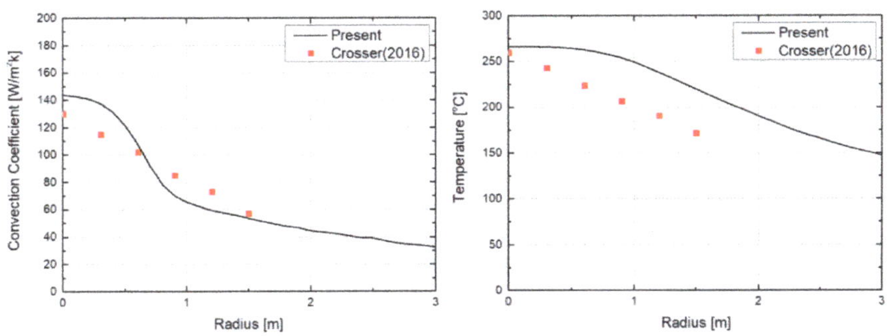

Figure 18. Comparison of experimental and numerical convection coefficient (**left**) and temperature (**right**).

The result of the convection coefficient was approximately 15% higher in the stagnation point than the experimental result, but the experimental result was higher after the stagnation region. Furthermore, both the numerical and the experimental temperature results are approximately 250 °C at the stagnation point. However, as r increases, the numerical results are larger than the experimental data. The heat flux was calculated to analyze the thermal load of convection heat transfer; the heat flux results are depicted in Figure 19. As shown in Figure 19, the convective heat transfer coefficient and temperature distribution are different from the experimental values. The maximum value at the stagnation point has a difference of about 10%, but the further away from the stagnation point, the smaller the difference between the experimental value and the numerical result. Nonetheless, the heat flux distribution showed an acceptable level of difference.

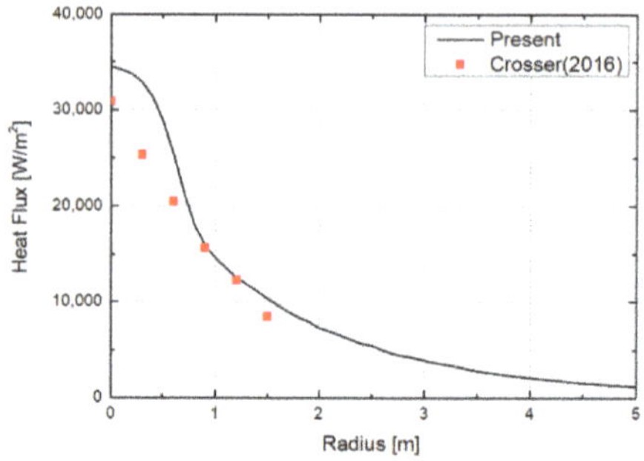

Figure 19. Comparison of experimental and numerical heat flux.

4.5. Heat Transfer Analysis

The temperature history and distribution were calculated using heat transfer analysis to estimate the thermal stress and deformation of a deck by the flame jet. The results of thermal flow analysis were applied as the thermal load conditions, and the maximum temperature history of a deck was compared with the results of a real-scale experiment [4]. The size of a deck is 15 × 15 m and it is considered a dangerous area for propeller downwash, as defined by the Naval Air System Command [41]. The geometry of the deck is as depicted in Figure 20, and the spacing and shape of the longitudinal frame and transverse web frame were defined from the study by Wadley et al. [42]. For efficiency of time cost, a symmetric condition was applied to define a quarter model. The deck was modeled by the shell element. Information on boundary conditions and material properties are summarized in Tables 6 and 7. Forced convection by jet was applied on the upper side of the deck, and natural convection was applied to the others. The deck material was defined as HY100 steel, which is commonly used in a naval ship, and the effects of coating, painting, and insulation were not considered.

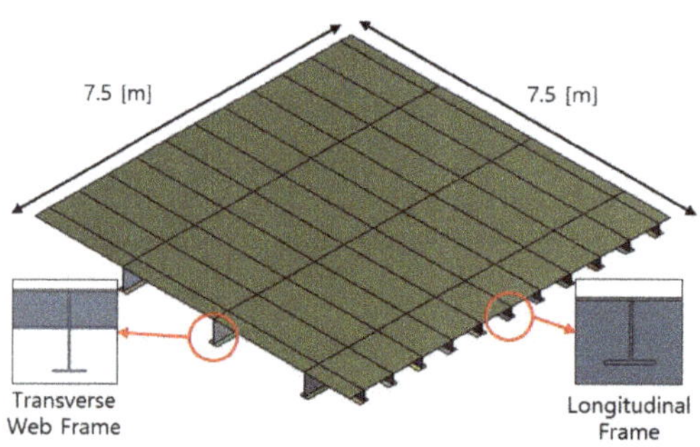

Figure 20. Simulation model of the heat transfer and thermal structure (deck).

The distribution of temperature and history of maximum temperature at the deck and frame are depicted in Figure 21. The maximum temperature is approximately 231 °C on the center of a deck after 1800 s. The temperature increased to 203 °C at the longitudinal frame and 150 °C at the transverse web frame by conduction. The history of the maximum temperature on the deck is compared with the experimental results [4] from the real-scale structure, as depicted in Figure 22. These results are reasonable considering natural convection conditions and paint insulation under real-scale experimental conditions.

Table 6. Summary of heat transfer analysis.

Specification	Value
Thickness of deck	0.013 (m)
Initial Temperature	25 (°C)
Forced convection by Impinging Jet	$h_{ex}(r)$, $T_{ex}(r)$ depicted in Figure 15
Natural convection by Ambient (h_a)	5 (W/m^2K)
Ambient Temperature (T_∞)	22 (°C)
Analysis time	1800 (s)

Table 7. Material properties of HY 100.

Material Property	Value
Density	7744 (kg/m^3)
Specific Heat	407 (J/kg·K)
Thermal Conductivity	34 (W/m·K)

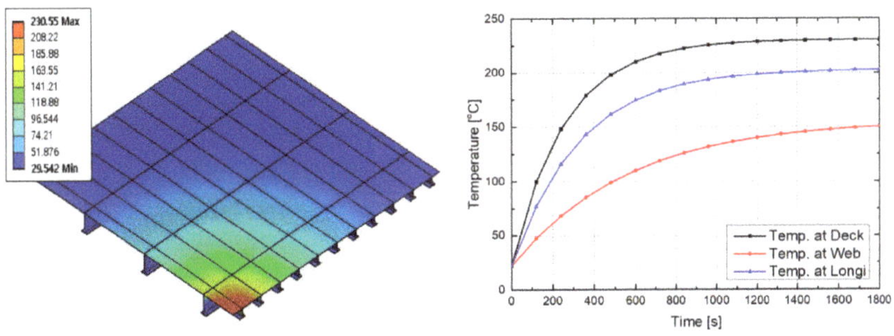

Figure 21. Distribution (**left**) and history (**right**) of temperature on the deck.

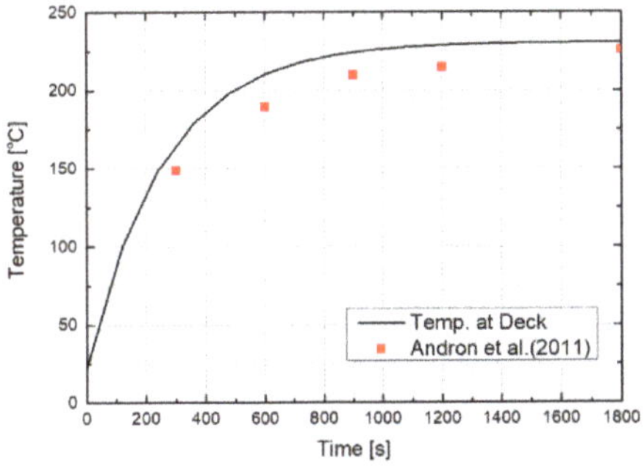

Figure 22. Comparison of experimental and numerical maximum temperature history on the deck.

4.6. Transient Thermal-Structural Analysis

The thermal-structural analysis was performed with transient analysis, and a safe operation time was suggested based on the results. The thermal-structural analysis model is depicted in Figure 23 (left); the fixed condition was applied at the side edge. The temperature gradient was imported from heat transfer analysis to thermal-structural analysis. The distribution and history of deck temperature (Figure 23, right) are applied to the thermal-structural analysis. The material properties of a deck used to calculate the thermal stress are summarized in Table 8.

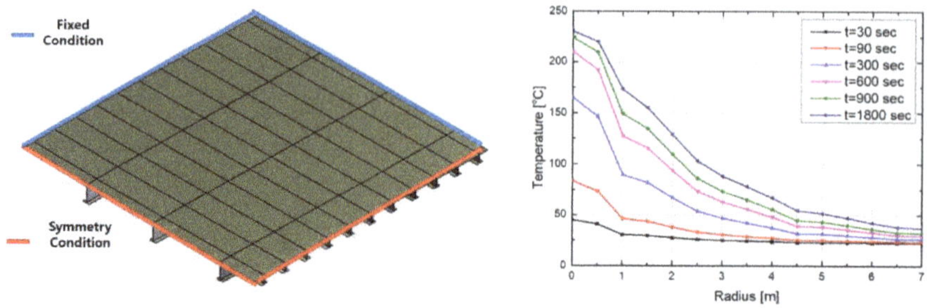

Figure 23. Boundary conditions of thermal-structural analysis (**left**) and temperature distribution and history (**right**).

Table 8. Deck material properties.

Material Property	Value
Elastic Modulus	207 (GPa)
Poisson's Ratio	0.3
Yield Stress	690 (MPa)
Thermal Expansion Coefficient	1.4×10^{-5} (1/K)

In this study, only the elastic properties were considered because the yield stress of the material was defined as a safety criterion. The contour and history of thermal stress and deformation results of the deck are depicted in Figures 24 and 25. The yield stress was reached in approximately 450 s (7 min 30 s) at the center of a deck and the longitudinal frame in Table 9. As mentioned in the second section, the present study assumed that the exhaust gas was sprayed onto the deck while the VTOL was fixed in the landing state. Although this assumption overestimates the thermal stress of the deck, it is a conservative design condition to ensure the structural safety of the deck. In an actual VTOL operation, the time to reach the yield stress is expected to be more than 450 s because the height changes dynamically during take-off and landing. Therefore, it is necessary to estimate the exact time to reach the yield stress by considering the dynamically changing VTOL height through future study.

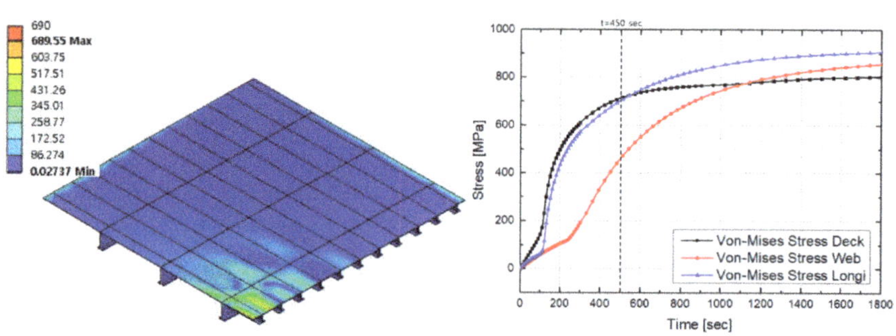

Figure 24. Stress distribution (**left**) and history (**right**) of the deck structure.

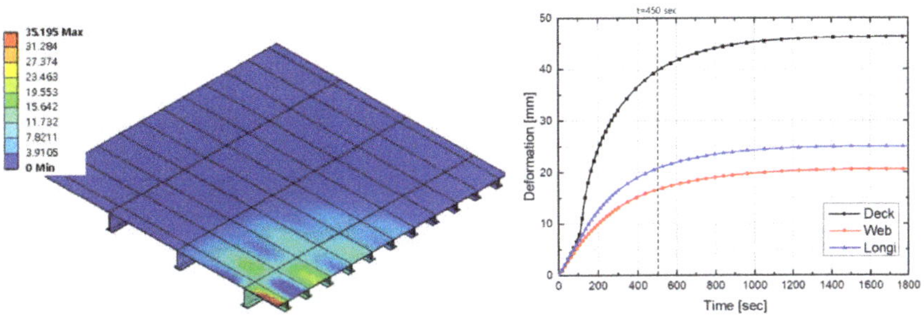

Figure 25. Deformation distribution (**left**) and history (**right**) of the deck structure.

Table 9. Thermal-structural analysis results.

Structural Member	Arrival Time at Yield Stress
Deck	450 (s)
Web frame	810 (s)
Longitudinal	450 (s)

5. Conclusions

In this study, thermal flow analysis was performed to calculate the thermal load on the deck structure from the high-speed and high-temperature exhaust gas during landing and take-off. However, it was a physically unreasonable steady-state result because the coefficient of convection has a low value in the stagnation region. A reasonable convection coefficient was calculated using equivalent velocity based on the law of energy conservation considering the velocity and pressure. The heat transfer analysis was applied based on the thermal load conditions obtained from thermal flow analysis, and the temperature distribution and history were calculated. The temperature and convection coefficient distribution of exhaust gas were compared and verified with a real-scale experiment. Subsequently, the history of maximum temperature on the deck was verified again with the result. The results of thermal load were reasonable. Therefore, the distribution and history of temperature on the deck were applied to the thermal-structural analysis. Finally, this study suggested the time to reach the yield of the deck by the thermal stress and deformation from the thermal-structural analysis.

Author Contributions: J.-H.L. suggested the concept of thermal flow and stress analysis of processing VTOL system; S.-Y.H. analyzed the data and suggested an analysis procedure; H.-S.J. and S.-Y.H. performed a case study based on finite element analysis; H.-S.J. and J.-H.L. wrote the paper. All authors have read and agreed to the published version of the manuscript.

Funding: This research was funded by Defense Industry Technology Center (DITC) of Korea (Project code UC200005D).

Institutional Review Board Statement: Not applicable.

Informed Consent Statement: Not applicable.

Data Availability Statement: Not applicable.

Acknowledgments: This research was supported by Defense Industry Technology Center (DITC) of Korea (Project code UC200005D).

Conflicts of Interest: The authors declare no conflict of interest.

References

1. Jambunathan, K.; Lai, E.; Moss, M.A.; Button, B.L. A review of heat transfer data for single circular jet impingement. *Int. J. Heat Fluid Flow* **1992**, *13*, 106–115. [CrossRef]
2. Katti, V.; Prabhu, S.V. Experimental study and theoretical analysis of local heat transfer distribution between smooth flat surface and impinging air jet from a circular straight pipe nozzle. *Int. J. Heat Mass Transf.* **2008**, *51*, 4480–4495. [CrossRef]
3. Crosser, K.E. Heat Transfer Assessment of Aluminum Alloy Corrugated Naval Ship Deck Panels under VTOL Aircraft Thermal Loads. Ph.D. Dissertation, Virginia Polytechnic Institute and State University, Blacksburg, VA, USA, 2016.
4. Andron, P.; Conley, J.; Wells, B.; Erick, A. *Land Based Testing of V-22 Osprey Aircraft for Deck Heating Effects—Test Results and Load Standardization*; Naval Surface Warfare Center: Carderock, MD, USA, 2011.
5. Zuckerman, N.; Lior, N. Jet Impingement Heat Transfer: Physics, Correlations, and Numerical Modeling. *Adv. Heat Transf.* **2006**, *39*, 565–631.
6. Pattamatta, A.; Singh, G.; Mongia, H. Assessment of turbulence models for free and confined impinging jet flows. In Proceedings of the 42nd AIAA Thermophysics Conference, Honolulu, HI, USA, 27–30 June 2011; p. 3952.
7. Barata, J.M. Fountain flows produced by multiple impinging jets in a crossflow. *AIAA J.* **1996**, *34*, 2523–2530. [CrossRef]
8. Matsumoto, R.; Ishihara, I.; Yabe, T.; Ikeda, K.; Kikkawa, S.; Senda, M. Impingement heat transfer within arrays of circular jets including the effect of crossflow (AJTE99-6386). In Proceedings of the 5th ASME/JSME Thermal Engineering Joint Conference, San Diego, CA, USA, 14–19 March 1999; pp. 1–8.
9. Gauntner, J.W. *Survey of Literature on Flow Characteristics of a Single Turbulent Jet Impinging on a Flat Plate*; National Aeronautics and Space Administration, Lewis Research Center: Washington, DC, USA, 1970.
10. Martin, M.J.; Boyd, I.D. Stagnation-point heat transfer near the continuum limit. *AIAA J.* **2009**, *47*, 283–285. [CrossRef]
11. Rao, S.R.; Srinath, S.; Bhargavi, V.; Dere, B.R. Analysis of Vertical Take-off Landing Aircraft using CFD. *Int. J. Eng. Res. Technol.* **2014**, *3*, 250–263.
12. Annaswamy, A.; Choi, J.J.; Sahoo, D.; Egungwu, O.; Lou, H.; Alvi, F. Active-adaptive control of acoustic resonances in supersonic impinging jets. In Proceedings of the 33rd AIAA Fluid Dynamics Conference and Exhibit, Orlando, FL, USA, 23–26 June 2003; p. 3565.
13. Choi, J.J.; Annaswamy, A.; Egungwu, O.; Alvi, F. Active noise control of supersonic impinging jet using pulsed microjets. In Proceedings of the 43rd AIAA Aerospace Sciences Meeting and Exhibit, Reno, NV, USA, 10–13 January 2005; p. 798.
14. Beltaos, S.; Rajaratnam, N. Impinging circular turbulent jets. *J. Hydraul. Div.* **1974**, *100*, 1313–1328. [CrossRef]
15. Beltaos, S.; Rajaratnam, N. Impingement of axisymmetric developing jets. *J. Hydraul. Res.* **1977**, *15*, 311–326. [CrossRef]
16. Rajaratnam, N. *Turbulent Jets*; Elsevier: Amsterdam, The Netherlands, 1976.
17. Nawani, S.; Subhash, M. A review on multiple liquid jet impingement onto flat plate. *Mater. Today Proc.* **2021**, *46*, 11190–11197. [CrossRef]
18. O'Donovan, T.S.; Murray, D.B. Jet impingement heat transfer–Part I: Mean and root-mean-square heat transfer and velocity distributions. *Int. J. Heat Mass Transf.* **2007**, *50*, 3291–3301. [CrossRef]
19. O'Donovan, T.S.; Murray, D.B. Jet impingement heat transfer–Part II: A temporal investigation of heat transfer and local fluid velocities. *Int. J. Heat Mass Transf.* **2007**, *50*, 3302–3314. [CrossRef]
20. Han, B.; Goldstein, R.J. Jet-impingement heat transfer in gas turbine systems. *Ann. N. Y. Acad. Sci.* **2001**, *934*, 147–161. [CrossRef] [PubMed]
21. Maghrabie, H.M. Heat transfer intensification of jet impingement using exciting jets-A comprehensive review. *Renew. Sustain. Energy Rev.* **2021**, *139*, 110684. [CrossRef]
22. Vlachopoulos, J.; Tomich, J.F. Heat transfer from a turbulent hot air jet impinging normally on a flat plate. *Can. J. Chem. Eng.* **1971**, *49*, 462–466. [CrossRef]
23. Olsson, E.E.M.; Ahrne, L.M.; Trägårdh, A.C. Heat transfer from a slot air jet impinging on a circular cylinder. *J. Food Eng.* **2004**, *63*, 393–401. [CrossRef]
24. Parida, P.R.; Ekkad, S.V.; Ngo, K. Experimental and numerical investigation of confined oblique impingement configurations for high heat flux applications. *Int. J. Therm. Sci.* **2011**, *50*, 1037–1050. [CrossRef]
25. Angioletti, M.; Di Tommaso, R.M.; Nino, E.; Ruocco, G. Simultaneous visualization of flow field and evaluation of local heat transfer by transitional impinging jets. *Int. J. Heat Mass Transf.* **2003**, *46*, 1703–1713. [CrossRef]
26. Angioletti, M.; Nino, E.; Ruocco, G. CFD turbulent modelling of jet impingement and its validation by particle image velocimetry and mass transfer measurements. *Int. J. Therm. Sci.* **2005**, *44*, 349–356. [CrossRef]
27. Achari, A.M.; Das, M.K. Application of various RANS based models towards predicting turbulent slot jet impingement. *Int. J. Therm. Sci.* **2015**, *98*, 332–351. [CrossRef]
28. Coussirat, M.; Van Beeck, J.; Mestres, M.; Egusguiza, E.; Buchlin, J.M.; Escaler, X. Computational fluid dynamics modeling of impinging gas-jet systems: I. assessment of eddy viscosity models. *J. Fluids Eng.* **2005**, *127*, 691–703. [CrossRef]
29. Coussirat, M.; Van Beeck, J.; Mestres, M.; Egusquiza, E.; Buchlin, J.M.; Valero, C. Computational fluid dynamics modeling of impinging gas-jet systems: II. Application to an industrial cooling system device. *J. Fluids Eng.* **2005**, *127*, 704–713. [CrossRef]
30. Versteeg, H.K.; Malalasekera, W. *An Introduction to Computational Fluid Dynamics: The Finite Volume Method*; Pearson Education: London, UK, 2007.
31. Hwang, C.B.; Lin, C.A. Improved low-reynolds-number ke model based on direct numerical simulation data. *AIAA J.* **1998**, *36*, 38–43. [CrossRef]

32. Bolkcom, C. *V-22 Osprey Tilt-Rotor Aircraft*; Library of Congress Washington DC Congressional Research Service: Washington, DC, USA, 2004.
33. Benson, T. Turboprop Thrust. Retrieved from NASA. 2006. Available online: http://www.grc.nasa.gov/www/K-12/airplane/turbprp.html (accessed on 24 December 2021).
34. Rotaru, C.; Todorov, M. *Helicopter Flight Physics*; Flight Physics-Models, Techniques and Technologies, Defense Acquisition University: Fort Belvoir, VA, USA, 2018.
35. Newman, S. *Foundations of Helicopter Flight*; Elsevier: Amsterdam, The Netherlands, 1994.
36. Ansys Inc. *Ansys®CFX Theory Guide*; Ansys Inc.: Canonsburg, PA, USA, 2017.
37. Hannat, R.; Morency, F. Numerical Validation of Conjugate Heat Transfer Method for Anti-/De-Icing Piccolo System. *J. Aircr.* **2014**, *51*, 104–116. [CrossRef]
38. Park, T.H.; Choi, H.G.; Yoo, J.Y.; Kim, S.J. Streamline upwind numerical simulation of two-dimensional confined impinging slot jets. *Int. J. Heat Mass Transf.* **2003**, *46*, 251–262. [CrossRef]
39. Heyerichs, K.; Pollard, A. Heat transfer in separated and impinging turbulent flows. *Int. J. Heat Mass Transf.* **1996**, *39*, 2385–2400. [CrossRef]
40. Bejan, A. *Convection Heat Transfer*; John Wiley & Sons: Hoboken, NJ, USA, 2013.
41. Commander, Naval Air Systems Command. *NATOPS Flight Manual Navy Model MV-22B Tiltrotor*; Change 58; NAVAIR: Patuxent River, MD, USA, 2006.
42. Wadley, H.N.G.; Haj-Hariri, H.; Zok, F.; Norris, P.M. Multifunctional Thermal Management System and Related Method. USA Patent No. 10,107,560, 23 October 2018.

Journal of
Marine Science and Engineering

MDPI

Article

Numerical Simulation Study of the Horizontal Submerged Jet Based on the Wray–Agarwal Turbulence Model

Bo Hu [1,2], Chuan Wang [1,3], Hui Wang [3,4], Qian Yu [3,*], Jinhua Liu [5], Yong Zhu [6], Jie Ge [7], Xinxin Chen [6] and Yang Yang [3,4,*]

1 College of Mechatronics Engineering, Hainan Vocational University of Science and Technology, Haikou 571126, China
2 Department of Energy and Power Engineering, Tsinghua University, Beijing 100084, China
3 State Key Laboratory of Simulation and Regulation of Water Cycle in River Basin, China Institute of Water Resources and Hydropower Research, Beijing 100038, China
4 College of Hydraulic Science and Engineering, Yangzhou University, Yangzhou 225009, China
5 International Shipping Research Institute, Gongqing Institute of Science and Technology, Jiujiang 332020, China
6 National Research Center of Pumps, Jiangsu University, Zhenjiang 212013, China
7 SHIMGE Pump Co., Ltd., Taizhou 317525, China
* Correspondence: yuqian@iwhr.com (Q.Y.); yang_yang@yzu.edu.cn (Y.Y.)

Abstract: The horizontal submerged jet (HSJ), as a special form of jet, is widely used in aerospace, food and drug, water engineering, and other industries. In order to further understand the mechanism of the HSJ, the Wray–Agarwal turbulence model was used to predict the HSJ with different incidence heights H/D and Reynolds number (*Re*) conditions in this paper. The results show that the jet horizontal height H/D has a large influence on the flow field structure. The unsteady flow within the flow field is dominated by vortexes. In addition, their distribution is relatively independent of H/D. Under different H/D conditions, the axial velocity distribution of the jet has very high similarity, all of them have an obvious velocity inflection point at $x = 10D$. When H/D is small, the wall attachment effect of the jet and the boundary layer effect generated at the bottom of the fluid domain have a certain role in maintaining the velocity of the jet near the wall, resulting in a significantly higher axial velocity than other H/D conditions, up to 1.29 times. In this paper, we thoroughly investigated the structure of the internal flow field and velocity distribution of the submerged horizontal jet. The results have a guiding significance for engineering practice and academic research.

Keywords: horizontal submerged jet; Wray–Agarwal turbulence model; flow field structure; speed distribution

Citation: Hu, B.; Wang, C.; Wang, H.; Yu, Q.; Liu, J.; Zhu, Y.; Ge, J.; Chen, X.; Yang, Y. Numerical Simulation Study of the Horizontal Submerged Jet Based on the Wray–Agarwal Turbulence Model. *J. Mar. Sci. Eng.* **2022**, *10*, 1217. https://doi.org/10.3390/jmse10091217

Academic Editors: Carlos Guedes Soares and Constantine Michailides

Received: 29 May 2022
Accepted: 26 August 2022
Published: 31 August 2022

1. Introduction

With a relatively simple flow structure, jet streams are widely used in engineering practice, especially in the fields of water conservancy and marine engineering [1]. The difference between an ordinary jet and a submerged jet is that the submerged jet will form a velocity shear layer at the jet boundary during the flow process, and the destabilization of the shear layer will generate vortexes. The vortexes will continuously deform, break, and merge during the movement and finally develop into turbulence. They will entrain the surrounding fluid into the jet area, triggering the exchange of lateral momentum, heat, and mass between the jet and the ambient fluid. The submerged jet is widely used in scenarios such as underwater dredging and ship propulsion due to the above-mentioned characteristics, and, in view of this, the research related to the submerged jet was initiated earlier. As early as 1953, Rouse [2] made a detailed analysis of the cavitation phenomenon in the region of submerged jets by means of high-speed photography and found that the maximum intensity of jet turbulence decreases with increasing mean velocity; the authors concluded that the origin of this phenomenon lies in the velocity fluctuations caused by vortexes. Law et al. [3] studied the submerged shock jet in two dimensions in 1983 and found that the submerged shock jet flow field is mainly controlled by the recirculating vortex near the confinement plate and smaller vortexes on the impingement plate. Over time, the research focus has gradually changed from the analysis of macroscopic flow and flow field characteristics of submerged jets to the analysis of internal flow structure and dynamics. Wen et al. [4] studied the interaction between the submerged jet and the free surface at the spatial scale based on visualization experiments and found that the high Reynolds number jet is detrimental to the development of large-scale vortex structures in the flow field. Liu et al. [5] studied the characteristics of high-pressure submerged water jets using the Particle Image Velocimetry (PIV) technique, focusing on their cavitation phenomenon, and came to similar conclusions as Rouse [2]; they summarized the cavitation evolution process into three features: cavitation cloud extending, cavitation cloud preserving, and cavitation cloud shrinking. Yang et al. [6] considered the effect of shear stress in the study of high-pressure submerged jet cavitation using numerical simulations and experiments. They found that shear stress plays an important role in the formation of high-pressure submerged jet cavitation.

The HSJ, i.e., a submerged jet with an incidence angle of 0°, is a special form of the submerged jet. Unlike oblique or forward inundation jets, horizontal jets do not have the limitation of walls when developing horizontally, and turbulence can develop more fully, possessing a higher coiling and diffusion rate than impact jets under the same initial conditions. The HSJ has also been studied theoretically to some extent by domestic and foreign scholars. Ma et al. [7] focused on the motion trajectory and flow law of the submerged horizontal jet in a stationary, ambient fluid and established an accurate mathematical model of the jet trajectory which provides some reference for the flow prediction of the HSJ. Yang et al. [8] qualitatively analyzed the flow characteristics of a gas-liquid two-phase horizontal jet in a water tank with annular injection by high-speed photography and found that at high entrainment ratios, the two-phase HSJ flows as a thin film around the mixing tube wall to form an annular flow [9]. Shao et al. [10] used Planar Laser Induced Fluorescence (PLIF) to explore the flow patterns and stability of horizontal buoyant jet diffusion for different densities and Froude numbers (Fr) and identified three flow patterns with unstable small-scale structural features. At the same time, the engineering applications of horizontal inundation jets have received equal attention, especially in the areas of hydraulic scouring and energy dissipation. Chatterjee et al. [11], in 1994, studied the local scouring and sediment transport phenomena caused by horizontal inundation jets produced by sluices and found that the maximum scouring depth at equilibrium can be expressed as Fr based on the jet development thickness and is related to the particle size of the scoured media. Immediately after, Chiew et al. [12] investigated the local scouring caused by circular jets under deep submergence using comparative experiments. It was found that the scouring effect gradually decreases with the increase in the horizontal jet offset distance. Additionally, under the same conditions, the scouring pits produced by horizontal air jets are always larger than those produced by submerged horizontal jets. With the rapid development of computer technology and the iterative update of algorithms [13,14], Abdelaziz et al. [15] added the open channel sediment transport calculation module to FLOW-3D and studied the effect of horizontal jet flow on the calculation of stream-bed deformation in the flume. At the same time, Li et al. [16] used a combination of model tests and numerical simulations to evaluate the effectiveness of multi-stream multilayer horizontal jets in hydraulic energy dissipation projects and found that multi-stream multilayer horizontal inundation jet energy dissipation workers can better solve the flood dissipation problem in actual projects.

In summary, HSJs, as a special form of submerged jets, have been more comprehensively studied. Based on the extremely strong engineering application background of horizontal inundation jet, experts and scholars are more inclined to explore its macroscopic characteristics in practical engineering applications, especially in the field of horizontal inundation jet scouring and cavitation [17], and a large number of research results have built a more complete theoretical system. It is worth noting that, in the case of HSJ, the engineering application theory is more perfect; the study of its own flow structure and coupled flow characteristics with the environment fluid is less so, the submerged flow field in the fine flow structure is not yet clear, and in the development process of the jet flow vortex and other unsteady flow field derivation and transfer law has not yet been mastered. In this paper, the Wray–Agarwal turbulence model is used to numerically simulate the HSJ under different incidence heights H/D and Re conditions, focusing on capturing the typical flow structure and flow characteristics during the development of the jet, and dissecting the interaction mechanism and coupling law between the jet and the ambient fluid which can not only provide a basis for implementing flow control and improving the flow characteristics but also provide some theoretical support for actual engineering practice and optimization.

2. Modeling and Numerical Methods

2.1. Model Building

The horizontal jet fluid is injected into a square sink through a circular jet pipe parallel to the bottom wall, as shown in Figure 1. In order to investigate the general law of the flow structure of the HSJ, the geometrical parameters are made non-dimensional by dividing the inner pipe diameter D (D = 20 mm), given in Table 1. While ensuring the accuracy of the article presentation, the spatial right-angle coordinate system $oxyz$, with the center of the jet pipe exit circle as the coordinate origin, is established to better analyze the flow characteristics of the HSJ. The coordinate origin is set at the center of the circle at the outlet of the jet pipe.

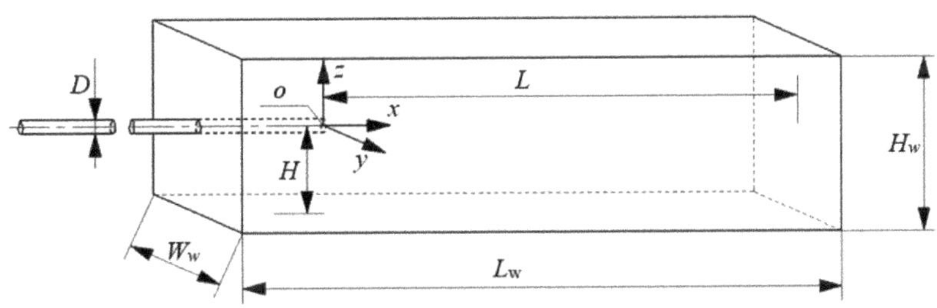

Figure 1. Schematic diagram of the three-dimensional horizontal submerged jet.

Table 1. Detailed dimensions of the computational model.

Abbreviations	Meaning	Dimensionless Length	Actual Length
H	Vertical distance of the horizontal jet axis from the bottom	-	-
L	Vertical distance of the horizontal jet outlet from the right wall	-	-
L_j	Length of jet pipe	50 D	1000 mm
L_w	Length of sink	55 D	1100 mm
W_w	Width of sink	15 D	300 mm
H_w	Height of sink	12 D	240 mm

2.2. Numerical Methods and Boundary Conditions

In this paper, ANSYS FLUENT 14.5 is used to perform numerical simulations. The wall roughness of the sink and jet pipe are set to 0.010 mm and 0.025 mm, respectively. A Pressure-Based solver is used to better predict the flow of incompressible fluids [18]. In addition, the finite volume method is used in the computational region to control the discrete equations, and the coupled pressure-velocity equations are solved using the SIMPLEC algorithm [19]. During the iterative calculation using the SIMPLEC algorithm, the residuals of each convergence parameter are set to 10^{-6} to ensure the accuracy of the calculation. The under-relaxation factors for Pressure, Density, Body Forces, and Momentum are set to 0.8, 1.0, 1.0, and 0.5, respectively. The spatial discretization of the gradient is chosen as "Least Squares Cell Based". Meanwhile, the accuracy of divergence operators is set as the second order, the precision of the solver is set as double precision, and the standard wall function method is chosen for near-wall fluid flow simulation.

The boundary conditions are set as follows:

Inlet: The working fluid is incompressible, so the velocity inlet is chosen and the inlet velocity V_b is calculated with Equation (1). The flow rate Q can be calculated with the following equation:

$$Q = \frac{V_b \pi D^2}{4}, \tag{1}$$

Outlet: The outlet boundary is farther downstream, and the turbulent flow is considered to reach relative equilibrium, so the outlet is set as a pressure outlet with a reference pressure of 1 atm [20].

Wall: The solid walls in this numerical simulation are set as smooth non-slip walls.

Free surface: In the numerical simulation study of this paper, considering the small influence of the HSJ on the ambient fluid flow field in the vertical direction, the jet has limited influence on the free liquid surface to decrease the size of the simulation domain. The rigid-lid hypothesis is adopted, such as the symmetric boundary conditions on the free liquid surface. The convective and diffusive fluxes on the symmetric plane are set as 0.

2.3. *Grid-Independence Analysis*

The quality of the grid is crucial to the numerical simulation calculation [21,22]. Since all internal points of the structured grid have the same adjacent cells, it can better achieve the fitting of regional boundaries and is suitable for the calculation of fluid and stresses on the surface. Hence, this paper adopts ANSYS ICEM for the structured meshing of the calculation domain [23]. A reasonable choice of the number of grids can not only ensure the accuracy of numerical simulation calculation but also shorten the computation time [24,25]. In this grid-independence analysis, the computational fluid domain is meshed in three scales, the number of grids is shown in Table 2.

Table 2. The number of grids for the different schemes.

Case	N1	N2	N3
Number of grids	2,504,627	4,258,458	7,308,570

The distribution of the axial velocity V/V_b of the HSJ under the condition of $H/D = 2$ (Re = 35,100) was calculated separately using the above-mentioned grids. The results are compared in Figure 2. The distribution of V/V_b with the different numbers of grids is in good agreement with each other with an error of up to 7%. The results almost remain the same when the grid number increases from N1 to N2. On the other hand, there is a large deviation in the numerical results in the jet development section ($10 \leq l/D \leq 30$) when the number of grids increases from N1 to N2. The results from the numerical simulation are processed using the grid-convergence index (GCI) proposed by Roache [26], using:

$$GCI_{1\text{-}2} = F_s \left| \frac{r^p\left(\frac{V_1}{V_b} - \frac{V_2}{V_b}\right)}{1 - r^p} \right|, \tag{2}$$

$$GCI_{2\text{-}3} = F_s \left| \frac{r^p\left(\frac{V_2}{V_b} - \frac{V_3}{V_b}\right)}{1 - r^p} \right|, \tag{3}$$

where F_s is the safety factor; r is the linear grid refinement factor; p is the formal order of accuracy; V_1 is the axial velocity of the jet calculated using N1; V_2 is the axial velocity of the jet calculated using N2; and V_3 is the axial velocity of the jet calculated using N3.

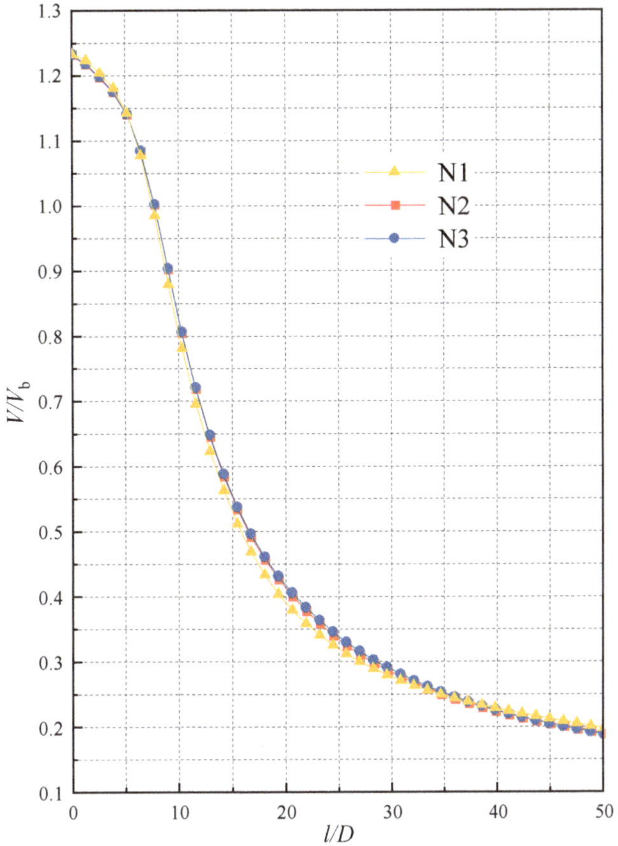

Figure 2. The distribution of the axial velocity V/V_b of the jet for the different numbers of grids.

The following is a detailed description of the values of the relevant variables; F_s generally takes a range of 1.25–3.00 and 1.25 is taken in the paper [27]; p is assigned a value of 2 here; r should theoretically take a value of 2 [28], but Celik et al. [29] point out that reasonable values of r are at least 1.3 and 1.1. Considering the accuracy of grid-independence analysis and the applicability of the grid, r is taken as 1.7 in this paper.

Figure 3 shows the distributions of $GCI_{1\text{-}2}$ and $GCI_{2\text{-}3}$ with the different numbers of grids. The values of $GCI_{1\text{-}2}$ are significantly larger than the $GCI_{2\text{-}3}$ which indicates that the refinement of the grid significantly improves the accuracy of the numerical simulation. The $GCI_{1\text{-}2}$ forms a crest in the region $10 \leq l/D \leq 20$, indicating that the accuracy of flow prediction in this region needs to be supported by a high grid density, while the relative error of the calculation results is significantly reduced after grid encryption. The mean values of $GCI_{1\text{-}2}$ and $GCI_{2\text{-}3}$ for the jet axial velocity V/V_b were 2.2% and 0.5%, and the maximum values of $GCI_{1\text{-}2}$ and $GCI_{2\text{-}3}$ are obtained at l/D = 10.3 and 23.1, and 4.7% and 1.0%, respectively. In summary, the simulation results with the basic grid show good accuracy. It is considered as the reasonable number of grids and selected for numerical simulation. Meanwhile, the accurate prediction of the boundary layer flow is also an important part of the numerical simulation process that cannot be neglected. Section 2.2 of the paper has already mentioned the proposed use of the standard wall function method for the near-wall flow in this numerical calculation. The standard wall function method uses empirical formulas to approximate the flow near the wall rather than directly solving for the complex flow in the corresponding region. Hence, the wall function method does not require high fineness of the boundary layer mesh, and it is usually sufficient to keep the y^+ value of the first layer mesh between 11.5 and 400. A too large or too small y^+ will make the nodes of the first mesh of the boundary layer out of the core region of the turbulence, thus making the wall function unavailable. It is verified that after the boundary layer of the mesh is encrypted, the minimum value of y^+ for the first layer of the mesh in the computational domain is 13.8 and the mean value is 30.4, which meets the requirement of using the standard wall function. The mesh is depicted in Figure 4.

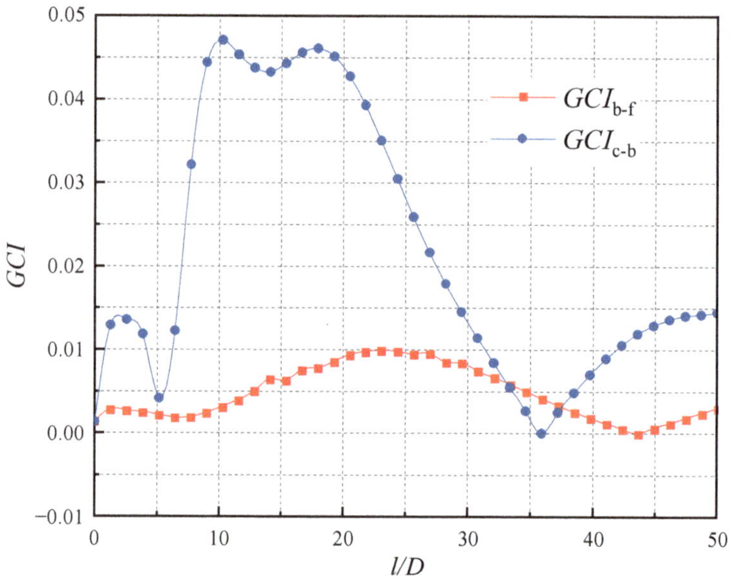

Figure 3. *GCI* distribution of numerical simulation results.

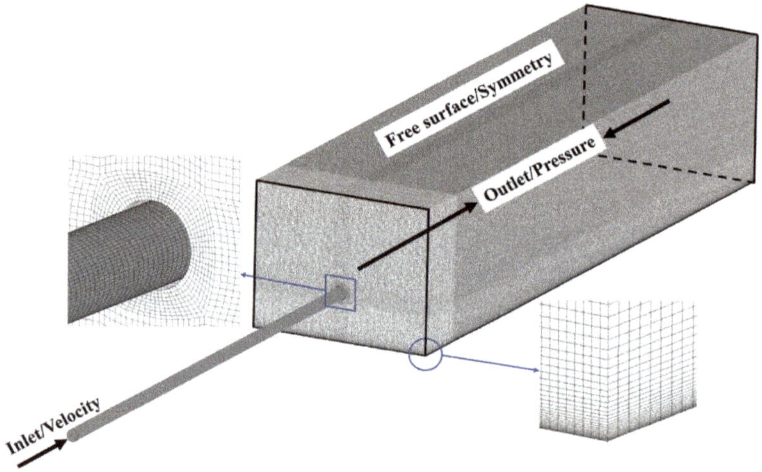

Figure 4. Schematic diagram of the computational model grid.

2.4. Turbulence Model Validation

In the numerical simulation, different turbulence models have different conditions of applicability. The selection of a suitable turbulence model not only allows for better coordination of computational resources but also for higher computational accuracy.

For the selection of computational models, this paper proposes to compare Wray–Agarwal, Standard k-ε, RNG k-ε, Realizable k-ε, Standard k-ω, and SST k-ω. All the above turbulence models are widely used in related research in the field [30,31] and have good simulation effects on complex flows such as secondary flows and cyclonic flows. Among them, the Wray–Agarwal [32,33] model is a one-equation linear eddy viscosity model that was derived from two-equation k-ω closure. It combines the most desirable characteristics of the one-equation k-ε model and the one-equation k-ω model. It uses the ω equation with a cross-diffusion term to increase the ability to solve for nonequilibrium flows, aiming to improve the accuracy of predicting the equilibrium flows [34–36].

Eddy viscosity:

$$\mu_T = f_\mu \rho R, \tag{4}$$

$$R = \frac{k}{\omega}, \tag{5}$$

R transport equation:

$$\begin{gathered}\frac{\partial \rho R}{\partial t} + \frac{\partial \rho u_j R}{\partial x_j} = \frac{\partial}{\partial x_j}\left[(\sigma_R \mu_T + \mu)\frac{\partial R}{\partial x_j}\right] + \rho C_1 R S \\ +\rho f_1 C_{2k\omega}\frac{R}{S}\frac{\partial R}{\partial x_j}\frac{\partial S}{\partial x_j} - (1-f_1)\rho C_{2k\varepsilon}\left(\frac{R\frac{\partial S}{\partial x_j}}{S}\right)\end{gathered} \tag{6}$$

where f_μ is the damping function; f_1 is the switching function; k is the turbulent pulsation kinetic energy, J; ω is the specific dissipation rate; t is the time, s; S is the average strain, 1/s; and the modulation constants C_1, $C_{2k\omega}$, $C_{2k\varepsilon}$, and σ_R are determined from empirical data and the wall constraint law:

$$\begin{gathered}C_{1k\omega} = 0.0829,\ C_{1k\varepsilon} = 0.1127,\ C_1 = f_1(C_{1k\omega} - C_{1k\varepsilon}) + C_{1k\varepsilon} \\ \sigma_{k\omega} = 0.72,\ \sigma_{k\varepsilon} = 1.0,\ \sigma_R = f_1(\sigma_{k\omega} - \sigma_{k\varepsilon}) + \sigma_{k\varepsilon} \\ C_{2k\omega} = \frac{C_{1k\omega}}{\kappa^2} + \sigma_{k\omega},\ C_{2k\varepsilon} = \frac{C_{1k\varepsilon}}{\kappa^2} + \sigma_{k\varepsilon},\ \kappa = 0.41\end{gathered} \tag{7}$$

Average strain:

$$S = \sqrt{2S_{ij}S_{ij}}, \tag{8}$$

The damping function f_μ is given by:

$$f_\mu = \frac{\chi^3}{\chi^3 + C_W^3}, \tag{9}$$

$$\chi = \frac{R}{v}, \tag{10}$$

It is important to note that more detailed information and descriptions of the Wray–Agarwal turbulence model can be found on the NASA Turbulence Modeling Resource (TMR) website (https://turbmodels.larc.nasa.gov/wray_agarwal.html, accessed on 29 July 2022).

HSJs have better conditions for turbulence development. On the other hand, they lack fast-changing parameters, namely the velocity gradients to characterize the accuracy of turbulence models for flow prediction. Based on the group's previous experimental study of PIV for oblique submerged impact jets [37], the distribution of axial dimensionless velocity V/V_b along the jet axis under fixed impact angle (θ = 45°) and impact height (H/D = 3) conditions are calculated using the above turbulence models as shown in Figure 5a. The predicted results of different turbulence models are in good agreement with each other in the prediction of the jet flow trend, i.e., the axial velocity remains basically constant in the free jet region ($0 \leq l/D \leq 3$) and decreases rapidly in the impact region ($3 \leq l/D \leq 4$). The absolute errors between the results from either numerical simulation or PIV experiments in the effective monitoring range are shown in Figure 5b. Combining Figure 5a with Figure 5b, in the initial stage of the free jet ($0 \leq l/D \leq 0.8$), the predicted results of conventional turbulence models such as SST k-ω show high accuracy, while the predicted results of the Wray–Agarwal model are up to 3.3 % larger. With the free development of the jet, both the simulation results and the PIV experimental data show a slightly increasing trend. The results from Wray–Agarwal turbulence model fit better with the measurement results. At the same time, it is noted that as the velocity gradient and pressure gradient at the near-wall surface increase, the numerical results derived from each turbulence model are less and less different from the experimental data, showing a better prediction accuracy. The differences between the predicted results of each turbulence model gradually decrease and all agree well with the experimental data. Overall, the predicted results of the Wray–Agarwal turbulence model are larger. Meanwhile, the Wray–Agarwal model fits well with the results from experiments in most of the regions. It is noteworthy that the predicted results deviate from those from experiments in the jet exit region where the l/D is small. The focus of this paper is on capturing and analyzing the flow field structure within the HSJ, rather than focusing on the microscopic flow in the inlet region of the jet. Therefore, the prediction error in the jet exit area is considered acceptable. Through the above consideration, this paper decided to use the Wray–Agarwal model for numerical simulation.

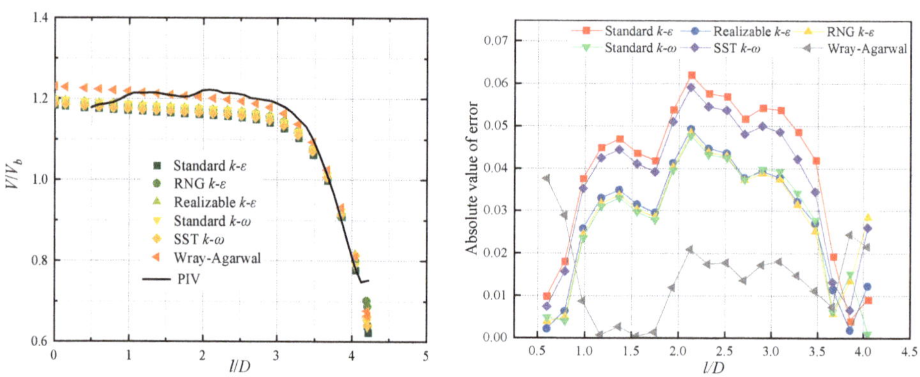

(a) (b)

Figure 5. Results and analysis of V/V_b obtained from different turbulence models: (**a**) Distribution of axial velocity V/V_b along the axis of the jet; (**b**) Absolute error values of numerical results and PIV experiments.

3. Results and Discussion

3.1. Flow Field Analysis

Figure 6 depicts the distributions of V/V_b with streamlines on the mid-section (*oxz* plane) of the jet at different horizontal heights (*Re* = 35,100). The parameter h is the vertical height from the bottom of the sink. As the surrounding fluid is continuously sucked into the jet, the jet gradually diffuses with decreasing axial velocity. When the horizontal height H/D equals 0.5, the HSJ structure is similar to a wall jet, where the jet diffusion near the wall side (along $-z$ direction) is blocked. The jet diffuses freely away from the wall side (along +z direction). When the horizontal height is in the range of $0.5 < H/D \leq 2$, the jet spreads to the bottom wall in a relatively short time. As the H/D is larger than 2, the main stream of the jet moves towards the bottom wall. The mass on the side away from the wall is more than the side near the wall which is due to the unequal mass distribution of the ambient fluid affected by entrainment on both sides of the jet at the same time. With the turbulent mixing effect on the boundary of the jet, the energy transferred from the jet to the ambient fluid on both sides is basically equal. Hence, the entrainment velocity in the $-z$ direction is larger, while the pressure is smaller, resulting in a shift of the main stream of the jet along the $-z$ direction. The jet attachment effect then occurs. In addition, a vortex (E_1) appeared downstream of the jet exit. The vortex affected about $1/3$ of the entire length of the flume. As the horizontal height of the jet increases, the position of the vortex core at the outlet gradually moves along the $+z$ direction and the area of influence becomes smaller (E_5–E_7) until it disappears. This suggests that the associated backflow is more likely to occur in areas outside the main stream diffusion zone of the jet. The back flow needs some space for further development. At the same time, when H/D is small, the distributions of the entrainment area and the scope of the vortex in the flow field are highly consistent. The vortex areas (E_1–E_6) are both located in the upper right of the calculation domain and the boundary between them is clear and located near $x/D = 30$. Due to the entrainment effect, the jet will absorb a large amount of ambient fluid into the main stream. With a smaller H/D, there is not enough space for entrainment below the jet, resulting in a large amount of fluid being sucked into the area above the jet. With the full development of the jet, the fluid in the upper part of the computational domain at $x/D = 30$ reaches a certain entrainment velocity in the $-z$ direction and mixes with the ambient fluid, forming a large reflux zone. As the H/D increases, the volume of entrainment on both sides of the jet gradually becomes equal. Meanwhile, the unsteady flow is relatively smaller, and the flow field in the sink is more regular.

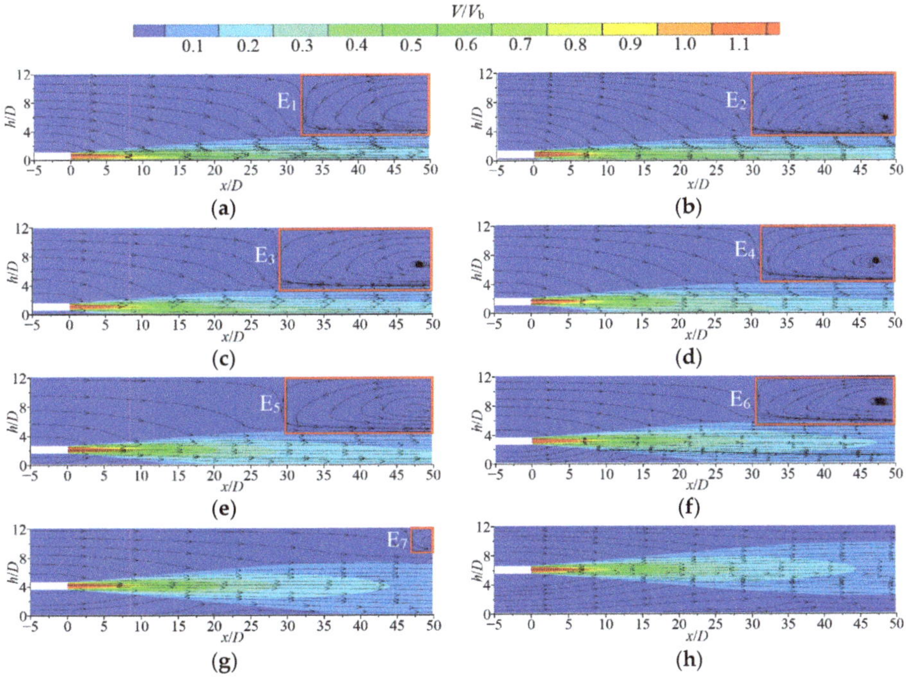

Figure 6. Contours with streamline diagrams of cross-sectional V/V_b distribution (*oxz* plane) in the jet under different H/D conditions: (**a**) $H/D = 0.5$; (**b**) $H/D = 0.75$; (**c**) $H/D = 1$; (**d**) $H/D = 1.5$; (**e**) $H/D = 2$; (**f**) $H/D = 3$; (**g**) $H/D = 4$; and (**h**) $H/D = 6$.

Figure 7 shows the contours of the V/V_b velocity distribution with streamlines on the jet cross section (*oxy* plane) at different horizontal heights (*Re* = 35,100). As for $0.5 \leq H/D \leq 1.5$, both the velocity contours and diagrams of streamlines show good symmetries; no obvious vortexes appear in the flow field. All of them have vector rotation points (VRP) which are distributed within the parameter range of $25 \leq x/D \leq 30$. They also show high symmetry, which is highlighted in the figure with a red circle. The explanation appears to be that when H/D is small, the jet is fully developed in the *oxy* plane. In the $-z$ direction of the *oxz* plane, however, the development of the jet is inhibited by the bottom wall. Hence, the turbulent kinetic energy contained in the jet cannot be adequately exchanged with the ambient fluid in the *oxy* plane, such that there is not enough energy in the direction perpendicular to the main stream velocity to engage in entrainment action with the surrounding fluid. Only VRPs without intact vortex structures appear. At $H/D = 2$, the suppression effect of the bottom wall of the jet becomes weak. The vortexes begin to appear in the flow field of the jet and are symmetrically distributed on both sides of its main stream of the jet near the wall. With the increase in H/D, the vortex near the side wall continues to develop and gradually moves in the $-x$ direction. The inhibitory effect of the wall on jet development basically disappears at larger H/D when the turbulence is fully developed and the high-speed zone of the jet is significantly reduced. At the same time, the momentum exchange phenomenon with the ambient fluid is significant. This results in a significant increase in unstable three-dimensional flow in the flow field, creating an asymmetric distribution of the cross-sectional flow field in the jet. More unsteady flow is observed on the main stream side of the jet and a relatively flat flow on the other side.

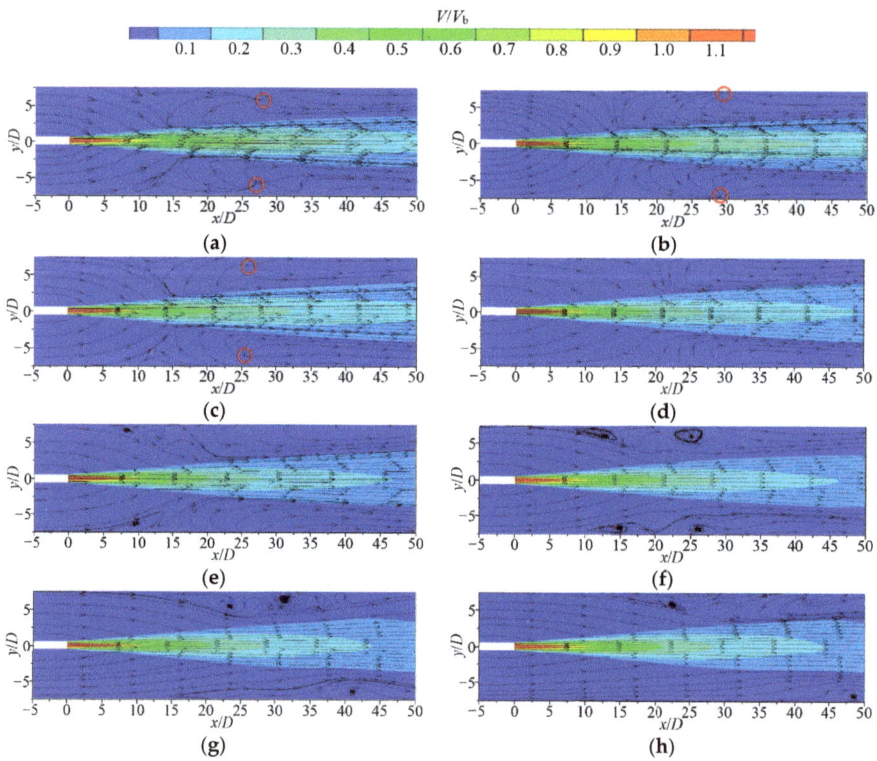

Figure 7. Contours with streamline diagrams of V/V_b distribution in the jet cross-section (*oxy* plane) under different H/D conditions: (**a**) $H/D = 0.5$; (**b**) $H/D = 0.75$; (**c**) $H/D = 1$; (**d**) $H/D = 1.5$; (**e**) $H/D = 2$; (**f**) $H/D = 3$; (**g**) $H/D = 4$; and (**h**) $H/D = 6$.

To understand the process of vortex generation and its development in the sink, the contours of V/V_b and streamlines for different vertical cross sections (*oxz* plane) at different horizontal heights ($Re = 35{,}100$) are compared and analyzed, as shown in Figure 8.

Comparing Figure 7c with Figure 8a, it is not difficult to find that the formation of VRP originates precisely from the sinking of the fluid at the edges of the vortex caused by the spatial vortex structure consisting of $E_{1\text{-}0}$ to $E_{1\text{-}6}$. When H/D is 1, a vortex with a large influence range ($E_{1\text{-}0}$) appears downstream of the jet outlet in the middle section. The scope of the vortex gradually decreases ($E_{1\text{-}0}$~$E_{1\text{-}6}$) during the development of this vortex to both sides. A vortex ($G_{1\text{-}4}$) occurs near the bottom of the cross-section at $y/D = 4$. On the cross-section at $y/D = 6$, the core position of this vortex was almost unchanged. The vortex developed significantly, and the influence range gradually increased. At $H/D = 3$ and the same value of y, the scope of vortexes generated at the jet downstream outlet ($E_{3\text{-}0}$ and $E_{3\text{-}2}$) are smaller than those ($E_{1\text{-}0}$ and $E_{1\text{-}2}$) generated in $H/D = 1$. The reduction in space above the jet and the increase in the volume of entrainment below contribute to this phenomenon. The position of the vortex core moves along the +z direction as well. Meanwhile, the vortexes $G_{3\text{-}4}$ and $G_{3\text{-}6}$ move gradually in the $+x$ direction. Therefore, the incidence height H/D affects the size and location of the vortex in the flow field. There are obvious wall vortexes that are generated and well developed on both sides along the y-direction.

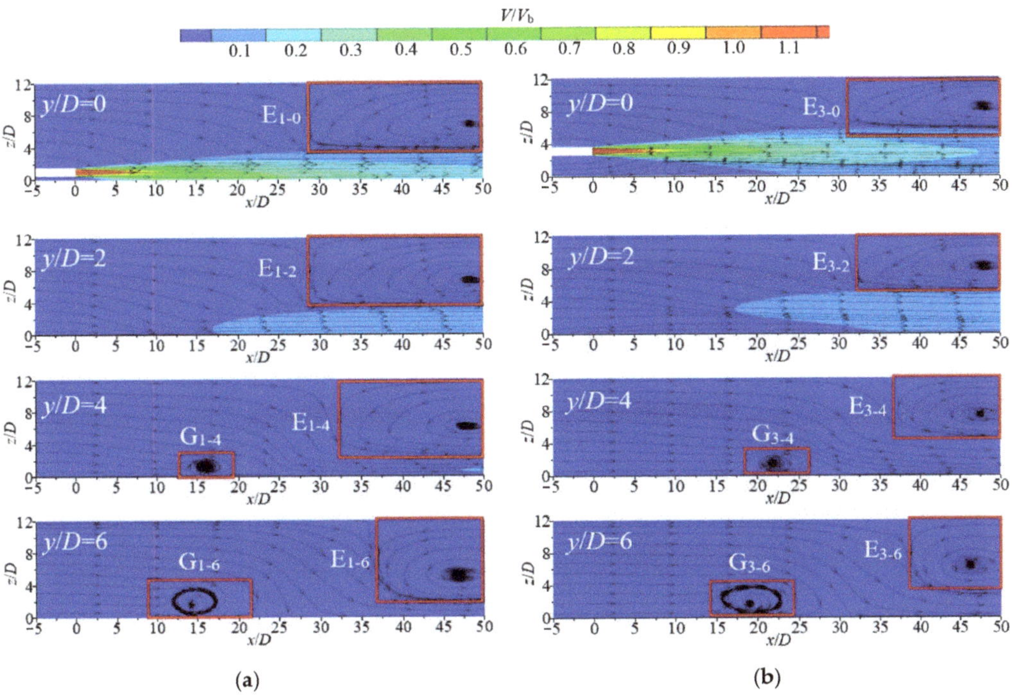

Figure 8. Contours with streamline diagrams of V/V_b distribution for different shot vertical sections: (**a**) $H/D = 1$; (**b**) $H/D = 3$.

With different longitudinal sections x/D, the contour of V/V_b on the longitudinal section of the jet (oyz plane) at different horizontal heights at Re = 35,100 is depicted in Figure 9. The values of V/V_b for each longitudinal section of the jet are distributed in a circular pattern for different H/D conditions. The dimensionless velocity decreases gradually from the jet center to the edge. The entrainment effect during the development of the jet makes the jet flow increase along the range; the section is expanding, and the velocity gradually decreases. At $H/D = 0.5$, the jet flows along the bottom wall and diffuses more obviously along the lateral direction y than in the vertical direction z, indicating that the bottom wall limits the diffusion of the jet in the vertical direction and enhances the lateral diffusion. As H/D increases, the influence of the bottom wall on the jet becomes weak. At the same time, the velocity of the jet arriving at the cross-section of $x/D = 50$ becomes smaller with the increase in H/D. The relative high-speed region in the center of the V/V_b distribution contours gradually disappears. Indicating that with the increase in H/D, the energy exchange between the jet and the ambient fluid becomes more adequate, leading to the rapid dissipation of its own energy.

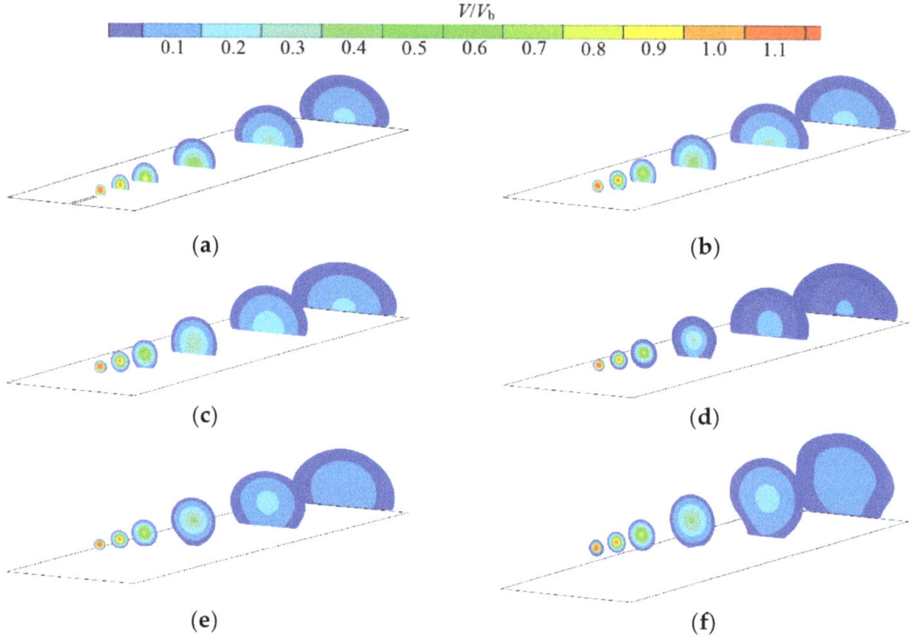

Figure 9. V/V_b distribution contours of longitudinal section (*oyz* plane) of jet under different H/D conditions: (**a**) $H/D = 0.5$; (**b**) $H/D = 1$; (**c**) $H/D = 1.5$; (**d**) $H/D = 2$; (**e**) $H/D = 3$; and (**f**) $H/D = 4$.

Figure 10 shows the variations of V/V_b at different horizontal heights at $Re = 35{,}100$. The cyan dashed line with arrows in the figure indicates the variation of the position of the maximum axial velocity point in the *z*-direction. As shown in the figure, the jet gradually spreads, and the maximum axial velocity gradually decreases as the jet flow x increases at $H/D = 0.5$. Due to the influence of the bottom wall, the parameter V/V_b distributes asymmetrically. The values of V/V_b near the bottom wall increase rapidly from 0 to the maximum. On the side of the HSJ away from the bottom, the values of V/V_b distribution are similar to the free jet velocity distribution. As the jet flow x increases, the point where V/V_b reaches the maximum value is gradually deviated from the central axis of the jet (x-axis) and gradually approaches the bottom wall. At $H/D = 1$, the maximum value of V/V_b is obtained on the centerline at $x/D = 10$. As for $H/D = 0.75$, however, the maximum value of V/V_b at the same position has deviated from the axis. This indicates that the distance between the position where the deflection of the jet occurs and the jet outlet increases with the growth of the horizontal height. As for $H/D = 6$, the bottom wall has almost no effect on the velocity distributions of the cross-section in the jet. The velocity distribution at each section is almost symmetric to the axis line. In summary, the bottom wall restricts the diffusion of the horizontal jet in the $-z$ direction, increasing the axial velocity of the jet in the near-wall area. It then forms a low-pressure area below the main stream of the jet, which makes the jet deflect downward and leads to the wall attachment effect of the jet.

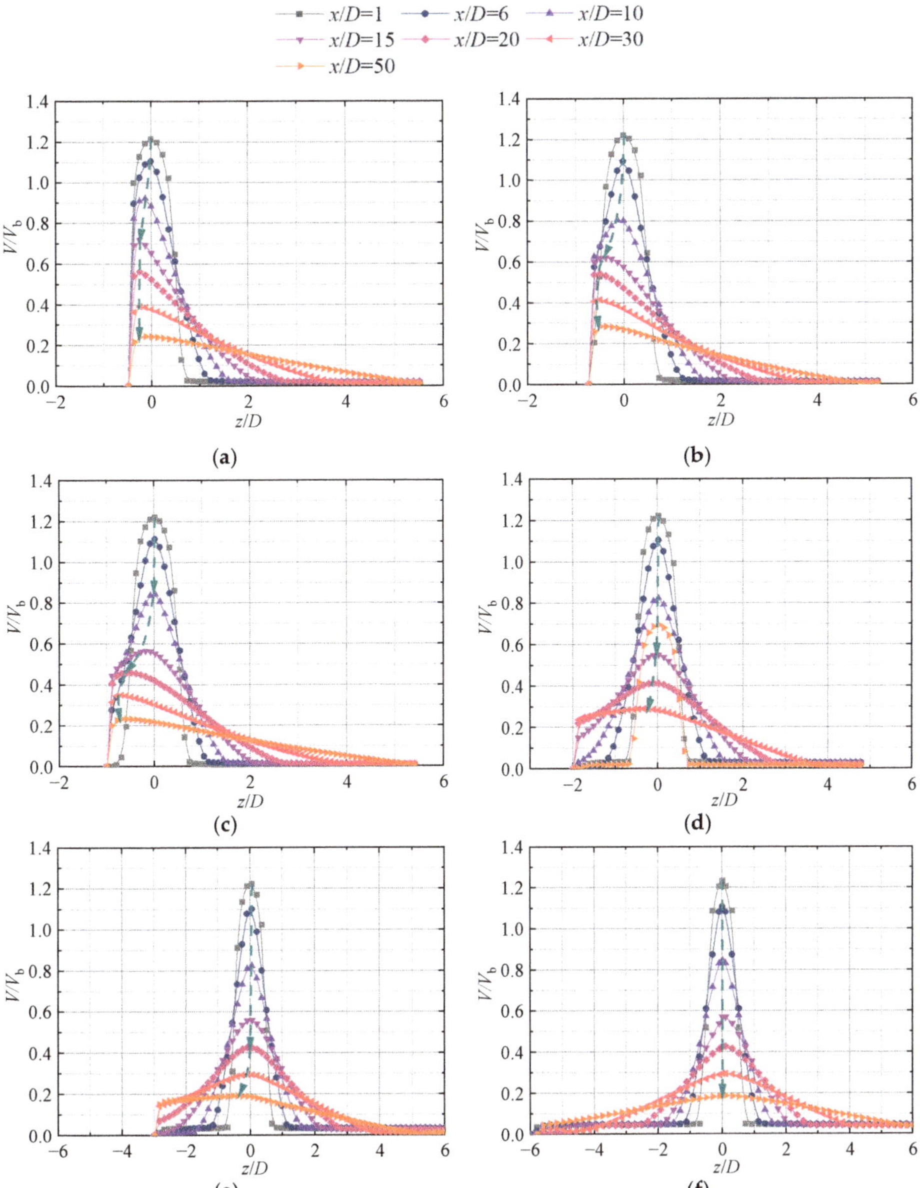

Figure 10. Axial V/V_b distribution in the mid-section (*oxz* plane) of the jet under different H/D conditions: (**a**) $H/D = 0.5$; (**b**) $H/D = 0.75$; (**c**) $H/D = 1$; (**d**) $H/D = 2$; (**e**) $H/D = 3$; and (**f**) $H/D = 6$.

Figure 11 shows the distribution of V/V_b along the axis of the horizontal jet at different horizontal heights ($Re = 35{,}100$). The trends are similar at different horizontal heights. In the initial range ($0 \leq x/D \leq 6$), the values of V/V_b decrease slowly from 1.23 to 1.15. When the jet further develops, it decays rapidly to about 0.4 in the region of $6 \leq x/D \leq 25$. As x/D continues to increase, the rate of decrease in V/V_b slows down and gradually fluctuates around 0.2. At the same time, due to the influence of the bottom wall, the mixing effect with the ambient fluid below the jet is greatly weakened. This is why the values of V/V_b of the horizontal jet with $H/D = 0.5$ are up to 1.29 times larger than those of other horizontal heights in the region of $x/D > 6$.

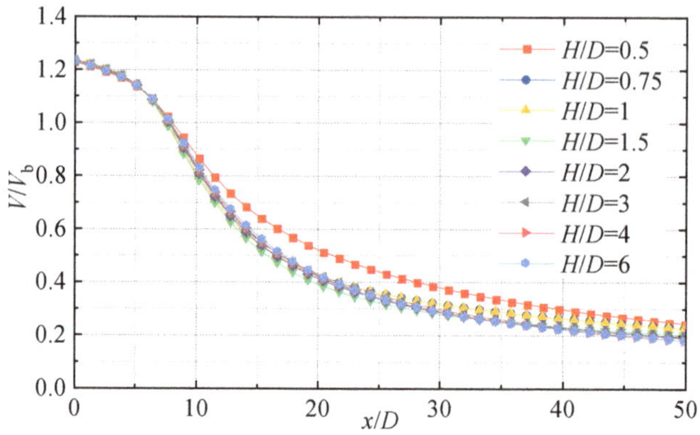

Figure 11. Distribution of horizontal jet axial dimensionless velocity V/V_b at different horizontal heights.

The distributions of each axial velocity have obvious inflection points. The horizontal inundation jet is divided into two zones. In zone I, the velocity drops sharply. In zone II, the drop of axial velocity slows down. To explore the general law of the inflection point, its first-order derivative image is made, as shown in Figure 12. The minimal value points of the V/V_b distribution curves under different H/D conditions are more concentrated and are all located near $x/D = 10$, suggesting that the inflection points of each axial velocity distribution curve have high consistency. The initial velocity of the jet is very fast after it is ejected from the outlet and mixed with the ambient fluid. During the process, there is a strong energy exchange and the axial velocity decreases along the course. As mentioned above, the cross-sectional area increases while the jet flow increases along the range, due to the entrainment effect, which leads to increased resistance of the main stream of the jet during its development, thus leading to an increase in the decay rate of the jet axial velocity and reaching a maximum at $x/D = 10$. As for $x/D > 10$, the jet gradually develops completely, and the energy exchange and entrainment effect then reach a relative equilibrium. The axial velocity decay rate remains at a low level and decreases along the range, which leads to an obvious inflection point on the V/V_b distribution curve. Further exploration reveals that although the extreme points (inflection points) of the first-order derivative images of each curve are consistent, there are large differences in their distributions of extreme value. When the horizontal height H/D is taken as 0.5 and 1.5, respectively, the extreme values of the corresponding axial velocity distribution curves obtain the maximum or minimum values. This indicates that at $H/D = 0.5$ and 1.5, the absolute value of the jet axial velocity gradient at the inflection point ($x/D = 10$) achieves the minimum and maximum values, respectively, due to the wall effect generated at the bottom wall. With the full development of the jet, a situation diametrically opposite to that described earlier emerges: the absolute value of the velocity gradient of V/V_b is maximum when the horizontal height $H/D = 0.5$, and the absolute value of the velocity gradient of V/V_b is minimum when the horizontal height $H/D = 1.5$.

Figure 13 shows the variations of V/V_b at different oxy and oxz cross-sections at different horizontal heights (Re = 35,100). At $H/D = 0.5$, $z/D = \pm 0.25$ and $y/D = 0.25$, the values of V/V_b almost remain as a constant in the range of $0 \leq x/D \leq 2$. The values of V/V_b for each operating condition decrease gradually with the increase in the jet flow distance. The values of V/V_b in the range of $2 \leq x/D \leq 40$ is located at $z/D = -0.25$, $y/D = 0.25$, and $z/D = 0.25$ cross-sections from the largest to the smallest. This indicates that the velocity of the jet fluid on the side close to the wall is increased due to the influence of the wall. The values of V/V_b at $z/D = 0.5$ and $y/D = 0.5$ rapidly increase from 0 to 0.6 in the region of $0 \le x/D \le 2$. Meanwhile, the values of V/V_b at $z/D = 0.5$ and $y/D = 0.5$ gradually decrease. For $H/D \ge 2$, the values of V/V_b at $z/D = \pm 0.25$ and $y/D = 0.25$ are almost equal, and the trend of velocity distribution is the same. Combined with Figure 8, the overall dimensionless velocity V/V_b of the jet is less and less influenced by the wall as H/D increases. In summary, the influence of the bottom wall on the jet is mainly manifested by limiting the diffusion in the $-z$ direction, increasing the velocity in the near-wall area, and the influence of the bottom wall on the jet decreases with the increase in the horizontal height until it disappears.

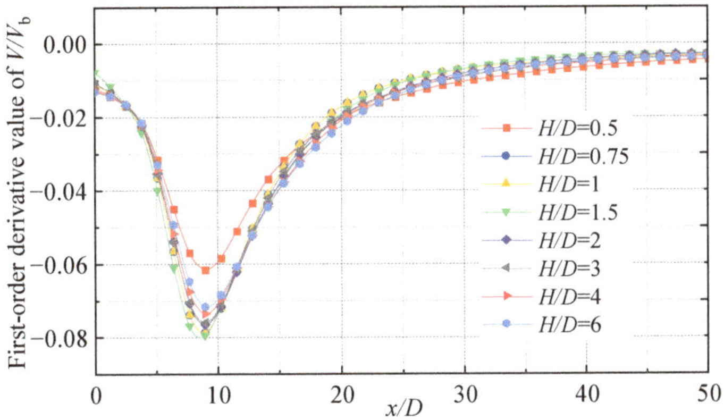

Figure 12. First-order derivative curve of V/V_b.

3.2. Jet Velocity Distribution in the Near-Wall Region

Figure 14 shows the contours of dimensionless velocity with the surface streamlines in the mid-section of the horizontal jet under the condition of horizontal height $0.5 \le H/D \le 1.5$ in the parameter ranges of $0 \le x/D \le 10$ and $0 \le H/D \le 2$. The red rectangular block indicates the point where the pressure on the bottom wall is zero and is called the jet contact point (JCP). The JCP parameter gradually moves downstream of the jet with the increase in H/D. At $H/D = 0.5$, the jet flows along the bottom wall. At $H/D = 0.75$, the jet diffuses to the bottom wall at the position around $x/D = 1.5$, after which, the bottom wall restricts the vertical downward diffusion of the jet, resulting in a continuous backward movement of the JCP. On the side away from the bottom, however, the diffusion of the jet upward is almost unaffected, which leads to an asymmetry in the jet velocity distribution and vector diagram.

In order to explore the general law of JCP distribution, the horizontal distance of the JCP from the jet pipe outlet is denoted by L_p. Figure 15 depicts the variation of the L_p/D with H/D for different Reynolds number conditions. As shown in the figure, at a lower Reynolds number, the jet contains less turbulent kinetic energy and its ability to mix the ambient fluid is diminished, resulting in a small L_p/D in most of the H/D range. In particular, in the smaller H/D case ($H/D = 0.75$), the wall attachment effect of the HSJ is very weak because there is very little ambient fluid below the main stream of the jet. At the same time, the lower jet velocity further weakens the wall attachment effect, and the main stream of the jet is no longer deflected towards the bottom wall. In addition, the fluid below the main stream of the jet has a large velocity gradient in the z-direction, which produces a clear boundary layer effect. The slip phenomenon occurs when the jet main stream diffuses to the boundary layer, resulting in a large L_p/D for a small H/D. Overall, the distribution pattern of L_p/D under different Reynolds number conditions has a high consistency and shows a relatively independent property from the Reynolds number when the Reynolds number of the jet increases.

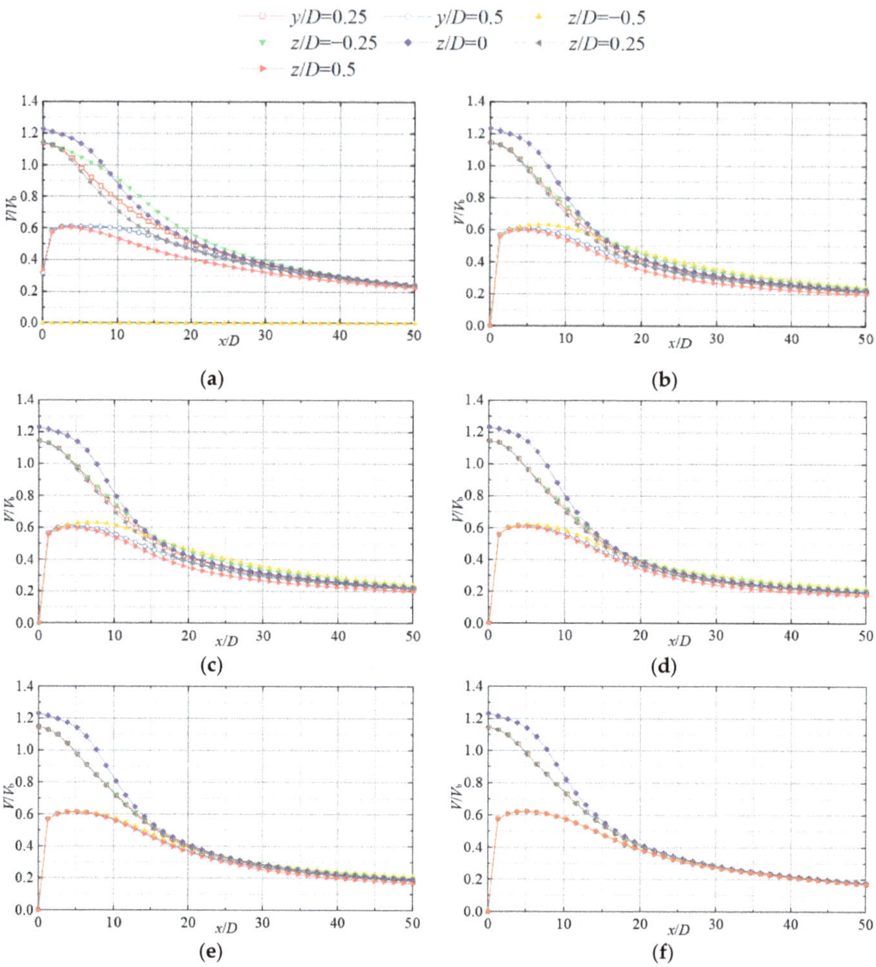

Figure 13. Distribution of V/V_b at different y/D and z/D positions at different horizontal heights H/D: (**a**) H/D = 0.5; (**b**) H/D = 0.75; (**c**) H/D = 1; (**d**) H/D = 1.5; (**e**) H/D = 3; and (**f**) H/D = 4.

Figure 16 depicts the distribution of V/V_b at different values of z in the near-wall region in the parameter range of $0 \leq z/D \leq 0.5$. When H/D = 0.5, the jet structure is a typical wall jet. The values of V/V_b decrease gradually with the increase in the jet range. At H/D = 0.75, the distribution of V/V_b shows different trends for different z values. At h/D = 0.1 and 0.2, the values of V/V_b increase gradually as x increases in the region of $0 \leq x/D \leq 10$ due to the entrainment effect of the jet, and thereafter, the values of V/V_b decrease gradually. The velocity gradient when the jet decelerates is smaller than that when the flow is accelerated. As for increasing h/D, the values of V/V_b decrease gradually with the increase in the jet range, and is especially noted that when h/D = 0.3; a velocity smoothing zone appears near the outlet by the influence of the boundary layer of the main stream of the jet. In the zone, the values of V/V_b fluctuate around 0.72. When the H/D ranges from1.25 to 2, the velocity in the near-wall area shows a trend of gradual acceleration from 0. After that, a slow deceleration can be found. At the same time, V/V_b gradually approaches 0.2 as the horizontal height increases. In addition, the acceleration zone near the wall gradually expands.

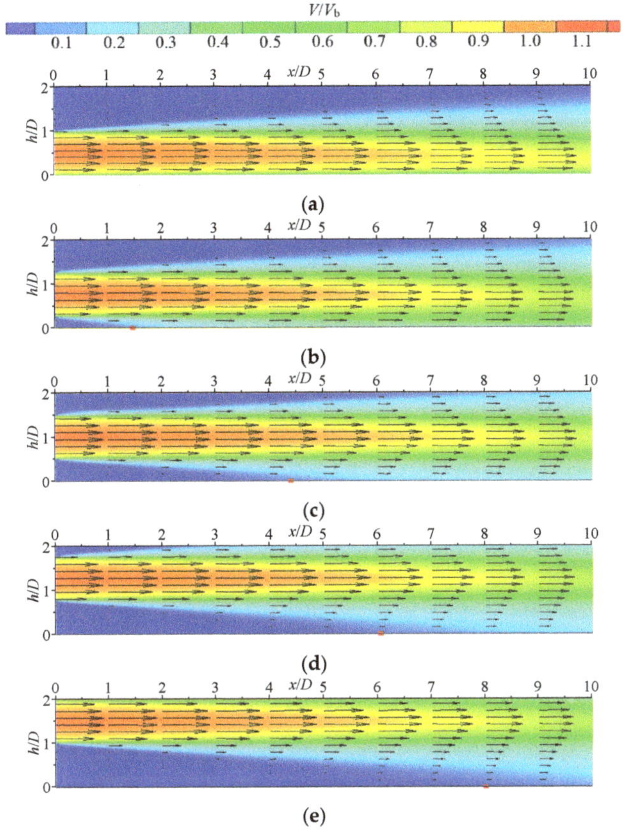

Figure 14. Contours with vector diagrams of jet velocity distribution at different incidence heights H/D: (**a**) $H/D = 0.5$; (**b**) $H/D = 0.75$; (**c**) $H/D = 1$; (**d**) $H/D = 1.25$; and (**e**) $H/D = 1.5$.

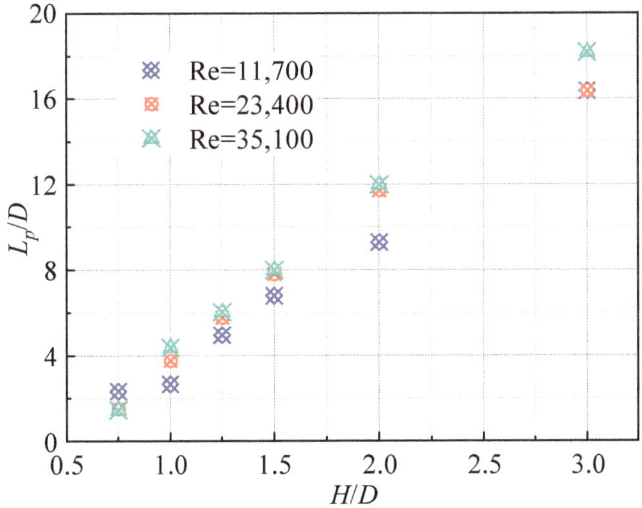

Figure 15. The plot of dimensionless distance $L_{\rm p}/D$ with H/D.

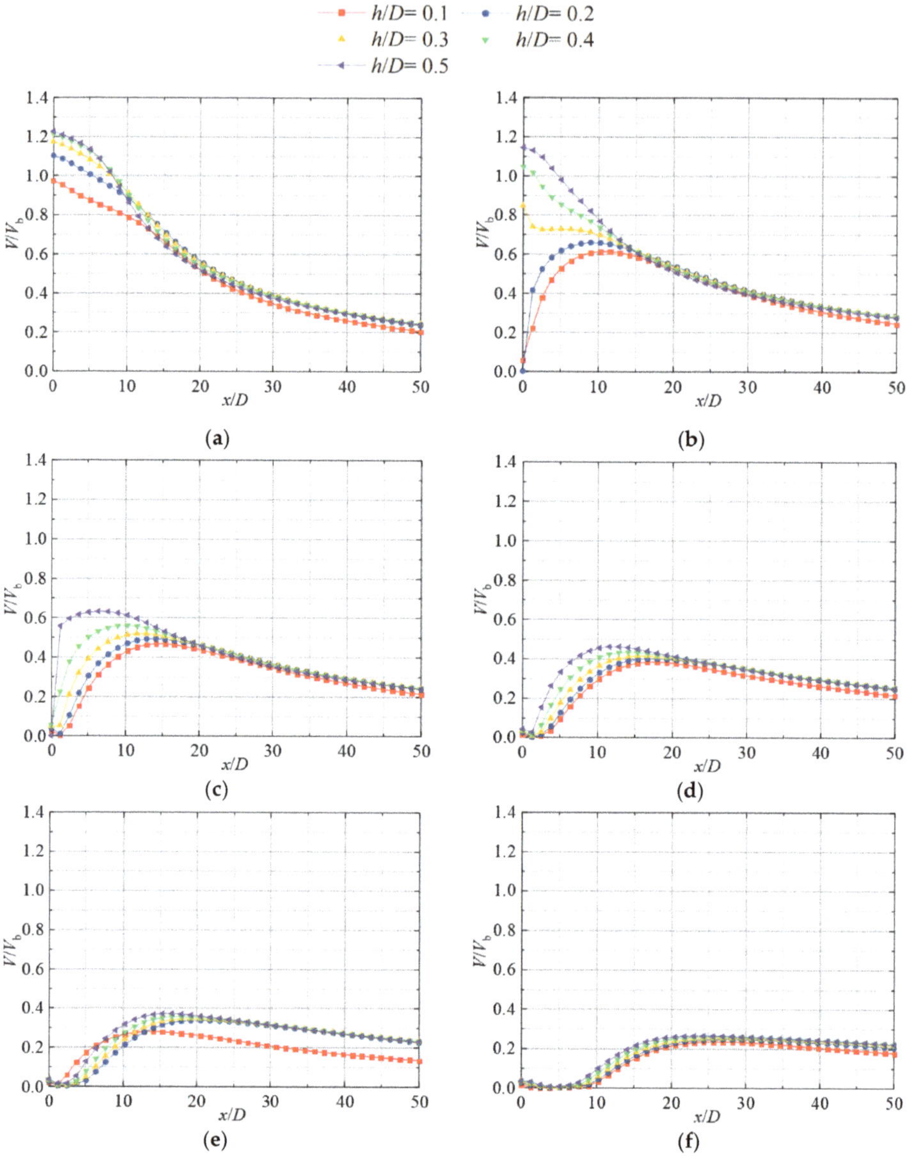

Figure 16. Distribution of near-wall dimensionless velocity V/V_b at different incidence heights H/D: (**a**) H/D = 0.5; (**b**) H/D = 0.75; (**c**) H/D = 1; (**d**) H/D = 1.25; (**e**) H/D = 1.5; and (**f**) H/D = 2.

4. Conclusions

In this paper, based on the Wray–Agarwal turbulence model, a numerical simulation study of HSJs with different incidence heights is carried out, and the main conclusions are as follows:

(1) The jet horizontal height H/D has a large effect on the flow field structure of HSJ. When H/D is small, the unsteady structure in the jet flow field is dominated by vortexes, and the distribution is more regular, all starting generation around x/D = 30. In the meantime, the vortex structure in the flow field has a clear boundary and is

insensitive to the variation of H/D. In addition, significant wall-attached vortexes are generated on both sides of the flow field. In the $+y$ direction, the wall-attached vortexes gradually develop. They are also observed to gradually move in the $+x$ direction as H/D increases.

(2) The distance between the JCP and the jet pipe outlet increases linearly with the increase in the incidence height H/D, but the distribution pattern of the JCP gradually becomes relatively independent of the Reynolds number with the increase in the jet Reynolds number.

(3) The velocity distribution of the jet axial velocity under different incidence heights H/D has very high similarity, and all of them have obvious velocity inflection points at $x = 10D$. The horizontal inundation jet is divided into two zones; in zone I, the velocity drops sharply, while in zone II, the drop of axial velocity slows down. In addition, when H/D is small, its axial speed is significantly higher than in other working conditions, up to 1.29 times. It can be seen that the wall attachment effect of the jet and the boundary layer effect generated at the bottom of the fluid domain have a certain role in maintaining the velocity of the jet near the wall. The flow velocity of the main stream of the jet near the wall is obviously large. This phenomenon disappears with the increase in H/D. In addition, this velocity retention effect is most significant at $H/D = 1$.

The above conclusions have a certain guiding significance for engineering practice. In marine engineering, horizontal inundation jets are often used for underwater sand flushing and dredging where the distribution law of the horizontal jet JCP can be fully utilized, and the position of the JCP can be adjusted by controlling the incidence height to achieve precise construction. At the same time, the incidence height can be sufficiently reduced to ensure a large axial velocity of the jet in order to achieve a better scouring effect. It is not negligible that the vortex structure in the flow field is fully developed and has a stable distribution pattern at a small incidence height, which also provides good conditions for the rapid transport of scouring materials. In addition, in water conservancy engineering, horizontal submerged jet energy dissipation as a new flood energy dissipation method is developing rapidly, and the research results of this paper lay the theoretical foundation for the study of its energy dissipation effect, which is of great practical significance.

Author Contributions: Data curation, B.H. and H.W.; formal analysis, J.L. and Q.Y.; writing—original draft, C.W. and Y.Z.; writing—review and editing, J.G. and X.C.; software, Y.Y. All authors have read and agreed to the published version of the manuscript.

Funding: This research was supported by the Open Research Fund of State Key Laboratory of Simulation and Regulation of Water Cycle in River Basin (China Institute of Water Resources and Hydropower Research) (Grant No: IWHR-SKL-201719), Science Center for Gas Turbine Project (P2021-A-I-003-002), the National Natural Science Foundation of China (Grant No: 51979240 and 51609105), the National Key R&D Program of China (Grant No: 2020YFC1512402), and the Natural Science Foundation of Jiangsu Province (Grant No: BK20220587).

Institutional Review Board Statement: Not applicable.

Informed Consent Statement: Not applicable.

Data Availability Statement: Not applicable.

Conflicts of Interest: The authors declare no conflict of interest.

References

1. Dong, Z. *Impinging Jets*; China Ocean Press: Beijing, China, 1997.
2. Rouse, H. Cavitation in the mixing zone of a submerged jet. *La Houille Blanche* **1953**, *1*, 9–19. [CrossRef]
3. Law, H.S. Numerical prediction of the flow field due to a confined laminar two-dimensional submerged jet. *Comput. Fluids* **1984**, *12*, 199–215. [CrossRef]
4. Wen, Q.; Kim, H.D.; Liu, Y.; Kim, K.C. Dynamic structures of a submerged jet interacting with a free surface. *Exp. Therm. Fluid Sci.* **2014**, *57*, 396–406. [CrossRef]

5. Liu, H.; Kang, C.; Zhang, W.; Zhang, T. Flow structures and cavitation in submerged waterjet at high jet pressure. *Exp. Therm. Fluid Sci.* **2017**, *88*, 504–512. [CrossRef]
6. Yang, Y.; Li, W.; Shi, W.; Zhang, W.; El-Emam, A.M. Numerical Investigation of a High-Pressure Submerged Jet Using a Cavitation Model Considering Effects of Shear Stress. *Processes* **2019**, *7*, 541. [CrossRef]
7. Ma, J.; Song, Y.; Zhou, P.; Cheng, W.; Chu, S. A mathematical approach to submerged horizontal buoyant jet trajectory and a criterion for jet flow patterns. *Exp. Therm. Fluid Sci.* **2018**, *92*, 409–419. [CrossRef]
8. Yang, H.C. Horizontal two-phase jet behavior with an annular nozzle ejector in the water tank. *J. Vis.* **2015**, *18*, 359–367. [CrossRef]
9. Xi, B.; Wang, C.; Xi, W.; Liu, Y.; Wang, H.; Yang, Y. Experimental investigation on the water hammer characteristic of stalling fluid in eccentric casing-tubing annulus. *Energy* **2022**, *253*, 124113. [CrossRef]
10. Shao, D.; Huang, D.; Jiang, B.; Law, A.W.K. Flow patterns and mixing characteristics of horizontal buoyant jets at low and moderate Reynolds numbers. *Int. J. Heat Mass Transf.* **2017**, *105*, 831–846. [CrossRef]
11. Chatterjee, S.S.; Ghosh, S.N.; Chatterjee, M. Local scour due to submerged horizontal jet. *J. Hydraul. Eng.* **1994**, *120*, 973–992. [CrossRef]
12. Chiew, Y.M.; Lim, S.Y. Local scour by a deeply submerged horizontal circular jet. *J. Hydraul. Eng.* **1996**, *122*, 529–532. [CrossRef]
13. Tang, S.; Zhu, Y.; Yuan, S. Intelligent Fault Diagnosis of Hydraulic Piston Pump Based on Deep Learning and Bayesian Optimization. *ISA Trans.* **2022**, in press. [CrossRef]
14. Zhu, Y.; Li, G.; Tang, S.; Wang, R.; Su, H.; Wang, C. Acoustic Signal-based Fault Detection of Hydraulic Piston Pump using a Particle Swarm Optimization Enhancement CNN. *Appl. Acoust.* **2022**, *192*, 108718. [CrossRef]
15. Abdelaziz, S.; Bui, M.D.; Rutschmann, P. Numerical simulation of scour development due to submerged horizontal jet. *River Flow* **2010**, *2010*, 1597–1604.
16. Li, H.; Zheng, T.; Dai, L.; Chen, X. Theoretical Research on New Energy Dissipation Type of Multi-horizontal Submerged Jets. *Water Resour. Power* **2010**, *20*, 74–76+166.
17. Li, W.; Zhang, W.; Shi, W.; Yang, Y.; Cao, W. Experiment study on cavitation cloud evolution law of cavitation jet of convergent-divergent nozzle. *J. Irrig. Drain. Eng.* **2020**, *38*, 547–552.
18. Kan, B.; Gao, Y.; Xu, S.; Ding, J. Numerical simulation of flow field characteristics of pressure controllable pipe with microporous walls. *J. Irrig. Drain. Eng.* **2020**, *38*, 183–187.
19. Qi, M.; Wang, L.; Chen, Q.; Zhao, J.; Ju, Y.; Fu, Q. Numerical simulation of cavitating jet in dual chamber self-oscillation pulse nozzle. *J. Irrig. Drain. Eng.* **2020**, *38*, 457–461.
20. Jiang, H.; Yan, H.; Xiao, Z.; Cheng, L.; Liu, H. Numerical simulation of hydraulic characteristics of three-side inlet of vertical shaft inlet channel of horizontal pumping station. *J. Irrig. Drain. Eng.* **2021**, *39*, 1008–1013.
21. Shi, G.; Yao, X.; Wang, S.; Li, H. Analysis of stress and strain of multiphase pump blade considering medium viscosity. *J. Irrig. Drain. Eng.* **2020**, *38*, 991–996.
22. Zhang, H.; Zheng, Y.; Zhang, Z.; Kan, K.; Mo, X. Influence of placement angle of sewage mixer on its hydraulic characteristics. *J. Irrig. Drain. Eng.* **2021**, *39*, 483–487.
23. Wang, H.; Long, B.; Wang, C.; Han, C.; Li, L. Effects of the impeller blade with a slot structure on the centrifugal pump performance. *Energies* **2020**, *13*, 1628. [CrossRef]
24. Liu, Y.; Pan, Z. Analysis of internal flow characteristics of marine water-jet propulsion. *J. Irrig. Drain. Eng.* **2022**, *40*, 55–61.
25. Quan, H.; Cheng, J.; Peng, G. Effect of screw centrifugal inducer on mechanical characteristics of vortex pump. *Drain. Irrig. Mach. Eng.* **2021**, *39*, 345–350.
26. Roache, P.J. A method for uniform reporting of grid refinement studies. *ASME-Publ.-Fed* **1993**, *158*, 109. [CrossRef]
27. Khayrullina, A.; van Hooff, T.; Blocken, B.; van Heijst, G. Validation of steady RANS modelling of isothermal plane turbulent impinging jets at moderate Reynolds numbers. *Eur. J. Mech. B/Fluids* **2019**, *75*, 228–243. [CrossRef]
28. Liu, H.; Liu, M.; Bai, Y.; Du, H.; Dong, L. Grid convergence based on GCI for centrifugal pump. *J. Jiangsu Univ. Nat. Sci. Ed.* **2014**, *35*, 279–283.
29. Celik, I.B.; Ghia, U.; Roache, P.J.; Freitas, C.J. Procedure for estimation and reporting of uncertainty due to discretization in CFD applications. *J. Fluids Eng. Trans. ASME* **2008**, *130*, 7.
30. Ying, J.; Yu, X.; He, W.; Zhang, J. Volume of fluid model-based flow pattern in forebay of pump station and combined rectification scheme. *J. Irrig. Drain. Eng.* **2020**, *38*, 476–480.
31. Feng, L.; Xu, R.; Feng, J. Numerical simulation of flow and heat transfer characteristics in helical pipe. *J. Irrig. Drain. Eng.* **2020**, *38*, 697–701.
32. Wray, T.J.; Agarwal, R.K. Low-Reynolds-number one-equation turbulence model based on k-ω closure. *AIAA J.* **2015**, *53*, 2216–2227. [CrossRef]
33. Han, X.; Rahman, M.; Agarwal, R.K. Development and Application of Wall-Distance-Free Wray-Agarwal Turbulence Model (WA2018). In Proceedings of the 2018 AIAA Aerospace Sciences Meeting, Kissimmee, FL, USA, 8–12 January 2018; p. 0593.
34. Hu, B.; Wang, H.; Liu, J.; Zhu, Y.; Wang, C.; Ge, J.; Zhang, Y. A Numerical Study of a Submerged Water Jet Impinging on a Stationary Wall. *J. Mar. Sci. Eng.* **2022**, *10*, 228. [CrossRef]
35. Zhang, D.; Wang, H.; Liu, J.; Wang, C.; Ge, J.; Zhu, Y.; Chen, X.; Hu, B. Flow Characteristics of Oblique Submerged Impinging Jet at Various Impinging Heights. *J. Mar. Sci. Eng.* **2022**, *10*, 399. [CrossRef]

36. Xu, W.; Wang, C.; Zhang, L.; Ge, J.; Zhang, D.; Gao, Z. Numerical study of continuous jet impinging on a rotating wall based on Wray–Agarwal turbulence model. *J. Braz. Soc. Mech. Sci. Eng.* **2022**, *44*, 433. [CrossRef]
37. Wang, C.; Wang, X.; Shi, W.; Lu, W.; Tan, S.K.; Zhou, L. Experimental investigation on impingement of a submerged circular water jet at varying impinging angles and Reynolds numbers. *Exp. Therm. Fluid Sci.* **2017**, *89*, 189–198. [CrossRef]

Journal of
Marine Science and Engineering

MDPI

Article

Numerical Study on the Effect of Port Orientation on Multiple Inclined Dense Jets

Seyed Ahmad Reza Saeidi Hosseini [1], Abdolmajid Mohammadian [1,*], Philip J. W. Roberts [2] and Ozeair Abessi [3]

[1] Department of Civil Engineering, University of Ottawa, 75 Laurier Ave E, Ottawa, ON K1N 6N5, Canada; ssaei041@uottawa.ca

[2] School of Civil and Environmental Engineering, Georgia Institute of Technology, Atlanta, GA 30332, USA; phil.roberts@ce.gatech.edu

[3] School of Civil Engineering, Babol Noshirvani University of Technology, Babol 4714871167, Iran; oabessi@nit.ac.ir

* Correspondence: majid.mohammadian@uottawa.ca

Abstract: Wastewaters are commonly discharged into the seas and oceans through multiport diffusers. Accurate prediction of the complex interactions of multiport diffusers with the receiving water bodies is significant for the optimal design of outfall systems and has yet to be fully illuminated. In the current study, the mixing and dilution characteristics of multiple inclined dense jets are studied using a three-dimensional numerical simulation. The Launder, Reece, and Rodi (LRR) turbulence model is employed to perform the simulations, and the predictions are compared against available experimental data. The results indicate that the LRR turbulence model is a promising tool for the study of inclined dense jets discharged from multiport diffusers, and it can provide more accurate predictions of the mixing behavior than standard and re-normalization group (RNG) k-ε turbulence models. The model is further employed to evaluate and compare the dispersion capabilities of multiport diffusers with uniform and non-uniform jet orientation to the horizontal, as a novel idea. The comparisons demonstrate the middle discharge may have a longer trajectory (7% and 5% increase in terminal rise height and impact point distance, respectively) and therefore a higher dilution rate (14% increase in impact dilution) when its adjacent jets are disposed with a different angle, compared to that of uniform discharges. The outcomes may be favorable for outfall systems applications involving dilution.

Keywords: numerical simulations; multiple discharges; inclined dense jets; mixing and dilution; discharge inclination

Citation: Saeidi Hosseini, S.A.R.; Mohammadian, A.; Roberts, P.J.W.; Abessi, O. Numerical Study on the Effect of Port Orientation on Multiple Inclined Dense Jets. *J. Mar. Sci. Eng.* **2022**, *10*, 590. https://doi.org/10.3390/jmse10050590

Academic Editors: Fuping Gao and Alessandro Antonini

Received: 22 March 2022
Accepted: 23 April 2022
Published: 26 April 2022

1. Introduction

Brine wastewaters from desalination plants are commonly discharged into the seas and oceans through diffusers. These effluents, which have a density higher than the surrounding water, sink to the sea-bottom and can increase salinity near the diffuser, which can result in adverse effects on the local natural environment if not properly discharged [1]. The mixing characteristics of the wastewater discharges are closely related to the properties of the outfall diffusers [2,3]. Thus, it is essential to predict the mixing behavior of the jets from different kinds of diffusers to enable an economically efficient design while complying with regulatory demands.

Discharges may be positively, neutral, or negatively buoyant. Positively buoyant jets have a density lower than the ambient water density and rise due to a positive buoyancy until they reach the water surface and then spread horizontally as a plume; the horizontal and vertical multiple discharges are typically positively buoyant or neutral. Negatively buoyant jets have a density higher than the surrounding water, and they are affected by a negative buoyancy which hinders the flow moving upward and leads the jets to move back to the sea bottom; the inclined jets are mainly negatively buoyant. Thus, the trajectory and therefore mixing behavior of inclined jets may differ from the horizontal and vertical discharges, which highlights the importance of the separate study of inclined jets. On the other hand, the diffusers may be either single- or multi-port. Single-port discharges can freely mix with the receiving water without any interactions. However, multiport diffusers are frequently applied around the world, e.g., Melbourne, Perth, Sydney, and Boston, to enable high dilutions with high effluent flow rates [4,5]. The mixing process of multiport discharges differs from single-port jets, and its study is much more complicated because of the complex mechanisms resulting from jet interactions and the Coanda effect. Although the mixing behavior of single inclined dense jets has been widely studied for many years, the dilution properties of multiple jets have received less attention.

The investigations on multiple buoyant jets were quite often experimental or analytical. Some prior experimental studies on multiple buoyant jets have focused on multiple horizontal [5–11] and some others on vertical [4,12] discharges. Some studies analytically investigated multiple buoyant jets and estimated the distance over which the interacting 3D buoyant jets behave as a plane jet [7,13]. Knystautas theoretically predicted the velocity field of multiple turbulent jets using the superposition technique based on the Reichardt hypothesis [6]. Knystautas' model was further improved to predict the concentration field based on an extension of the Reichardt hypothesis to the lateral transport of pollutants [14]. Wang and Davidson [5] argued that the models presented by Knystautas [6] and Hodgson et al. [14] did not capture the possible change in spreading rate due to jet merging, and extended their models by considering the changes in the spreading rates and concentration to velocity length scale ratio between the axisymmetric and 2D flow limits. Yannopoulos and Noutsopoulos analyzed interacting vertical buoyant jets in stagnant ambient conditions, and developed analytical expressions concerning the concentration and velocity distributions based on the entrainment restriction approach [12] and superposition method [15]. Lai and Lee [16] proposed a general semi-analytical model for the dynamic interaction of multiple buoyant jets, and showed that merging point, jet trajectories, as well as the centerline concentration and velocity of the buoyant jet group were perfectly matched with the measurements. Furthermore, multiple inclined discharges have been mainly studied experimentally [17–23].

Although experimental studies have significantly contributed to a better understanding of multiple jets, they are expensive and time-consuming options. Thus, supplementary methods such as integral entrainment models have always been of interest. These models are widely employed by designers to assess jet mixing behavior and are a basis for commercial mixing tools such as JetLag [24,25], CoreJet [26], and UM3 [27]. According to integral entrainment models, the velocity profiles of a jet is assumed to be Gaussian or top hat and axisymmetric without radial changes, resulting in simplified governing momentum and mass conservation equations [28]. These models are unable to resolve re-entrainment, boundary effects, and the Coanda effect [28,29]. The last limitation may shed doubt on the prediction capabilities of integral entrainment models for multiple jets, where the Coanda effect reduces jet dilution and terminal rise height [19,30]. Palomar et al. [31] evaluated the prediction capabilities of the JetLag, CoreJet, and UM3 tools for inclined dense jets. They found a lot of discrepancies for the prediction of terminal rise height, and showed more than 50% under-prediction for dilution. This underestimation has roots in the simplified mass and momentum equations of the integral models resulting from the simplex assumptions for the jet velocity profiles.

Thanks to the advancements in computational resources over the past two decades, the application of computational fluid dynamics (CFD) modeling in the prediction of jet-mixing behavior has dramatically increased [32]. CFD includes fewer simplifying assumptions, compared to integral entrainment models, and is currently a promising approach to simulate desalination outfalls [29]. The application of CFD modeling, including the Reynolds Averaged Navier Stokes (RANS) model [33–38] and large eddy simulation (LES) [39–41], has been reported for single port discharges in recent works. However, there have been only a limited number of reports covering the predictive capabilities of CFD models for multiple jets. Some studies have numerically investigated multiple buoyant jets. Xue et al. [42] indicated that the standard k-ε model could reasonably predict the mixing behavior of multiple horizontal jets in cross-flow. Tang et al. [43] applied a buoyancy-corrected eddy viscosity model (EVM) to predict the primary mixing of buoyant jets from multiple ports, and compared the results to the CORMIX model predictions. They showed that the CFD model could capture the changes of the plumes' depth along the trajectory, which the CORMIX model was incapable of capturing. Yan et al. [44] investigated the prediction capabilities of four different turbulence closures including the standard k-ε, re-normalization group (RNG) k-ε, k-ω, and k-ω shear stress transport (SST) for multiple vertical jets and showed that (RNG) k-ε resulted in more accurate outcomes. As the mixing behavior of multiple inclined discharges may differ from buoyant jets due to the negative buoyancy, some studies have attempted to validate the application of CFD for those jets. Yan and Mohammadian [3] validated the performance of the standard and re-normalization group (RNG) k-ε turbulence models for the prediction of multiple inclined dense jets and showed that both models could provide predictions of the terminal rise height and the impact point dilution with errors lower than 15%. Moreover, the predictions for the impact distance based on the standard and (RNG) k-ε closures were found to be accurate with errors of 18.6% and 13.3%, respectively. A recent study [29] conducted numerical CFD simulations based on k-ω (SST) closure to investigate the Gold Coast Desalination Plant offshore inclined multiport brine diffuser while affected by different ambient velocity conditions, in the range of 0.03–0.26 m/s, and validated impact dilutions and plume trajectory against physical experimental data.

Experimental and numerical models are adopted to predict the mixing behavior of discharges, which are significantly affected by the properties of outfall diffusers [2,3]. For single dense discharges, inclined jets result in a higher dilution rate than vertical discharges since vertical jets tend to fall back on themselves and it significantly impairs the dilution process [45–47]. It is widely accepted that the 60° inclined discharges provide the longest trajectory for entrainment and therefore the highest dilution rate [48–51]. The effects of flow orientation on the flow field have been also investigated in other hydraulic conditions [52–54]. The main issue with multiple inclined dense discharges arises when the ports are not sufficiently spaced apart, and the jets are closely or moderately spaced. The dynamic interaction and the Coanda effect cause these jets to have a reduced terminal rise height and to bend inward, and thus to have a shorter trajectory compared to a single jet [19]. The shortened trajectory further decreases dilution since it diminishes the available surface for entrainment. To overcome this reduction in dilution, it has been recommended to space the jets sufficiently apart in multiport diffusers [19], which lets the jets freely entrain the surrounding water without any interactions, as single jets do. The effect of port spacing on the mixing process of multiport diffusers has been the focus of some prior research. Yan et al. [44] focused on multiple vertical buoyant jets and indicated that as the port spacing was reduced, the jets' centerline dilution was lower, and the plume-like region was placed farther from the port. Moreover, they proposed a novel formula for concentration profiles that considers the influence of port spacing. Abessi and Roberts [19] conducted laboratory experiments on the influences of port spacing on multiple inclined dense jets and indicated that the port spacing can have a negative effect on the rise height and other geometrical characteristics of discharges. They highlighted the minimum port spacing that avoids dynamic interaction of jets.

Based on the above explanations, two main research gaps need to be further elaborated when focusing on the study of multiport diffusers. First evaluating the prediction capabilities of less validated modeling approaches, from the CFD simulation point of view, and secondly investigating the influence of non-uniform port orientation on the mixing behavior, from the optimal design point of view. The available numerical studies have mainly investigated the prediction capabilities of the eddy viscosity models (EVMs). The EVMs consider that the Reynolds stresses are proportional to mean rates of deformation and the stresses are obtained based on the Boussinesq assumption [55]. These models provide good results for simple flows and some recirculating flows [55], but some drawbacks have been reported for them when employed for complex flow fields [56,57]. The main drawback of most EVMs can be the consideration of isotropic character for the eddy viscosity, which causes them to be insensitive to the orientation of the turbulence structure and its mixing and transport processes [35]. However, the stress anisotropy plays a significant role in changing in the streamline trajectory and stress-induced secondary motions [57]. The RSMs solve transport equations for the Reynolds stresses and consider different values for viscosity in each direction, which may result in more accurate predictions compared to the EVMs for the complex flow fields [55]. Gildeh et al. [35] applied five different RANS models, including the re-normalization group (RNG) k-ε, realizable k-ε, nonlinear k-ε, Launder, Reece, and Rodi (LRR), and Launder-Gibson, to predict the mixing behavior of 30° and 45° inclined dense jets. They showed that the two RSMs could provide more accurate predictions for the velocity fields compared to the other applied models. Additionally, the LRR model could better capture secondary flows. It is likely that the RSMs will begin to be more widely applied for industrial projects once a consensus has been achieved about the best numerical solution strategies for their convergence problems [55]. Additionally, the advances in computational resources in the early future may lead these models to be more employed by industrial users. Therefore, this study simulates multiple inclined dense discharges employing the Launder, Reece, and Rodi (LRR) model as a new approach for multiple jets, and compares the model predictions of terminal rise height, impact point distance, and impact dilution to existing experimental measurements.

Furthermore, previous studies have mainly focused on the influence of port spacing on the mixing characteristics of multiport diffusers. However, the influence of non-uniform port orientation on the merging process of multiport diffusers is rarely reported. Jet merging in multiport diffusers leads the jets to have shorter trajectories and lower dilution compared to single jets. It is expected that the jet merging and the Coanda effect diminish when the jets are discharged with non-uniform orientation to the horizontal, compared to uniform discharged diffusers. A change in port orientation can be of interest in practical projects, such as brackish water discharges into water bodies emitted by osmotic power plants, since it may provide an inexpensive way to enhance the mixing performance. Thus, as another contribution of this study, the effect of non-uniform port orientation on the merging process of multiple inclined dense jets is also evaluated using the validated LRR model.

The rest of this paper is devoted to the description of research methodology as well as presenting and discussing the results and conclusions. Section 2 describes methodology where dimensional analysis, governing equations, flow and mesh configurations, as well as model setup and numerical algorithms are discussed. Section 3 presents the results where the performance of the model and the effect of non-uniform port orientation on merging process are discussed. Section 4 also presents the conclusion and proposed future works.

2. Methodology

2.1. Analysis

A schematic view of multiple inclined dense jets is illustrated in Figure 1. The jets are discharged into the receiving water bodies at an initial angle θ to the horizontal with jet velocity U_0 and jet diameter D. Upon discharging, the jets move upward due to the initial vertical momentum flux. The negative buoyancy and jet entrainment continuously decrease this momentum flux until it almost vanishes at the point where the jets reach their maximum height, the terminal rise height y_t. Subsequently, the jets turn downward and impinge the seabed at the location of impact point, x_i. The dilution at the impact point S_i can be defined as $S_i = (\rho_0 - \rho_a)/(\rho - \rho_a)$, where ρ_0 is effluent density, ρ_a is ambient density, and ρ represents local density.

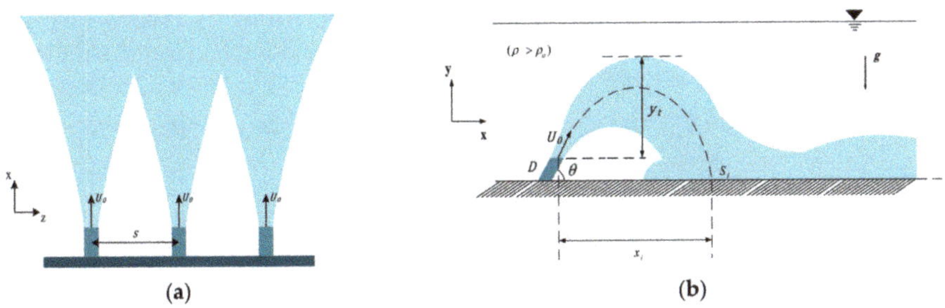

Figure 1. Schematic (**a**) plan and (**b**) side views of multiple inclined dense jets.

The mixing properties of the inclined dense jets are mainly characterized by discharge volume flux Q, kinematic momentum flux M, and buoyancy flux B, which can be defined by the following equations, respectively [28,45,58,59]:

$$Q = \frac{\pi D^2}{4}U_0;\ M = U_0 Q = \frac{\pi D^2}{4}U_0^2;\ B = g_0' Q \tag{1}$$

where g_0' is reduced gravitational acceleration and expressed as $g_0' = g(\rho_0 - \rho_a)/\rho_a$; where g is gravitational acceleration. In the case of dense discharges $\rho_0 > \rho_a$.

The most important length-scale of the flow is the momentum length scale, L_M, which refers to the distance beyond which the buoyancy produces momentum approximately equal to the initial momentum and is defined as [46,60]

$$L_M = \frac{M^{3/4}}{B^{1/2}} \tag{2}$$

which is commonly expressed as DF_d where F_d is the jet densimetric Froude number, which measures the ratio of inertia to buoyancy and can be defined as

$$F_d = \frac{U_0}{\sqrt{g_0' D}} \tag{3}$$

The effect of port spacing in dimensional analysis is expressed by the non-dimensional parameter s/DF_d where s is the port spacing. If the ports are widely spaced, $s/DF_d > \sim 2$, the jets behave as single jets and freely mix with the ambient water. If a single jet is fully turbulent and F_d is higher than approximately 20, the dynamic effect of Q becomes insignificant. Therefore, the dependent variables of y_t, x_i, and S_i are functions of M and B only. Following a dimensional analysis and experimental measurements [46]

$$\frac{y_t}{DF_d} = 2.2;\ \frac{x_i}{DF_d} = 2.4;\ \frac{S_i}{F_d} = 1.6 \tag{4}$$

For the case that the jets are not widely spaced, $s/DF_d < \sim 2$, the right-hand sides of Equation (4) are not constant anymore. The geometrical characteristics and impact dilution are functions of s/DF_d, as follows

$$\frac{y_t}{DF_d} = f\left(\frac{s}{DF_d}\right);\ \frac{x_i}{DF_d} = f\left(\frac{s}{DF_d}\right);\ \frac{S_i}{F_d} = f\left(\frac{s}{DF_d}\right) \tag{5}$$

Following the experiments conducted by Abessi and Roberts [19], the normalized terminal rise height and impact distance for one- and two-sided multiport discharges can be obtained by the empirical equations

$$\frac{y_t}{DF_d} = 1.9\left(\frac{s}{DF_d}\right)^{\frac{1}{2}};\ \frac{x_i}{DF_d} = 2.0\left(\frac{s}{DF_d}\right)^{\frac{1}{2}} \tag{6}$$

The impact dilution for one-sided multiport diffusers can be achieved by

$$\frac{S_i}{F_d} = 0.9\left(\frac{s}{DF_d}\right) \tag{7}$$

while for two-sided discharges it is about 20% lower.

2.2. Governing Equations

The mass and momentum conservation equations for an incompressible multiphase flow are as follows [61,62]:

$$\nabla \cdot U = 0 \tag{8}$$

$$\frac{\partial \rho U}{\partial t} + \nabla \cdot (\rho U U) = -\nabla \cdot \left(P_{rgh}\right) - gh\nabla\rho + \nabla \cdot (\rho T) \tag{9}$$

where

$$\rho = \alpha_1\rho_1 + \alpha_2\rho_2 = \alpha_1\rho_1 + (1-\alpha_1)\rho_2 \tag{10}$$

$$\mathrm{T} = -\frac{2}{3}\overline{\mu}_{eff}\nabla \cdot \mathrm{UI} + \overline{\mu}_{eff}\nabla \mathrm{U} + \overline{\mu}_{eff}(\nabla \mathrm{U})^{\mathrm{T}} \tag{11}$$

$$\overline{\mu}_{eff} = \alpha_1\left(\mu_{eff}\right)_1 + \alpha_2\left(\mu_{eff}\right)_2 \tag{12}$$

$$\left(\mu_{eff}\right)_i = (\mu - \mu_t)_i \tag{13}$$

where U is velocity, t indicates time, ρ is density, P_{rgh} represents the difference between static and hydraulic pressure, g is gravitational acceleration, h denotes height of fluid column, α is volume fraction, μ and μ_t are dynamic and turbulent viscosity, respectively, script i shows either the effluent ($i = 1$) or surrounding water ($i = 2$). The diffusion equation for the volume fraction of a fluid can be written as:

$$\frac{\partial \alpha_1}{\partial t} + \nabla \cdot (U\alpha_1) = \nabla \cdot \left(\left(D_{ab} + \frac{\nu_t}{S_C}\right)\nabla\alpha_1\right) \tag{14}$$

where D_{ab} indicates molecular diffusivity, ν_t represents turbulent eddy viscosity, and S_C denotes turbulent Schmidt number.

In CFD modeling, the three approaches of direct numerical simulation (DNS), LES, and RANS models can be used to simulate turbulent flows. The first two requires excessive computational cost for the simulation of multiple jets. The RANS models can be classified into the RSMs and EVMs. The RSMs usually need more computational resources. One study applied three different EVMs and two RSMs to predict the mixing behavior of single jets and showed that the RSMs were about 20% more expensive than the other models, regarding the computational costs [35]. However, the RSMs may be more accurate when the flow field is complex, which has roots in dealing with stress anisotropy [55]. The main difference between the RSMs and EVMs is that RSMs directly solve transport equations for the Reynolds stress, as presented in Equation (15), while the EVMs compute the stresses using the Boussinesq assumption. This difference may lead to more accurate predictions of the mixing behavior of multiple jets using an RSM compared to an EVM. In the current study, the LRR model, as an RSM, is adopted to solve the governing equations. The LRR model was developed by Launder, Reece, and Rodi in 1975 [63] and is expressed as:

$$\frac{D\overline{u_i u_j}}{\mathrm{Dt}} = \frac{\partial \overline{u_i' u_j'}}{\partial t} + \overline{u_k}\frac{\partial \overline{u_i' u_j'}}{\partial x_k} = D_{ij}^T + D_{ij}^\nu + P_{ij} + \phi_{ij} + \varepsilon_{ij} \tag{15}$$

where $\overline{u_i' u_j'}$ is the Reynolds stress, and D_{ij}^T, D_{ij}^ν, P_{ij}, ϕ_{ij}, and ε_{ij} represent the turbulent diffusion, viscous diffusion, production, pressure–strain, and turbulence dissipation rate parameters, respectively. The viscous and turbulent diffusion terms are calculated as

$$D_{ij}^\nu = \nu \frac{\partial^2 \overline{u_i' u_j'}}{\partial x_k x_k} \tag{16}$$

$$D_{ij}^T = \frac{\partial}{\partial x_l}\left(C_s \frac{\varepsilon}{k} \overline{u_i u_k} \frac{\partial \overline{u_i u_j}}{\partial x_k}\right) \tag{17}$$

where $C_s = 0.22$, and the pressure–strain term is expressed as

$$P_{ij} = -\overline{u_i' u_k'}\frac{\partial \overline{u_j}}{\partial x_k} - \overline{u_j' u_k'}\frac{\partial \overline{u_i}}{\partial x_k} \tag{18}$$

The dissipation term is calculated as

$$\varepsilon_{ij} = -2\nu \overline{\frac{\partial u_i'}{\partial x_k}\frac{\partial u_j'}{\partial x_k}} \tag{19}$$

$$\frac{\partial \varepsilon}{\partial t} + \overline{u_k}\frac{\partial \varepsilon}{\partial x_k} = C_{\varepsilon 1}\frac{P_k \varepsilon}{k} - C_{\varepsilon 2}\frac{\varepsilon^2}{k} + \frac{\partial}{\partial x_j}\left(C_\varepsilon \frac{k}{\varepsilon}\overline{u_i' u_j'}\frac{\partial \varepsilon}{\partial x_j}\right) \tag{20}$$

where $C_{\varepsilon 1} = 1.44$, $C_{\varepsilon 2} = 1.90$.

2.3. Flow and Mesh Configurations

As the purpose of the current study was to evaluate the predictive capabilities of RSM for multiple inclined dense jets and investigate the effects of non-uniform port orientation to the horizontal on the mixing behavior, previous experiments were first simulated to validate the LRR model predictions. The experiments conducted by Abessi and Roberts [19] were selected as a basis for model validation, and the results of their measurements and LRR simulations were compared. Their measurements were appropriate for model validation since concentration fields were mapped in three dimensions by laser-induced fluorescence, and a comprehensive dataset on multiport inclined dense effluents was provided. Afterward, the validated model was utilized for further investigations on the effects of non-uniform port angle on the mixing process. According to the experiments, the port spacing (s) was 57 mm, the jet diameter (D) was 1.93 mm, and the nozzle height above the lower wall (y_0) was 30 mm. The effluent and tank water density were 1013.8 kg/m^3 and 999.8 kg/m^3, respectively.

To model a long diffuser with many nozzles, three ports were considered in the computational domain as the representatives of all possible jets on a diffuser. Overall, five cases were considered for the simulations, see Table 1. Cases C1, C2, C3, and C4 were utilized for model validation, and were exactly the same as the experimental measurements conducted by Abessi and Roberts [19]. In these cases, s/DF_d varies between 0.47–0.85 to cover moderately spaced ports, which was the focus of this study, and all F_d values were more than 20 to assure fully turbulent flow. In the simulations, the three considered ports were discharged with an angle of 60° to the horizontal, which represents the first configuration (60°–60°–60°). In case S1, the second configuration (45°–60°–45°) is simulated where the central jet is discharges with an angle of 60° to the horizontal and the adjacent jets are discharged with an angle of 45°. Case S1, which has similar conditions to case

C2 except port inclinations, and case C2 were compared to evaluate the effects of port orientation on the mixing process of multiple inclined dense discharges.

Table 1. Characteristics of numerical cases.

Case	Diameters (mm)	Effluent Density (kg/m^3)	Ambient Density (kg/m^3)	Velocity (m/s)	F_d	s/DF_d	θ_1-θ_2-θ_3 (°)
C1	1.93	1013.8	999.8	1.03	63.1	0.47	60-60-60
C2	1.93	1013.8	999.8	0.92	56.8	0.52	60-60-60
C3	1.93	1013.8	999.8	0.85	52	0.57	60-60-60
C4	1.93	1013.8	999.8	0.57	34.7	0.85	60-60-60
S1	1.93	1013.8	999.8	0.92	56.8	0.52	45-60-45

The three considered jets could reflect the mixing behavior of the jets in the whole system using appropriate boundary conditions. The simulations were conducted in a rectangular domain with the dimensions of 600 mm length (along x-direction), 400 mm depth (along y-direction), and 171 mm width (along z-direction); similar domains have been used in many hydraulic simulations [64–66]. The computational domain was surrounded by four types of planes, including wall, outlet, free surface, and symmetry planes. Furthermore, the nozzles were defined by two sections, including nozzle tube and inlet (Figure 2).

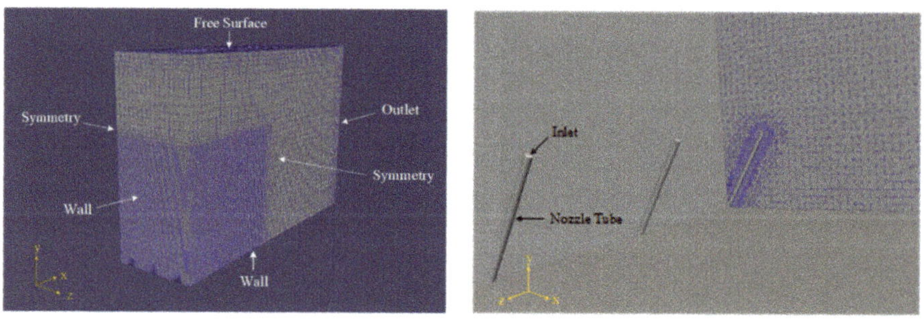

(a) (b)

Figure 2. (**a**) Computational domain and mesh setup; (**b**) detailed grids near the nozzles.

The accuracy and stability of numerical calculations of fluid flows are significantly affected by boundary conditions. As the wall planes and nozzle tubes are solid surfaces, a no-slip boundary condition was imposed on them to assume zero velocity at those surfaces. The boundary condition at the free surface and outlet planes was set to pressure–inlet outlet velocity. This boundary condition assigns a zero gradient condition for flow into the domain and a velocity, based on the flux in the path-normal direction, for flow out of the domain [67]. A fixed value of velocity was assigned to the inlets, and the initial k and ε values at the inlets were estimated based on [37], $k = 0.06u^2$ and $\varepsilon = 0.06u^3/D$. To simulate the system, the domain was discretized using the structured mesh shown in Figure 2. The mesh was modified by snappyHexMesh tool in OpenFOAM, which can divide the original grids into multiple sub-grids and snap the sub-grid boundaries onto the surfaces of interest. The employment of snappyHexMesh significantly decreased the required number of cells and therefore simulation time since there was no need to have small cells at the regions far from the area of interest. A double refinement was performed for the area in which jets move and merge, and a triple refinement was considered for the grids close to the inlets and nozzle tubes, which could perfectly determine the circular shape of the nozzles. The sensitivity of the simulation outcomes to the cell size was evaluated based on the procedures described and applied by Yan and Mohammadian [68,69]. Simulations were conducted based on four different meshes, including coarse and fine grids. The preliminary mesh structure was a relatively coarse mesh with 114,000 cells, and the difference between the predictions based on that mesh structure and experimental measurements were found to be about 10%. The finer mesh structures, Mesh 2, Mesh 3, and Mesh 4 included 311,000 cells, 385,000 cells, and 528,000 cells, respectively. The difference between the obtained results using Mesh 2 and Mesh 1 was about 10.5%. However, the difference between the predictions based on Mesh 2 and Mesh 3 was about 1.5%. The predictions of Mesh 3 and Mesh 4 were also very close, with about 2% difference between the results. Therefore, Mesh 4 was selected for simulations, and the final mesh structure included 528,000 cells.

2.4. Model Setup and Numerical Algorithms

The governing equations were numerically solved using the finite volume method. The simulations were performed using an open source code OpenFOAM [70], and the solver twoLiquidMixingFoam was employed to perform the implementation in OpenFOAM. This solver is a transient solver for multiphase fluids and has been widely applied and validated in prior research [2,3,39,41,44,71]. The Gauss linear scheme was adopted for the discretization of the gradient and Laplacian terms, the Euler scheme was utilized to discretize the temporal terms, and the Gauss upwind, Gauss vanleer, and Gauss linear schemes were employed to discretize the divergence terms. The adopted algorithms to solve the coupled pressure and momentum equations were PIMPLE algorithms, see Figure 3. The preconditioned conjugate gradient (PCG) and preconditioned bio conjugate gradient (PBiCG) methods were employed for the pressure field and the other fields, respectively.

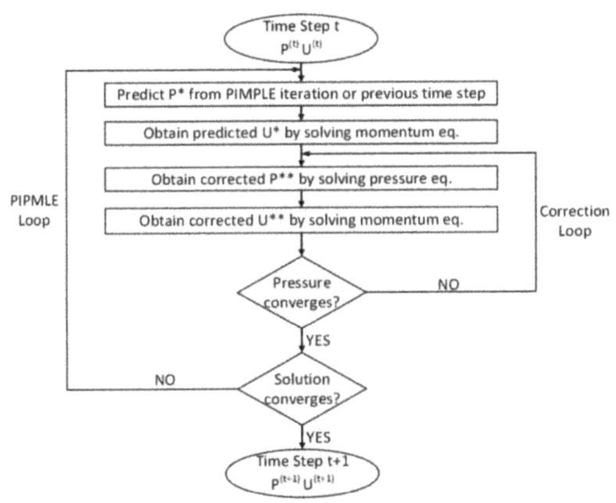

Figure 3. Flowchart of the PIMPLE algorithm; * and ** correspond to predicted and corrected fields, respectively.

The Courant number, C, can be defined as $C = U\Delta t/\Delta x$, where U is flow velocity, Δt represents time step, and Δx indicates cell size. The time step was adjusted during the simulations, and the maximum courant number was set to be 0.5 to provide small time steps. The simulations were run for 120 s, and the results obtained by time-averaging over the period of 70 s to 120 s to ensure fully-developed flow and stable results. Figure 4 shows the time-series graphs for the velocity field along the trajectory of the central jet of case C2. To accelerate the simulation process, the numerical domain was decomposed into 16 regions, 4 parts in x- and 4 part in y-directions, and then each region was processed by a processor. Afterward, the entire domain was reconstructed and the results were extracted. All the computations were performed in the advanced research computing (ARC) systems of Compute Canada, and the real-time computing duration was ranged from 3 to 4 days.

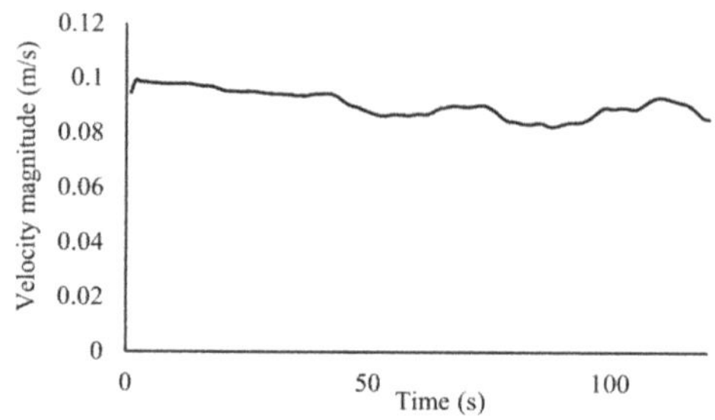

(**a**)

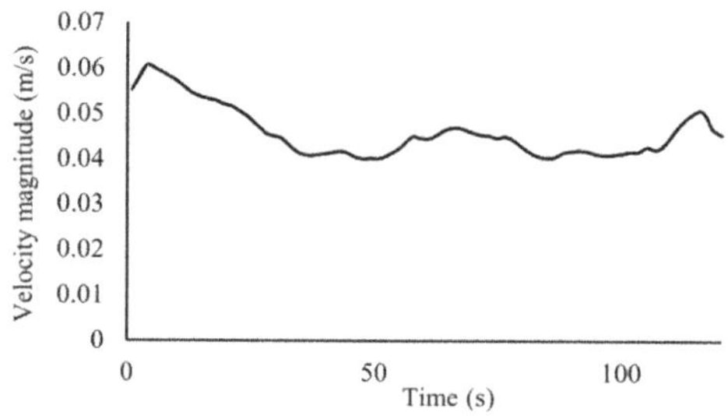

(**b**)

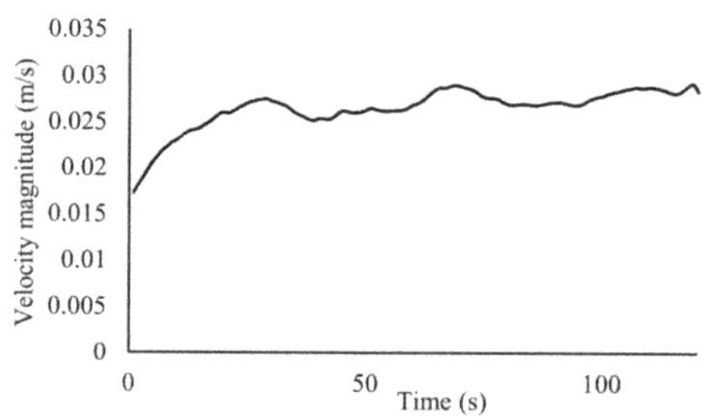

(**c**)

Figure 4. Time-series graphs for the velocity magnitude along the trajectory of the central jet in case C2 at (**a**) $x/(D.F_d) = 0.4$, (**b**) $x/(D.F_d) = 0.75$, and (**c**) $x/(D.F_d) = 1.2$.

To discuss the numerical results in the context of the literature, the accuracy of LRR predictions were compared with that of standard and (RNG) k-ε models adopted for the simulation of the same experimental work in a previous study [3]. Table 2 demonstrates the simulation characteristics of the previous and current study. It is worth mentioning that the numerical results obtained in both studies were mesh-independent, and adopting finer mesh does not necessarily lead to more accurate results. Therefore, the comparison of the accuracy of outcomes can provide a good insight into the performance of the turbulence models.

Table 2. Simulation characteristics of the previous [3] and present study.

	[3]	Current Study
Boundary conditions		
Bottom wall	Slip *	No-slip
Symmetry planes	Symmetric	Symmetric
Inlets	Fixed velocity inlet	Fixed velocity inlet
Atmosphere	Inlet–outlet	Pressure–inlet outlet
Outlet	Inlet–outlet	Pressure–inlet outlet
Number of cells	113,400	528,000
Software	OpenFOAM	OpenFOAM
Solver	twoLiquidMixingFoam	twoLiquidMixingFoam
Discretization schemes		
Temporal term	Euler	Euler
Gradient term	Linear	Linear
Laplacian term	Linear	Linear
Divergence term	Linear, vanLeer, and upwind	Linear, vanLeer, and upwind

* The results were the same when using slip or no-slip boundary condition for the bottom wall [3].

3. Results and Discussion

3.1. Model Validation

The four cases of C1, C2, C3, and C4, moderately-spaced discharges, $0.47 < s/DF_d < 0.85$, with different Froude numbers were numerically simulated by the LRR model, and the model predictions for the terminal rise height, impact distance, and impact dilution were compared to the measurements conducted by [19].

Terminal rise height is determined as where the vertical momentum flux almost vanishes along the jet trajectory [35]. However, the y_t value is not defined exactly the same in previous studies. For instance, Lai and Lee [59] defined it as where the concentration was 25% of the local maximum height concentration, C, while the integral model CorJet applied two cut-off levels, 3% and 25%, for the visual boundary [26]. Shao and Law [72] also adopted the cut-off level of 3%. The definition of y_t in the current simulations was considered to be the locus of 10% of the transverse maximum concentration at the location of jet maximum height, as it was defined in the experimental data [19]. Figure 5 demonstrates the results of normalized terminal rise height, $y_t/(DF_d)$, versus port spacing values, s/DF_d, including cases C1–C4. The value of $y_t/(DF_d)$ for single jets were also included in this figure [46]. The numerical results of terminal rise height were in good agreement with the experimental data. The forecasting accuracy of the model was quantified using mean absolute percentage error (MAPE), the same as in the study conducted by [3], where it was found to be 4.5%. The comparison of $y_t/(DF_d)$ values for cases C1–C4 and single ports showed that the terminal rise height of multiple inclined jets was lower than that of single discharges, which highlights the effect of jet merging. The LRR results for terminal rise height showed slight over prediction compared to the empirical equation, Equation (6), obtained based on the measurements for one- and two-sided multiport diffusers. However, the prediction of terminal rise height at $s/(DF_d) = 0.85$ was matched with Equation (6).

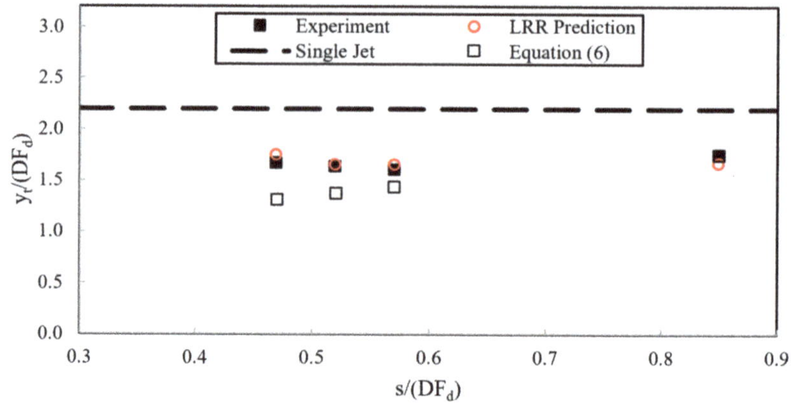

Figure 5. Experimental measurements [19] and numerical results of the normalized terminal rise height.

Figure 6 illustrates the results of normalized impact distance, $x_i/(DF_d)$, versus port spacing values. The model predictions of x_i/D demonstrated a good match with the measurements as the prediction error based on MAPE was less than 4%. The values of $x_i/(DF_d)$ increased with increasing values of port spacing, and all values were lower than the constant value presented for single discharges, $x_i/(DF_d) = 2.4$ [46]. Jet merging causes multiple inclined jets to bend inward, which results in lower impact distance compared to that of single discharges. However, single jets mix with the surrounding water without any intervention. As $s/(DF_d)$ decreases, this effect is more pronounced, and the value of $x_i/(DF_d)$ decreases. The LRR predictions showed higher values for impact distance compared to the general empirical equation for one- and two-sided multiple dense jets.

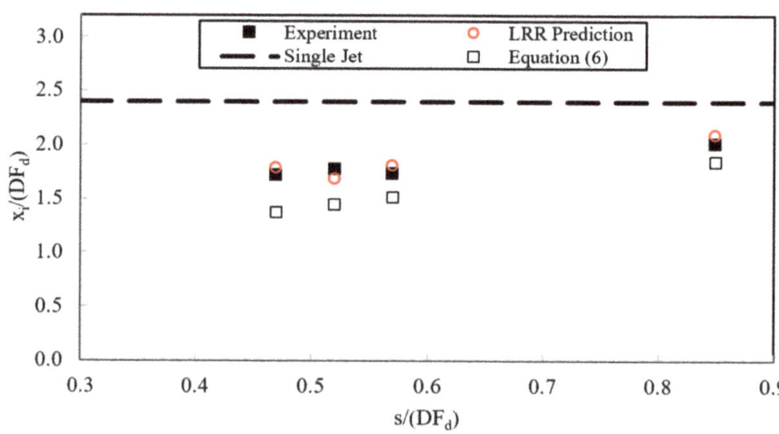

Figure 6. Experimental measurements [19] and numerical results of the normalized impact point distance.

Figure 7 indicates the results of the normalized impact dilution, S_i/F_d, against $s/(DF_d)$. The predictions were in an acceptable range of accuracy, with a MAPE value of 7%. As can be seen, $s/(DF_d)$ had a positive effect on S_i/F_d since the jet merging effect decreased for higher values of $s/(DF_d)$. Furthermore, the values of S_i/F_d were much lower for multiple discharges compared to single jets. The LRR results were in good agreement with the empirical equation presented for one-sided multiport diffusers, Equation (7).

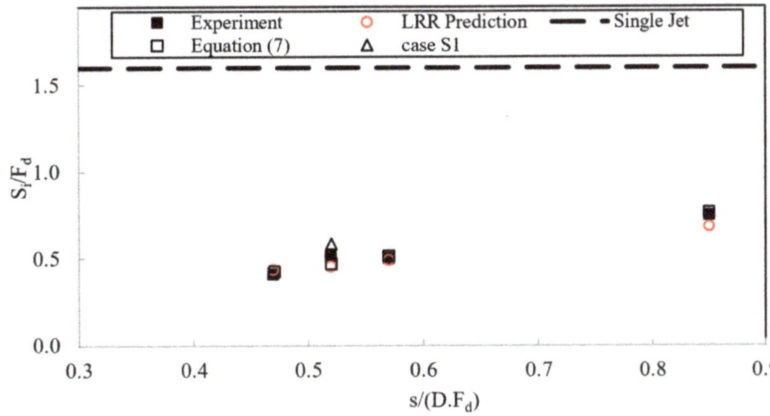

Figure 7. Experimental measurements [19] and numerical results of the normalized impact dilution.

Therefore, the numerical results demonstrated that LRR model, as an RSM, may be a reliable tool for the prediction of mixing behavior of multiple inclined dense jets, as it predicted the geometrical characteristics and impact dilution with mean absolute percentage errors of less than 5% and 7%, respectively. This model provided more accurate predictions compared to the EVMs employed for the simulation of same experimental work [3], see the simulation characteristics in Table 2. It has been shown that RNG k-ε closure may provide results with errors within the range of 15%, and the predictions based on this model were more accurate compared to the standard k-ε closure [3]. The reason behind the higher accuracy of LRR predictions may have its roots in the main difference between RSMs and EVMs: RSMs directly solve the Reynolds stresses, while EVMs model them. In particular, EVMs consider that Reynolds stresses are proportional to mean rates of deformation [55]. It has been also mentioned that RSMs may perform better relative to EVMs when the flow fields are sophisticated [55]. Although the employment of the LRR model for the simulation of multiple jets has rarely been reported, one study focusing on 30° and 45° single inclined dense jets highlighted the outperformance of the LRR model over the RNG k-ε and nonlinear k-ε models [35].

3.2. Effect of Non-Uniform Port Orientation on Jet Merging Process

As a novel contribution to the literature, the effect of non-uniform port orientation on jet merging process was investigated by comparing cases C2 and S1. Case C2 refers to a diffuser with uniform port orientation to the horizontal, including three jets which are discharged with an angle of 60°. In case S1, the central jet is discharged with an angle of 60°, and the adjacent jets with an angle of 45°. The volume fraction, α, in the governing equations is the normalized concentration, C/C0, obtained in post-processing, where C is the local concentration and C0 is the initial concentration. Figure 8 demonstrates the general shape of the jets for cases C2 and S1, and also the vertical slices where the data were extracted. In both cases, the jets moved upward due to the vertical component of the initial momentum, and this momentum continuously decreased due to jet entrainment and negative buoyancy until it equaled zero. Afterwards, the dense jets moved downward and subsequently impacted the bottom boundary. Unlike case C2, the central jet of the diffuser with non-uniform jet orientation to the horizontal, case S1, had higher y_t and lower x_i compared to its adjacent jets, which were discharged with an angle of 45° to the horizontal. This occurred because of the difference between the components of initial momentum of 45° and 60° inclined jets.

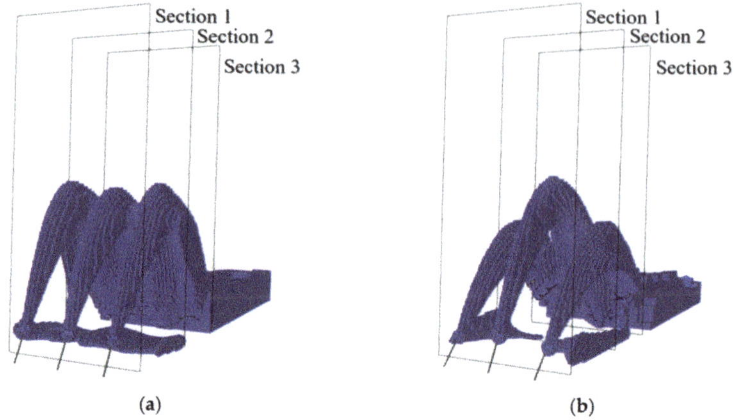

Figure 8. Visualization of (**a**) uniform (case C2) and (**b**) non-uniform (case S1) inclined dense jets, using a velocity isosurface (U = 0.015 m/s).

To investigate the jet merging process for the two different cases, concentration profiles at cross-sections close to and far from the nozzles were extracted, see Figure 9. Close to the ports, e.g., Figure 9a, the jets in both cases seemed to be completely separate, like single jets, with high concentration at the jet centers, and they were not wide enough to interact with each other. Along the trajectory, the jets became more diluted due to the entrainment of the surrounding water, due to shear induced entrainment at the jets' edges. Farther away, e.g., Figure 9b, the maximum concentrations were significantly decreased compared to Figure 9a, and the jets were more diffused. The distance between the jet's centers in case S1 are more than that of case C2 because of the difference between the jets' elevation in case S1. Different orientations of the ports provide more space for the central jet to expand before interacting with other jets. As shown in Figure 9b, at $x/(D.F_d) = 1.5$, the jets in case C2 were almost merged, and there was no space between them or under the central jet. This means the entrainment of fresh water into the jet was restricted in the z-direction and y-direction, from the lower edge. However, there were still some areas with almost zero concentration around the central jet in case S1. It is worth mentioning that the entrainment was restricted in both cases and, therefore, dilution was less than that of single jets. Eventually, in case C2, the individual jets were indistinguishable after the impact point of the central jet, see Figure 9c. However, there was still a trace of 45° inclined dense jets in case S1.

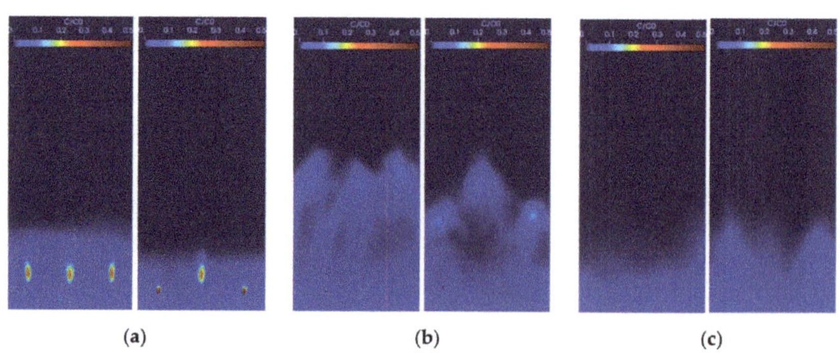

Figure 9. Concentration fields at different cross sections of uniform (case C2) and non-uniform (case S1) multiple inclined dense jets; (**a**) Section 1 ($x/(D.F_d) = 0.25$); (**b**) Section 2 ($x/(D.F_d) = 1.5$); (**c**) Section 2 ($x/(D.F_d) = 2.5$).

Jet merging may be the main source of difference between the mixing characteristics of single and multiple discharges. The Coanda interaction between the rising and descending parts of a dense jet causes the rising jet to attempt to re-entrain itself. As the jets are closely- or moderately-spaced, jet merging results in a virtually impenetrable wall that prevents the entrainment of ambient water to inner surfaces and intensifies the Coanda effect. The lack of clear water on inward faces causes the jets to bend inward more sharply in search of clear water [19]. Thus, multiple dense jets usually have shorter trajectories and therefore lower dilution rates compared to single jets. This difference between single and multiple jets can be also observed between the diffusers with uniform and non-uniform port orientation. Figure 10 indicates the contours of normalized concentration fields at the central planes for cases C2 and S1. The contours in both cases showed that the lower part of the jet was more expanded compared to the upper part. As shown in Figure 10, the central jet in case C2 bent inward more sharply, and there was less available clear water ($C < 0.02\ C0$) under the central jet when the discharges were uniform, compared to that of non-uniform jets. It was found that the impact dilution for case S1 was 14% higher than that of case C2, see Figure 7. The availability of more clear water at the inner edge of the jet in case S1, compared to that of C2, decreased the entrainment restriction and thus caused higher dilution rates.

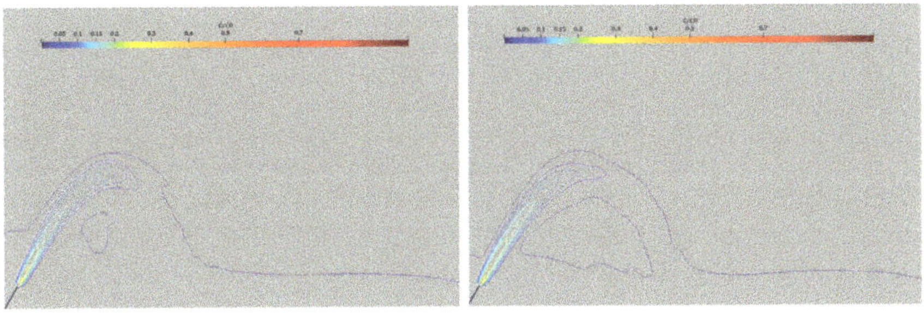

(a) (b)

Figure 10. Concentration contours at the central plane for (**a**) case C2 and (**b**) case S1.

A comparison of central jet trajectories more clearly highlights the effect of adjacent jets on the geometrical characteristics of central jets. As shown in Figure 11a, the trajectory of the central jet in case S1 was at a lower elevation, compared to case C2, before the terminal rise height. However, it reached higher elevations after the terminal rise height and did not bend inward as sharply as the other case, demonstrating the reduced effect of jet merging and the Coanda interaction. The terminal rise height and impact point distance were found to be 7% and 5% higher for case S1 relative to case C2. A small deviation between the trajectories before the terminal rise height may be attributed to the Coanda interaction between the jets. The jets mutually attempted to entrain one another because of a pressure force, resulting from a change in the entrained flow pattern, that moves a jet to the other jet [19]. This effect can be seen in a study focusing on multiple positive buoyant jets [73] where the end plumes bend inwards as a result of unbalanced inward pressure force on the plumes, while the central jets move vertically because of the equally balanced forces on each side of them. Similarly, in case S1, as the adjacent jets are at a different elevation, the induced low-pressure field differs from that of case C2. As the central jet tries to entrain into the adjacent ones, which are at a lower elevation, it becomes somewhat deflected. Figure 11b illustrates how the concentration changes along the trajectory of the central jet in case S1 and case C2. As shown, before the terminal rise height, the concentration of the central jet in case C2 was lower than that for case S1. However, beyond this point, where the jets in case C2 interacted with each other and there was limited fresh water around the jets, its concentration became lower compared to case S1.

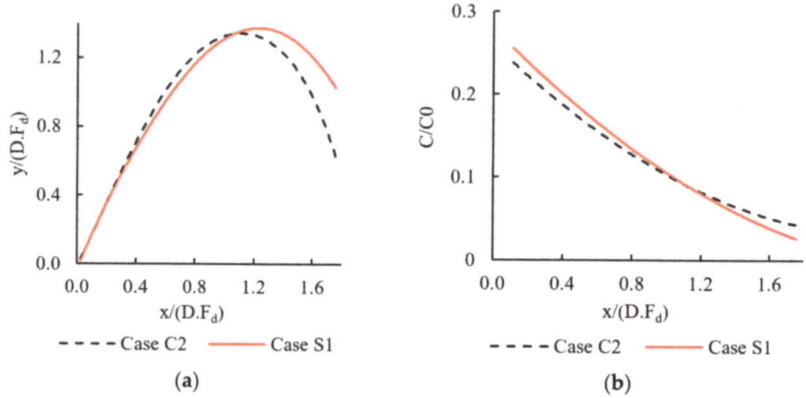

Figure 11. (**a**) Trajectory and (**b**) concentration of the central jet of cases C2 and S1.

To evaluate the overall effect of non-uniform port orientation, the concentration downstream of the impact point of all jets were investigated. Particularly, the concentration at $x/(D.F_d)$ = 3.4 and the elevation of nozzle tip on the centerline planes of central and adjacent jets are shown in Figure 12. According to Figure 12, the concentration beyond the impact point on the central jet of case S1 was 3.7% lower than that on the central jet of case C2. Additionally, the concentration on the planes of adjacent jets for the non-uniform case was 8.4% lower compared to that for the uniform case. Therefore, the results show some improvements in dilution when adopting the multiport diffuser with non-uniform port orientation compared to the uniform diffuser.

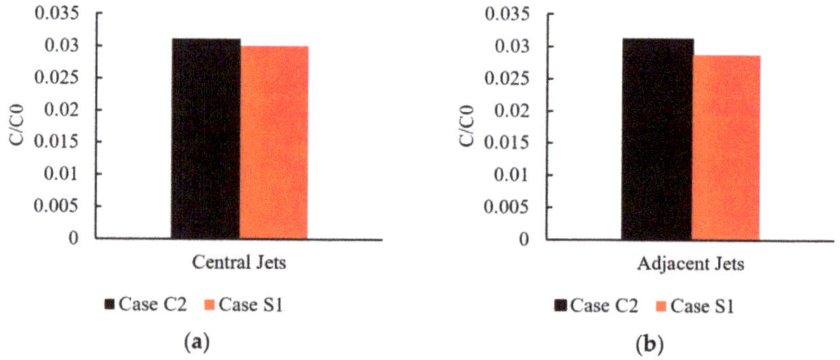

Figure 12. Concentration at $x/(D.F_d)$ = 3.4 and the elevation of nozzle tip on the centerline planes of (**a**) central and (**b**) adjacent jets.

4. Conclusions

The first aim of the present study was to evaluate the prediction capabilities of RSMs for multiple inclined dense jets. The LRR turbulence model, as a new approach for multiple jets, was employed to perform the simulations. The predictions of geometrical characteristics and impact dilution were compared to the existing experimental measurements. Subsequently, the validated model was applied to perform further investigations on the effect of non-uniform port orientation on the merging process of multiple inclined dense jets, as a novel contribution. Particularly, diffusers with uniform, 60°–60°–60°, and non-uniform, 45°–60°–45°, port orientation were considered, and the effect of adjacent ports on the central ports was determined.

It was found that the LRR turbulence model, as an RSM, can be a reliable model for the study of the mixing behavior of multiple inclined dense jets, and can provide results of geometrical characteristics and impact dilution within the range of 5% and 7%, respectively. It also provides more accurate results compared to the EVMs including the standard and RNG k-ε turbulence models. Furthermore, the comparison of the diffuser with uniform and non-uniform port orientation showed that jet merging process is significantly affected by the port's orientation and the middle discharge may have a longer trajectory and therefore a higher dilution rate when its adjacent jets are discharged at a different angle, compared to that of uniform discharges. The terminal rise height, impact distance, and impact dilution of the central jet on a diffuser with uniform port inclination were found to be 7%, 5%, and 14% higher than those of the uniform case, respectively. It can be concluded that the behavior of the central jet on the diffuser with non-uniform port orientation is more like a single jet, compare to the one with uniform ports. Generally, the effects of jet merging and dynamic interactions on the central jet, including inward bending and a reduction in geometrical characteristics, are more significant when the nozzles are uniform. The evaluation of the overall effects of non-uniform port orientation showed 3.7% to 8.4% increment in dilution beyond the impact point.

This study showed how the trajectory and dilution of the middle discharge were affected by the orientation of its adjacent jets and how merging process was influenced by the orientation. The results of this study can be of interest in practical projects applying multiport diffusers to discharge brine or saline water as a by-product of osmotic power plants and municipal dense wastewaters into water bodies. A change in port orientation may provide an inexpensive way to augment dilution, compared to costly methods such as an increase in port spacing. This study highlighted the effects of non-uniform port orientation on merging process of multiple dense jets in near-field; however, the effect of non-uniform port orientation on intermediate and far fields may also be of interest. The outcomes have the potential to be considered by outfall designers. More investigations on the other parameters, such as flowing current, stratification, along with port orientation, are also recommended.

Author Contributions: Conceptualization, A.M. and S.A.R.S.H.; methodology, A.M.; software, S.A.R.S.H. and A.M.; validation, S.A.R.S.H.; formal analysis, S.A.R.S.H.; investigation, S.A.R.S.H.; resources, S.A.R.S.H.; data curation, S.A.R.S.H., A.M., P.J.W.R. and O.A.; writing—original draft preparation, S.A.R.S.H.; writing—review and editing, A.M., P.J.W.R. and O.A.; visualization, S.A.R.S.H.; supervision, A.M.; project administration, A.M.; funding acquisition, A.M. All authors have read and agreed to the published version of the manuscript.

Funding: This research was funded by Natural Sciences and Engineering Research Council of Canada (NSERC), grant number 210717. The research of S.A.R.S.H. was supported by Mitacs through the Mitacs Research Training Award.

Institutional Review Board Statement: Not applicable.

Informed Consent Statement: Not applicable.

Data Availability Statement: Not applicable.

Conflicts of Interest: The authors declare no conflict of interest.

References

1. Einav, R.; Lokiec, F. Environmental aspects of a desalination plant in Ashkelon. *Desalination* **2003**, *156*, 79–85. [CrossRef]
2. Lai, C.; Zhao, B.; Law, A.W.K.; Adams, E.E. A numerical and analytical study of the effect of aspect ratio on the behavior of a round thermal. *Environ. Fluid Mech.* **2015**, *15*, 85–108. [CrossRef]
3. Yan, X.; Mohammadian, A. Numerical Modeling of Multiple Inclined Dense Jets Discharged from Moderately Spaced Ports. *Water* **2019**, *11*, 2077. [CrossRef]
4. Lyu, S.; Seo, I.W.; Kim, Y.D. Experimental investigation on behavior of multiple vertical buoyant jets discharged into a stagnant ambient. *KSCE J. Civ. Eng.* **2013**, *17*, 1820–1829. [CrossRef]
5. Wang, H.J.; Davidson, M.J. Jet Interaction in a Still Ambient Fluid. *J. Hydraul. Eng.* **2003**, *129*, 349–357. [CrossRef]
6. Knystautas, R. The Turbulent Jet from a Series of Holes in Line. *Aeronaut. Q.* **1964**, *15*, 1–28. [CrossRef]

7. Liseth, P. Mixing of merging buoyant jets from a manifold in stagnant receiving water of uniform density. In *Advances in Water Pollution Research, Proceedings of the Sixth International Conference, Munich, Germany, 1 January 1973*; Pergamon Press: Oxford, UK, 1973; pp. 921–936.
8. Seo, I.W.; Yeo, H.K. Near-field dilution of rosette type multiport wastewater diffusers. *Water Eng. Res. Int. J. Kings Waste Recycl. Auth.* **2002**, *3*, 93–111.
9. Tian, X.; Roberts, P.J.W.; Daviero, G.J. Marine wastewater discharges from multiport diffusers. I: Unstratified stationary water. *J. Hydraul. Eng.* **2004**, *130*, 1137–1146. [CrossRef]
10. Tian, X.; Roberts, P.J.W. Experiments on Marine Wastewater Diffusers with Multiport Rosettes. *J. Hydraul. Eng.* **2011**, *137*, 1148–1159. [CrossRef]
11. Roberts, P.J.W.; Hunt, C.D.; Mickelson, M.J.; Tian, X. Field and Model Studies of the Boston Outfall. *J. Hydraul. Eng.* **2011**, *137*, 1415–1425. [CrossRef]
12. Yannopoulos, P.C.; Noutsopoulos, G.C. Interaction of vertical round turbulent buoyant jets—Part I: Entrainment restriction approach. *J. Hydraul. Res.* **2006**, *44*, 218–232. [CrossRef]
13. Koh, R.C.Y.; Fan, L.N. Mathematical models for the prediction of temperature distributions resulting from the discharge of heated water in large bodies of water. *EPA Water Pollut. Control. Res. Ser.* **1970**, 16130. Available online: https://aquadocs.org/bitstream/handle/1834/25305/EPA_Koh.PDF?sequence=1&isAllowed=y (accessed on 20 April 2022).
14. Hodgson, J.E.; Moawad, A.K.; Rajaratnam, N. Concentration field of multiple circular turbulent jets. *J. Hydraul. Res.* **1999**, *37*, 249–256. [CrossRef]
15. Yannopoulos, P.C.; Noutsopoulos, G.C. Interaction of vertical round turbulent buoyant jets—Part II: Superposition method. *J. Hydraul. Res.* **2006**, *44*, 233–248. [CrossRef]
16. Lai, A.C.H.; Lee, J.H.W. Dynamic interaction of multiple buoyant jets. *J. Fluid Mech.* **2012**, *708*, 539–575. [CrossRef]
17. Adams, E.E. Submerged Multiport Diffusers in Shallow Water with Current. Ph.D. Thesis, Massachusetts Institute of Technology, Cambridge, MA, USA, 1972.
18. Marti, C.L.; Antenucci, J.P.; Luketina, D.; Okely, P.; Imberger, J. Near-Field Dilution Characteristics of a Negatively Buoyant Hypersaline Jet Generated by a Desalination Plant. *J. Hydraul. Eng.* **2011**, *137*, 57–65. [CrossRef]
19. Abessi, O.; Roberts, P.J.W. Multiport Diffusers for Dense Discharges. *J. Hydraul. Eng.* **2014**, *140*, 04014032. [CrossRef]
20. Abessi, O.; Roberts, P.J.W. Multiport Diffusers for Dense Discharge in Flowing Ambient Water. *J. Hydraul. Eng.* **2017**, *143*, 04017003. [CrossRef]
21. Abessi, O.; Roberts, P.J.W. Rosette Diffusers for Dense Effluents in Flowing Currents. *J. Hydraul. Eng.* **2018**, *144*, 06017024. [CrossRef]
22. Shrivastava, I.; Adams, E.E. Mixing of Tee Diffusers in Shallow Water with Crossflow: A New Look. *J. Hydraul. Eng.* **2019**, *145*, 04019006. [CrossRef]
23. Xu, Z.; Otoo, E.; Chen, Y.; Ding, H. 2D PIV Measurement of Twin Buoyant Jets in Wavy Cross-Flow Environment. *Water* **2019**, *11*, 399. [CrossRef]
24. Lee, J.H.W.; Cheung, V. Generalized Lagrangian Model for Buoyant Jets in Current. *J. Environ. Eng.* **1990**, *116*, 1085–1106. [CrossRef]
25. Lee, J.H.W.; Chu, V.H. *Turbulent Jets and Plumes: A Lagrangian Approach*; Springer Science & Business Media: Berlin/Heidelberg, Germany, 2003; Volume 1.
26. Jirka, G.H. Integral Model for Turbulent Buoyant Jets in Unbounded Stratified Flows. Part I: Single Round Jet. *Environ. Fluid Mech.* **2004**, *4*, 1–56. [CrossRef]
27. Frick, W.E. Visual Plumes mixing zone modeling software. *Environ. Model. Softw.* **2004**, *19*, 645–654. [CrossRef]
28. Taherian, M.; Mohammadian, A. Buoyant Jets in Cross-Flows: Review, Developments, and Applications. *J. Mar. Sci. Eng.* **2021**, *9*, 61. [CrossRef]
29. Baum, M.J.; Gibbes, B. Field-Scale Numerical Modeling of a Dense Multiport Diffuser Outfall in Crossflow. *J. Hydraul. Eng.* **2020**, *146*, 05019006. [CrossRef]
30. Roberts, P.J.W. Near Field Flow Dynamics of Concentrate Discharges and Diffuser Design. *Subser. Environ. Sci.* **2015**, *149*, 369–396. [CrossRef]
31. Palomar, P.; Lara, J.; Losada, I. Near field brine discharge modeling part 2: Validation of commercial tools. *Desalination* **2012**, *290*, 28–42. [CrossRef]
32. Taherian, M.; Saeidi Hossein, S.A.R.; Mohammadian, A. Overview of outfall discharge modeling with a focus on turbulence modeling approaches. In *Advances in Fluid Mechanics: Modeling and Simulation*; Springer: Singapore, 2022; ISSN 2364-6748.
33. Ardalan, H.; Vafaei, F. CFD and Experimental Study of 45° Inclined Thermal-Saline Reversible Buoyant Jets in Stationary Ambient. *Environ. Process.* **2019**, *6*, 219–239. [CrossRef]
34. Gildeh, H.K.; Mohammadian, A.; Nistor, I.; Qiblawey, H. Numerical Modeling of Turbulent Buoyant Wall Jets in Stationary Ambient Water. *J. Hydraul. Eng.* **2014**, *140*, 04014012. [CrossRef]
35. Gildeh, H.K.; Mohammadian, A.; Nistor, I.; Qiblawey, H. Numerical modeling of 30° and 45° inclined dense turbulent jets in stationary ambient. *Environ. Fluid Mech.* **2014**, *15*, 537–562. [CrossRef]
36. Gildeh, H.K.; Mohammadian, A.; Nistor, I.; Qiblawey, H.; Yan, X. CFD modeling and analysis of the behavior of 30° and 45° inclined dense jets—New numerical insight. *J. Appl. Water Eng. Res.* **2016**, *4*, 112–127. [CrossRef]

37. Huai, W.-X.; Li, Z.-W.; Qian, Z.; Zeng, Y.-H.; Han, J.; Peng, W.-Q. Numerical Simulation of Horizontal Buoyant Wall Jet. *J. Hydrodyn.* **2010**, *22*, 58–65. [CrossRef]
38. Oliver, C.J.; Davidson, M.J.; Nokes, R.I. k-ε Predictions of the initial mixing of desalination discharges. *Environ. Fluid Mech.* **2008**, *8*, 617–625. [CrossRef]
39. Zhang, S.; Jiang, B.; Law, A.W.-K.; Zhao, B. Large eddy simulations of 45° inclined dense jets. *Environ. Fluid Mech.* **2015**, *16*, 101–121. [CrossRef]
40. Zhang, S.; Law, W.K.A.; Jiang, M. Large eddy simulations of 45° and 60° inclined dense jets with bottom impact. *J. Hydro-environ. Res.* **2017**, *15*, 54–66. [CrossRef]
41. Jiang, M.; Law, A.W.-K.; Lai, A.C.H. Turbulence characteristics of 45° inclined dense jets. *Environ. Fluid Mech.* **2018**, *19*, 27–54. [CrossRef]
42. Xue, W.; Huai, W.-X.; Qian, Z.; Yang, Z.; Zeng, Y. Numerical simulation of initial mixing of marine wastewater discharge from multiport diffusers. *Eng. Comput.* **2014**, *31*, 1379–1400. [CrossRef]
43. Tang, H.S.; Paik, J.; Sotiropoulos, F.; Khangaonkar, T. Three-Dimensional Numerical Modeling of Initial Mixing of Thermal Discharges at Real-Life Configurations. *J. Hydraul. Eng.* **2008**, *134*, 1210–1224. [CrossRef]
44. Yan, X.; Ghodoosipour, B.; Mohammadian, A. Three-Dimensional Numerical Study of Multiple Vertical Buoyant Jets in Stationary Ambient Water. *J. Hydraul. Eng.* **2020**, *146*, 04020049. [CrossRef]
45. Fischer, H.B.; List, E.J.; Koh, R.C.Y.; Imberger, J.; Brooks, N.H. *Mixing in Inland and Coastal Waters*; Academic Press: Amsterdam, The Netherlands, 1979.
46. Roberts, P.J.W.; Ferrier, A.; Daviero, G. Mixing in Inclined Dense Jets. *J. Hydraul. Eng.* **1997**, *123*, 693–699. [CrossRef]
47. Missimer, T.M.; Jones, B.; Maliva, R.G. *Intakes and Outfalls for Seawater Reverse-Osmosis Desalination Facilities: Innovations and Environmental Impacts*; Springer: New York, NY, USA, 2015.
48. Kikkert, G.A.; Davidson, M.J.; Nokes, R.I. Inclined Negatively Buoyant Discharges. *J. Hydraul. Eng.* **2007**, *133*, 545–554. [CrossRef]
49. Zeitoun, M.; McIlhenny, W. Conceptual Designs of Outfall Systems for Desalination Plants. In *Offshore Technology Conference*; OnePetro: Washington, DC, USA, 1970.
50. Zeitoun, M.A.; McHilhenny, W.F. *Model Studies of Outfall Systems for Desalination Plants; Part III. Numerical Simulation and Design Considerations*; Dow Chemical Co.: Freeport, TX, USA, 1972; Available online: https://www.tib.eu/de/suchen/id/ntis:sid~{}oai:ds2:ntis/5319e8d74fe2d5c13d04dc1c/Model-Studies-of-Outfall-Systems-for-Desalination?tx_tibsearch_search%5Bsearchspace%5D=tn&cHash=7fcf481ee07b3258c85f7bc313ad0c16 (accessed on 20 April 2022).
51. Abessi, O.; Roberts, P.J.W. Effect of Nozzle Orientation on Dense Jets in Stagnant Environments. *J. Hydraul. Eng.* **2015**, *141*, 06015009. [CrossRef]
52. Lim, A.; Lam, Y. Numerical Investigation of Nanostructure Orientation on Electroosmotic Flow. *Micromachines* **2020**, *11*, 971. [CrossRef] [PubMed]
53. Lim, A.E.; Lam, Y.C. Vertical Squeezing Route Taylor Flow with Angled Microchannel Junctions. *Ind. Eng. Chem. Res.* **2021**, *60*, 14307–14317. [CrossRef]
54. Roberts, P.J.W.; Toms, G. Inclined Dense Jets in Flowing Current. *J. Hydraul. Eng.* **1987**, *113*, 323–340. [CrossRef]
55. Versteeg, H.K.; Malalasekera, W. *An Introduction to Computational Fluid Dynamics: The Finite Volume Method, 2nd ed*; Pearson Education: London, UK, 2007.
56. Peyret, R.; Krause, E. (Eds.) *Advanced Turbulent Flow Computations*; Springer: Wien, Austria, 2000.
57. Hanjalic, K. *Closure Models for Incompressible Turbulent Flows*; Lecture Notes at Von Kármán Institute: Brussels, Belgium, 2004; p. 75.
58. Abessi, O.; Saeedi, M.; Davidson, M.; Zaker, N.H. Flow Classification of Negatively Buoyant Surface Discharge in an Ambient Current. *J. Coast. Res.* **2012**, *278*, 148–155. [CrossRef]
59. Lai, C.C.; Lee, J.H. Mixing of inclined dense jets in stationary ambient. *J. Hydro Environ. Res.* **2012**, *6*, 9–28. [CrossRef]
60. Roberts, P.J.W.; Toms, G. Ocean outfall system for dense and buoyant effluents. *J. Environ. Eng.* **1988**, *114*, 1175–1191.
61. Holzmann, T. *Mathematics, Numerics, Derivations and OpenFOAM®*; Holzmann CFD: Loeben, Germany, 2016.
62. Killingstad, P.E. A Study of Dead Water Resistance Reynolds Averaged Navier Stokes Simulations of a Barge Moving in Stratified Waters. Master's Thesis, University of Oslo, Oslo, Norway, 2018. Available online: https://www.duo.uio.no/bitstream/handle/10852/63458/Peter_Killingstad_masterThesis.pdf?sequence=1&isAllowed=y (accessed on 20 April 2022).
63. Launder, B.E.; Reece, G.J.; Rodi, W. Progress in the development of a Reynolds-stress turbulence closure. *J. Fluid Mech.* **1975**, *68*, 537–566. [CrossRef]
64. Lauria, A.; Alfonsi, G.; Tafarojnoruz, A. Flow Pressure Behavior Downstream of Ski Jumps. *Fluids* **2020**, *5*, 168. [CrossRef]
65. Tafarojnoruz, A.; Lauria, A. Large eddy simulation of the turbulent flow field around a submerged pile within a scour hole under current condition. *Coast. Eng. J.* **2020**, *62*, 489–503. [CrossRef]
66. Calomino, F.; Alfonsi, G.; Gaudio, R.; D'Ippolito, A.; Lauria, A.; Tafarojnoruz, A.; Artese, S. Experimental and numerical study of free-surface flows in a corrugated pipe. *Water* **2018**, *10*, 638. [CrossRef]
67. Palomar, P.; Lara, J.; Losada, I.; Rodrigo, M.; Alvárez, A. Near field brine discharge modelling part 1: Analysis of commercial tools. *Desalination* **2012**, *290*, 14–27. [CrossRef]
68. Yan, X.; Mohammadian, A. Numerical Modeling of Vertical Buoyant Jets Subjected to Lateral Confinement. *J. Hydraul. Eng.* **2017**, *143*, 04017016. [CrossRef]

69. Yan, X.; Mohammadian, A.; Chen, X. Three-Dimensional Numerical Simulations of Buoyant Jets Discharged from a Rosette-Type Multiport Diffuser. *J. Mar. Sci. Eng.* **2019**, *7*, 409. [CrossRef]
70. OpenFOAM. *The OpenFOAM Foundation*; OpenCFD Ltd.: Bracknell, UK, 2015.
71. Li, Y.; Geng, X.; Wang, H.; Zhuang, X.; Ouyang, J. Simulating the frontal instability of lock-exchange density currents with dissipative particle dynamics. *Mod. Phys. Lett. B* **2016**, *30*, 1650200. [CrossRef]
72. Shao, D.; Adrian, W.K.L. Mixing and boundary interactions of 30 and 45 inclined dense jets. *Environ. Fluid Mech.* **2010**, *10*, 521–553. [CrossRef]
73. Roberts, P.J.W.; Snyder, W.H.; Baumgartner, D.J. Ocean Outfalls. III: Effect of Diffuser Design on Submerged Wastefield. *J. Hydraul. Eng.* **1989**, *115*, 49–70. [CrossRef]

MDPI
St. Alban-Anlage 66
4052 Basel
Switzerland
Tel. +41 61 683 77 34
Fax +41 61 302 89 18
www.mdpi.com

Journal of Marine Science and Engineering Editorial Office
E-mail: jmse@mdpi.com
www.mdpi.com/journal/jmse

www.ingramcontent.com/pod-product-compliance
Lightning Source LLC
Chambersburg PA
CBHW061833210926
51850CB00006B/877